DIGITAL COMMUNICATIONS SYSTEMS
With Satellite and Fiber Optics Applications

Harold Kolimbiris
Seneca College

Prentice Hall

Upper Saddle River, New Jersey   Columbus, Ohio

Library of Congress Cataloging-in-Publication Data

Kolimbiris, Harold.
  Digital communications systems : with satellite and fiber-optics applications / by Harold Kolimbiris.
    p.   cm.
    ISBN 0-13-081543-8
    1. Digital communications.   2. Artificial satellites in telecommunication.   3. Optical fibers.
I. Title.
  TK5103.7.K64   2000
  621.382—dc21
                                                                                    99-32117
                                                                                    CIP

Cover photo: FPG International
Publisher: Charles E. Stewart, Jr.
Production Editor: Alexandrina Benedicto Wolf
Production Coordination: York Production Services
Design Coordinator: Karrie Converse-Jones
Cover Designer: Alice Shikina
Production Manager: Matthew Ottenweller
Marketing Manager: Ben Leonard

This book was set in Times Roman by York Graphic Services and was printed and bound by R. R. Donnelley & Sons Company. The cover was printed by Phoenix Color Corp.

© 2000 by Prentice-Hall, Inc.
Pearson Education
Upper Saddle River, New Jersey 07458

Printed in the United States of America

10 9 8 7 6 5 4 3 2

ISBN: 0-13-081543-8

Prentice-Hall International (UK) Limited, *London*
Prentice-Hall of Australia Pty. Limited, *Sydney*
Prentice-Hall of Canada, Inc., *Toronto*
Prentice-Hall Hispanoamericana, S. A., *Mexico*
Prentice-Hall of India Private Limited, *New Delhi*
Prentice-Hall of Japan, Inc., *Tokyo*
Prentice-Hall (Singapore) Pte. Ltd., *Singapore*
Editora Prentice-Hall do Brasil, Ltda., *Rio de Janeiro*

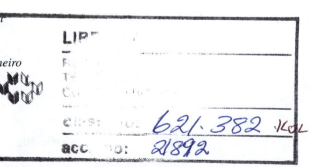

Preface

During the last two decades, the world has witnessed an ever-increasing demand for reliable and secure communications. These demands were satisfied with the implementation of digital transmission for short and long distances. In the early 21st century, the volume and clarity of voice, data, and video transmission will increase substantially. Educational institutions will offer more courses in the area of digital communications to satisfy the high demand for qualified graduates.

This book examines, in some detail, the three media utilized in digital transmission: line-of-sight, satellite, and optical fibers. Intended for senior college students in communications, practicing engineers, and managers and technical personnel in related industries, it provides practical examples of system design.

Chapter 1 deals with the important concept of electronic noise, its sources, and its effects on communications systems from the applications point of view. Numerous examples illustrate the competitiveness that exists between the signal power and noise power of a line-of-sight link and those of a satellite microwave link.

Chapter 2 presents the theory of quantization and the various methods of voice signal digitization. Emphasis is given to the two most commonly used techniques: pulse-code modulation and delta modulation. The North American, Japanese, and European digital hierarchies set forth by the International Telegraph and Telephone Consultative Committee are listed, and the concept of jitter, its sources, and its effect on digital transmission are examined.

Chapter 3 examines the fundamental components of a digital radio. The important concept of digital modulation/demodulation and various techniques such as Phase-shift keying, Differential Phase-shift keying, and Quadrature amplitude modulation are presented in detail. Particular emphasis is given to the important relationship between channel capacity, error power, and carrier-to-noise ratio.

Chapter 4 gives a complete analysis and design of both analog and digital microwave links, including the selection of the appropriate modulation scheme and calculations of path loss, system gain, carrier-to-noise ratio, bit error rate, and antenna tower height.

Chapter 5 discusses the unique features and components of geosynchronous communications satellites. Elements of planetary mechanics, the classification of satellites, space launching, and satellite subsystems are examined.

Chapter 6 examines the highly complex and sophisticated technology involved in the design of a satellite earth station. Important topics include elements of electromagnetic propagation, parabolic reflector antennas, and calculations of the effective isotropic radiated power (EIRP), figure-of-merit (G/T), and system noise temperature. It includes a detailed discussion of the building blocks of both the transmitter and the receiver:

high-power amplifiers, transmitter redundancy, parametric amplifiers, MMICs, and HEMT-LNA devices.

Chapter 7 presents the concepts and the various methods of satellite access, frequency-division multiple access, time-division multiple access, frame efficiency, and channel capacity.

Chapter 8 discusses the design and analysis of a complete digital satellite communications link. Calculations include up-link and down-link losses, EIRP, antenna gain, carrier-to-noise ratio, G/T, bit energy-to-noise ratio, receiver noise figure, and channel capacity.

Chapter 9 presents the principal concepts of optical fiber communications technology, including a review of optical laws, the classification and operating characteristics of optical fibers, the type description and operating characteristics of optical sources, and optical detectors.

Chapter 10 describes the various components of an analog link and a digital optical fiber communications link. Design examples of optical fiber communications systems to process voice and video signals are thoroughly examined.

Chapter 11 examines the various techniques of testing and measuring digital radio and digital communications links including satellites. Those discussed are modulation/demodulation sensitivity measurement, linearity and differential gain measurement, noise and interference measurement, jitter measurement, and receiver noise measurement. This chapter also presents important measurement techniques for link performance evaluation, especially the eye diagram and constellation display.

Chapter 12 examines the emerging technology of HDTV and fundamental concepts such as video compression, resolution, format comparison, video encoders and decoders, transport and voice digitization, and compression.

Since the field of digital communications is developing very rapidly, emphasis is given to those concepts and principles that are basic to the field.

Harold Kolimbiris

ACKNOWLEDGMENTS

I wish to express my deepest gratitude and appreciation to Dr. J. Moran of Seneca College, for reviewing the manuscript and providing me with very valuable suggestions. I would like to thank Hewlett Packard-Canada for their permission to use a number of pictures and graphs.

I would also like to thank Karen Ebden for the partial typing of the manuscript and my colleagues in the electronics department at Seneca College for their valuable suggestions and encouragements. Last but not least, I would like to express my deepest appreciation and gratitude to my graduate student Farshid Ghavami for the countless late nights he spent designing the artwork and typing the complex mathematical formulas. Without his assistance this project would not have been possible.

Contents

In memory of my father,
to my mother
and to my son Panagiotis Kolimbiris, whom I love very much.

Introduction

Electronic communications deals with the transmission and reception of information (voice, video, and data) through electronic means. A brief review of the history of communications technology shows that between the years 1800 and 1837, certain discoveries—such as the dry-cell battery by Volta, experiments in electromagnetism by Faraday and others, and the theoretical work in applied mathematics of Laplace and Fourier—set the foundation for the development of telegraphy by Morse, which eventually led to its commercial implementation in 1844.

Maxwell's theoretical work in electromagnetic propagation was verified by Hertz and Marconi, resulting in the first wireless system established in 1897. A few years earlier, Alexander Graham Bell had perfected his acoustic transducer, and in 1878 the first telephone exchange was fully operational. The early 20th century witnessed a dramatic acceleration in communications technology—the development of the diode tube by DeForest, followed by Bell's first transcontinental line in 1915, and the substantial improvement of the super-heterodyne receiver by Armstrong in 1918. In 1923, Baird and Jenkins' successful attempts at image reproduction set the foundation for the development of television transmission.

The period between 1923 and 1940 was characterized as the steady-state period; during this time there were no major breakthroughs in the field. With the advent of WWII, the high demand for reliable communications and the exclusive use of frequency modulation by the military resulted in a dramatic improvement of FM technology. The area of microwave applications benefited most from the war demands. The urgent need for an early warning system to protect Britain's coastal cities accelerated research and development in radar technology. The ever-increasing demand for high-power microwave tubes, microwave amplifiers, and directional antennas substantially enhanced microwave technology development.

The possibility of reflecting electromagnetic waves from a conducting medium had been considered. In 1903, Hulsmeyer, a German engineer, was experimenting with the reflection of electromagnetic waves from ships. His findings, although successful, were not well received because the military failed to realize their potential applications in military warfare.

In the late 1940s, the theoretical work of Rice, Shannon, and others in the area of statistical communications and information theory set the foundation for a new era in electronic communications. This was enhanced by perhaps the most important technological development of all time, that of the semiconductor transistor made by Bell-Lab scientists in 1948. In the 1950s color television made its appearance, and J. R. Pearce proposed that Arthur C. Clarke's fictional accounts of global communications via geosynchronous satellites could become reality. In the early 1960s, the mass production of integrated circuits opened new horizons for the telecommunications industry, and shortly thereafter the first communications satellite was launched into orbit.

In the microwave device area, Gunn improved his microwave oscillator in 1963, and in 1965 the Mariner spacecraft was able to transmit color from the surface of Mars. In the late 1960s, improvements in pulse-code modulation (PCM), digital signaling, and error control and coding greatly contributed to digital transmission. Without these technological advancements, the lunar landing in June 1969 would not have been possible. Since 1967, digital communications systems have continued to replace analog systems mainly due to their high reliability and the volume of information they can handle. It is estimated that by early 21st century, over 90 percent of all communications systems in the industrialized world will be digital.

A brief look at the history of information transmission capacity shows the following: In the 1930s the maximum transmission data rate through a telephone channel was approximately 600 bits per second (b/s). By the 1950s it was up to 1200 b/s, and in the 1970s it reached an impressive 56 Kb/s. It is projected that by the end of the century, digital voice channels transmitted via optical cables will be measured in the hundreds of thousands. An ever-increasing demand for voice, data, and high-resolution video transmission can very effectively and economically be accommodated by digital means. Today, the United States, Japan, Canada, and Italy are leading the research, development, and implementation of digital communications systems.

Now the questions arise: What is digital communications? How does it compare with analog systems? What are the major components of the system?

The basic block of a digital communications system is illustrated in Figure I.1. It shows that a digital communications system is composed of three main sections: the transmitter, the channel, and the receiver. The transmitter processes the information generated by the source and transmits it through the transmission medium (channel). The receiver demodulates the information, decodes it, and appropriately processes it to individual transponders. The characteristics of the information depend on the generating source and can be either analog or digital. An analog signal is defined as a signal whose amplitude is continuously varying with time; a digital signal is defined as a signal whose various states are discrete intervals of time.

The basic advantages of digital transmission are as follows:

Easy for multiplexing and signaling
Adaptive to the state-of-the-art solid-state technology
Easy for baseband generation

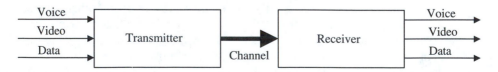

FIGURE I.1 Block diagram of a digital communications system.

Better signal-to-noise ratio (SNR)

Easier to scramble/de-scramble

A cost comparison between time-division multiplexing (TDM) and frequency-division multiplexing (FDM) shows that TDM systems are simpler and less expensive. Another advantage is that, because all control signals are digital, it can easily be accommodated within TDM systems. Furthermore, digital communications systems are using basically the same device technology used by computers, thus providing substantial design flexibility and cost reductions. On the other hand, special provisions are required for FDM signals. Bell's development of the common-channel-interface-signaling (CCIS) phone was an attempt to overcome problems related to signaling.

A voice channel requires a maximum bandwidth of 8 kHz for analog transmission. However, when the same channel is digitally transmitted, it requires 64 kHz of equivalent baseband bandwidth. This dramatic increase of bandwidth requirements for digital transmission is balanced by a significant increase in the signal-to-noise ratio. Advances in digital modulation techniques have improved the information-to-bandwidth ratio above the 3 b/s/Hz mark.

The last, but not least important, part of any communications system is the transmission medium, the link between the transmitter and the receiver. This medium can be (a) free space, (b) the ionosphere, (c) a transmission line, or (d) an optical fiber. The use of any of these depends on the system carrier frequency. Since communications systems cannot be considered without the transmission medium, the medium itself is a source of some of the most difficult problems encountered by the system designer. Such problems include transmission losses, signal degradation, and noise, among others. It is the transmitter's design specifications and performance that will offset all the problems induced into the system by the transmission medium. The role of the receiver is to distinguish the information transmitted by the transmitter, regardless of the degree of degradation the information has suffered through the transmission medium. In this book, all three fundamental components of a digital communications system will be thoroughly examined.

1

Noise in Communications Systems

Objectives

- Define and classify electronic noise.
- Explain the difference between external and internal noise.
- Stress the importance of thermal noise.
- Describe in detail noise temperature and phase noise.
- Compute the signal-to-ratio (SNR) and noise figure (NF).
- Derive relationships between the carrier-to-noise, carrier–to–noise power density, and noise power density ratios.
- Define figure of merit (G/T).

Key Terms

Electronic noise
Shot noise
Thermal or Johnson noise
Deterministic signals
Nondeterministic signals
External noise
Internal noise
Noise temperature
Thermal emission
Nonthermal emission
Popcorn noise
Flicker noise
Maximum power transfer
Bandwidth
Diode noise temperature
   factor

Figure of merit
Noise power
Noise voltage
Equivalent noise
   temperature
Quantization noise
Phase noise
Angular velocity
Bit error rate
NBS (National Bureau of
   Standards)
Amplitude fluctuation
Phase fluctuation
Signal-to-noise ratio (SNR)
Carrier-to-noise ratio (CNR)
Noise figure

Boltzmann's constant
Receiver noise figure
Noise factor
Internal noise power
Reference noise power
Attenuator noise figure
Mixer noise figure
Single-ended mixer
Balanced mixer
Mixer conversion loss
Bit energy
System bit rate
Bit-time duration
Noise power density
Noise bandwidth
Bit energy–to–noise ratio

INTRODUCTION

Electronic noise is defined as an unwanted signal present in a system that alters the amplitude, frequency, and phase of the information processed by that system in an unpredictable manner. In the early days of the development of communication systems, the presence of noise was interpreted as a form of degradation of the transmitted electronic signals. It was thought at that time that an improvement in the transmitter/receiver design would eliminate the problem of noise. Although substantial improvements were made in this area, the problem of noise was still evident in systems. In 1918, W. Schottky, while experimenting with vacuum tubes, observed an unavoidable irregularity in the dc current at the plate of the tubes. This type of irregularity was called **shot noise**, and it was attributed to the particle nature of free electrons composing the dc plate current of the tubes. In 1927, J. B. Johnson observed another type of electron noise, called in his honor **Johnson noise**. Today Johnson noise is commonly referred to as **thermal noise** and is attributed to thermally agitated free-electron motion within the conducting material. A closer study of the nature of thermal noise revealed that this unwanted and highly disturbing electronic signal has its source at the very heart of the matter, the atom. Therefore, a very brief review of the molecular and atomic structure of matter is essential for a better understanding of the noise phenomenon.

1.1 ELEMENTS OF ATOMIC STRUCTURE

In the middle of the 5th century B.C., the Greek philosophers Leucippus, Thales, and Democritus developed the hypothesis that all matter in nature is composed of discrete indivisible units called "atoms" (Greek: "single, nondivisible"). This philosophical interpretation of matter was carried on by the Epicurian School of the 3rd century B.C. For almost 17 centuries, these original ideas and philosophies were buried intentionally by the Romans. Later, they were covered by the shroud of medieval darkness until the 17th century A.D., when they were brought back to light by Galileo, Newton, Boyle, and others.

All matter in nature is composed of basic building blocks called molecules (Latin: "small objects"). Furthermore, each molecule is composed of a number of discrete units that are not further divisible by chemical means, called "atoms." If all the atoms composing the molecules are of same kind, then the resultant matter is called an "element." If the atoms composing the molecules are of different kinds, then the resultant matter is called a "substance." Therefore, the atom is the smallest unit in an element, retaining all the fundamental chemical characteristics of the element. The atom itself is also a microcosmic system, resembling in some ways the macrocosmic planetary system. It is composed of an extremely massive central core called the "nucleus," surrounded by a predetermined number of orbiting electrons. The number and arrangement of electrons within the atomic structure determine the behavior of atoms with each other and also determine the overall physical and chemical properties of these elements or substances. The nucleus of the atom is not homogeneous, but is composed of subatomic particles called "protons" (Greek: "first") and neutrons. Protons are positively charged particles with an electrical charge equal and opposite to that of the orbiting electron ($+1.602 \times 10^{-19}$ C), and a mass equal to 1.643×10^{-27} kg. The neutron is a particle carrying no electric charge and having a mass almost equal to that of the proton. In 1897, the English scientist J. J. Thomson discovered the electron, and Ernest Rutherford and James Chadwick discovered the proton and neutron in 1914 and 1932 respectively. Around 1911, Rutherford suggested that practically all the mass of an atom is concentrated in the nucleus. While the diameter of the atom is approximately 10^{-10} m, the diameter of the nucleus is equal to 10^{-15} m. Thus protons and neutrons in the nucleus are in virtual contact with each other. This observation defied the law of electromagnetic force, whereby like charges repel each other and unlike forces attract each other. Until 1935, the two known forces in nature were the electromagnetic force, with a relative strength equal to one, and the gravitational force, with a relative strength equal to 10^{-37}. Employing the principle of gravitational force to interpret why like charges such as protons are held together at the nucleus of an atom was unsuccessful because the gravitational force is too weak (weak force). On the other hand, the electromagnetic force, although strong, reacts only to electric charges in terms of repulsion or attraction and is also unable to provide a solution to the problem. In 1935, the Japanese scientist H. Yucawa introduced the theory of nuclear force. He proposed that when protons and neutrons are in very close proximity, almost in contact with each other, a nuclear force holds them together. The same nuclear force diminishes almost to zero for a distance of separation larger than 5×10^{-15} m. On the other hand, the electrons, separated from the protons by a relatively large distance and carrying opposite charges, are subject to electromagnetic field influence and thus are kept in orbit around the nucleus. For the outermost electron orbit in the atom of a conducting material, the electromagnetic force is quite weak. A relatively small amount of energy absorbed by this electron will be enough for it to break away from its predetermined orbit and become a "free electron." Free electrons and their random behavior are the main source of electronic noise.

1.2 CLASSIFICATION OF ELECTRONIC SIGNALS

Electronic signals are classified into two basic categories: deterministic and nondeterministic. **Deterministic signals** can be expressed as a direct function of time; their amplitude, frequency, and phase can be predicted. **Nondeterministic** (random) **signals** are those whose amplitude, frequency, and phase cannot be determined; their overall presence and effect can be statistically predicted. Electronic noise is therefore a nondeterministic signal.

1.3 CLASSIFICATION OF ELECTRONIC NOISE

Electronic noise is also divided into two categories: external noise and internal noise.

1.3.1 External Noise

External noise is the noise generated outside the electronic equipment used. Its sources can be terrestrial and extraterrestrial; such sources as the earth, the moon, the sun, the galaxies, and the universe as a whole. Of course, all these sources do not affect the entire communications frequency spectrum, but with various intensities, they affect certain frequencies at certain times and locations. The earth is a source of noise, with an average noise temperature of 254 K as seen from a satellite antenna with global average characteristics. The atmosphere is also a source of electronic noise. Lightning discharges and radiation absorption due to water vapor and oxygen molecules are the main causes of atmospheric noise. Figures 1.1 and 1.2 illustrate atmospheric noise in terms of operating frequencies and antenna elevations.

In 1932, K. G. Jansky of Bell labs proposed that noise sources outside the earth's atmosphere do exist and may have devastating effects on communications links if ignored. The strongest source of such an electronic noise outside the earth's atmosphere is the sun. This type of noise is referred to as "solar noise." The equivalent **noise temperature** (T_e) received by a directional antenna directed toward the sun is approximately equal to 10^5 K. Such a noise level completely blots out the transmitted signal. Figure 1.3 shows the relationship between noise temperature generated by the sun and system operating frequency at different solar activity states.

Stars, nova and supernova remnants, quasars, black holes, and ionized interstellar gases within our home galaxy are also sources of electronic noise. Their emissions are classified into two basic categories: thermal and nonthermal. **Thermal emission** is the result of electron-ion collision with an ionized gas cloud; **nonthermal emission** is the result of electron motion within a magnetic field. This type of electronic noise has almost no effect at frequencies above 1 GHz, but it affects systems operating at frequencies in the 40- to

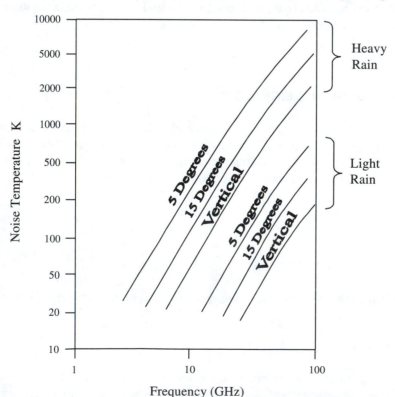

FIGURE 1.1 Noise temperature versus operating frequency resulting from oxygen content in the atmosphere and at different antenna elevations.

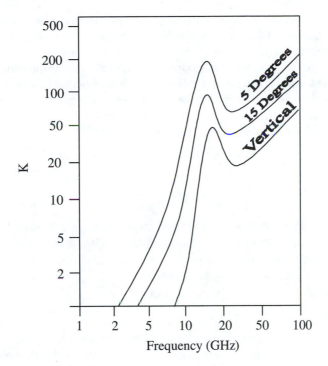

FIGURE 1.2 Noise temperature versus frequency as a result of water vapor content in the atmosphere at different antenna elevations.

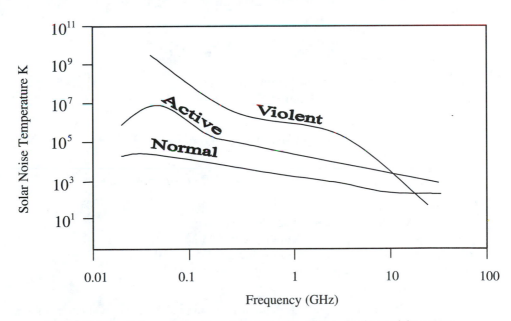

FIGURE 1.3 Solar noise versus frequency for different solar activity states.

Noise in Communications Systems

5

250-MHz range. The universe (cosmos) is a source of electronic noise as well. Cosmic noise is the noise generated by sources outside our home galaxy. This type of noise decreases very rapidly as the frequency increases, and at frequencies above 10 GHz maintains a minimum level of 3 K, with a maximum of 100 K. This minimum 3 K is thought to be the background-noise remnant of the Big Bang, which occurred some 16 billion years ago. Terrestrial noise is generated mainly in urban areas by electrical machine operations, and affects frequencies below 1 GHz. Satellite earth stations are susceptible to immediate environmental conditions such as rain, clouds, and fog.

1.3.2 Internal Noise

Internal noise is the electronic noise generated by the passive and active components incorporated in the designs of communications equipment. There exist several different types of internal noise, such as shot noise, popcorn noise, flicker noise, quantum noise, and the most important of all, thermal noise.

Shot Noise

W. Schottky, while experimenting with vacuum tubes, observed that certain spontaneous fluctuations of the dc anode current were occurring periodically. Further observations and studies concluded that these fluctuations were a result of the particle nature of free electrons moving randomly under the influence of a potential difference applied across the anode and the cathode of the tube. In semiconductor devices, shot noise is the result of electron–hole recombination and minority carrier random diffusion. The power spectral density of shot noise is proportional to the current passing through the device and is given by

$$P_n = 2 \cdot I_s \cdot q \tag{1.1}$$

where I_s = saturation current (A)
q = electron charge $(1.59 \times 10^{-19} \, \text{C})$

The mean square value of shot noise is given by

$$V_n = 2 \cdot I_s \cdot q \cdot \text{BW} \tag{1.2}$$

where V_n = noise voltage (V)
I_s = saturation current (A)
q = electron charge $(1.59 \times 10^{-19} \, \text{C})$
BW = operating bandwidth (Hz)

Shot noise is an important element incorporated into the evaluation of linear circuit performance by noise-temperature and noise-figure methods.

Popcorn Noise

Popcorn noise originates in semiconductor and integrated circuit devices. It occurs in bursts of several hundreds per second to very few pulses per minute, with a pulse width similar to and amplitude much larger than that of thermal noise. The power spectral density of popcorn noise is given by

$$P_n = \frac{K}{f^2 \cdot T^2 + 1} \tag{1.3}$$

where K = a constant
f = operating frequency
$T = \dfrac{1}{\text{bursts/s}}$

Equation (1.3) shows that at frequencies close to zero, the power spectral density becomes a constant (K) and progressively decreases at high frequencies.

Flicker Noise

Flicker noise or excess noise is thought to be related to dc current flow through imperfect conductors in semiconductor devices. The real nature of flicker noise is not yet fully understood. This type of noise is also referred to as $1/f$ noise. A more precise expression of flicker noise is given by $1/f^n$, where n takes values between 0.8 and 9.3. Flicker noise voltage is expressed as follows:

$$V_n = \int_{f_1}^{f_2} (K \cdot I^a \cdot f^b) df \tag{1.4}$$

where $f = f_1 - f_2$ = bandwidth (Hz)
K, a, b = constants, characteristic of the components used
I = saturation current (A)

Quantum Noise

Quantum noise is encountered at optical frequencies and has little or no effect at frequencies below optical range. The quantum noise power spectral density is expressed by

$$P_n = hf \tag{1.5}$$

where h = Planck's constant (6.63×10^{-36} J-s)
f = frequency (Hz)

Thermal Noise

Thermal noise is the result of thermally agitated free-electron motion within a conducting material. Thermally agitated electrons within a conductor collide with the molecules of that conductor, thus setting in motion a chain reaction with all the other free electrons. The average of thermal noise is zero, at which point a system is said to be in a thermal equilibrium; therefore there is a dc component. The voltage probability is Gaussian with a mean square voltage, given by

$$V_n = 4KT \cdot \int_{f_1}^{f_2} R(f)P(f)df \tag{1.6}$$

$P(f)$ is defined as follows:

$$P(f) = \frac{h \cdot f}{K \cdot T} \cdot (e^{-hf/kT} - 1)^{-1} \tag{1.7}$$

where

T = absolute temperature (290 K)
K = **Boltzmann's constant** (1.38×10^{-23} J-K)
f = frequency (Hz) (**bandwidth** of the observed signal)
h = Planck's constant (6.62×10^{-34} J-s)
R = impedance of the network

For a temperature $T = 290$ K and frequency $f \geq 100$ GHz, $P(f)$ takes a value between 0.992 and 1 ($0.992 < P(f) < 1$); therefore, $P(f) \cong 1$. Thus, Eq. (1.6) becomes

$$(V_n)^2 = 4KTBWR \quad \text{or} \quad V_n = \sqrt{KTBWR} \tag{1.8}$$

The power delivered to a load under matched conditions is calculated as follows:

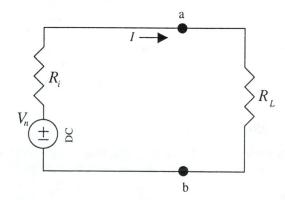

FIGURE 1.4

For the equivalent circuit of Figure 1.4, the current (I) is expressed by

$$I = \frac{V_n}{R_i + R_L} \tag{1.9}$$

The power dissipation at R_L is given by

$$P_n = I^2 \cdot R_L \tag{1.10}$$

Substituting Eq. (1.9) into Eq. (1.10) we have

$$P_n = \left(\frac{V_n}{R_i + R_L}\right)^2 \cdot R_L \tag{1.11}$$

Differentiating Eq. (1.11) gives

$$\frac{dP_n}{dR_L} = (V_n)^2 \frac{(R_i + R_L)^2 - 2R_L(R_i + R_L)}{(R_i + R_L)^2} \tag{1.12}$$

For maximum P_n, $dP_n/dt = 0$, or

$$V_n = \frac{(R_i + R_L)^2 - 2R_L(R_i + R_L)}{(R_i + R_L)^2} \tag{1.13}$$

or

$$V_n(R_i + R_L)^2 - 2R_L(R_i + R_L) = 0 \tag{1.14}$$

Since only the numerator is required to be zero in order for the expression to be equal to zero,

$$(R_i)^2 + (R_L)^2 + 2R_iR_L = 2R_iR_L + 2(R_L)^2$$
$$(R_i)^2 = (R_L)^2$$

Therefore,

$$R_i = R_L \tag{1.15}$$

For **maximum power transfer,** the load impedance must be equal to the Thevenin equivalent impedance. Thus the **noise voltage** across the load R_L is

$$V_{n(R_L)} = \left(\frac{R_L}{R_i + R_L}\right) \cdot V_n$$

$$= \frac{R_L}{2R_L} \times V_n$$

$$= \frac{V_n}{2}$$

Therefore:

$$V_{n(R_L)} = \frac{V_n}{2} \qquad \textbf{(1.16)}$$

To compute the **noise power** P_n at the load we have

$$P_n = \frac{(V_{n(R_L)})^2}{R}$$

$$= \frac{(V_n)^2/2}{R}$$

$$= \frac{(V_n)^2}{4R}$$

Since $(V_n)^2 = 4KTBWR$,

$$P_n = \frac{4KTBWR}{4R}$$

Therefore,

$$P_n = KTBW \qquad \textbf{(1.17)}$$

Noise power is proportional to noise temperature and operating bandwidth.

EXAMPLE 1.1

Determine the noise voltage V_n and noise power P_n for the circuit in Figure 1.5 with operating temperature of 330 K and bandwidth of 20 MHz.

Solution

$BW = 20$ MHz

$R_L = 100 \ \Omega$

$T = 300$ K

R_n = noiseless resistor

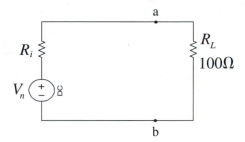

FIGURE 1.5

Noise Voltage (V_n)

$$V_n = \sqrt{4 \cdot K \cdot T \cdot \text{BW} R_L}$$
$$= (4 \times 1.38 \times 10^{-23} \times 20 \times 10^6 \times 100 \times 330)^{1/2}$$
$$= 6.03 \times 10^{-6} \text{ V}$$

Therefore, the noise voltage $V_n = 6.03 \ \mu\text{V}$.

Noise Power (P_n)

$$P_n = K \cdot T \cdot \text{BW}$$
$$= 1.38 \times 10^{-23} \times 330 \times 20 \times 10^6$$
$$= 9.1 \times 10^{-14}$$

Therefore, the noise power $P_n = 9.1 \times 10^{-14}$ W.

Note: The product KT is also referred to as "noise density," and at the ambient temperature of 290 K is calculated to be equal to:

$$KT_0 = 1.38 \times 10^{-23} \times 290$$
$$= 400.2 \times 10^{-23}$$

or, in dBm,

$$KT_0 = 10 \log(400.2 \times 10^{-23})$$
$$\cong -174$$

Therefore, $KT_0 = -174$ dBm.

EXAMPLE 1.2

Determine the noise voltage (V_n) for the circuits in Figure 1.6(a) and (b) with an operating bandwidth equal to 0.8 MHz and temperature of 310 K.

Noise in Communications Systems

FIGURE 1.6(a).

Solution

Circuit of Figure 1.6(a)

BW = 0.8 MHz

$K = 1.38 \times 10^{-23}$

$T = 310 \text{ K}$

$R_1 = 1 \text{ k}\Omega$

$R_2 = 1.5 \text{ k}\Omega$

$$V_n = \sqrt{4 \cdot K \cdot T \cdot \text{BW} \cdot R_t}$$
$$= \sqrt{4 \times 1.38 \times 10^{-23} \times 0.8 \times 10^6 \times 2.5 \times 10^3 \times 310}$$
$$= 18.5 \times 10^{-7}$$

Therefore, $V_n = 1.85 \, \mu\text{V}$.

Circuit of Figure 1.6(b)

BW = 0.8 MHz

$K = 1.38 \times 10^{-23}$

$T = 310 \text{ K}$

$R_1 = 2 \text{ k}\Omega$

$R_2 = 2 \text{ k}\Omega$

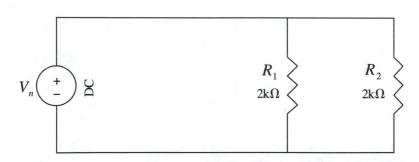

FIGURE 1.6(b).

$$R_t = R_1 \| R_2$$
$$= 1 \text{ k}\Omega$$
$$\therefore R_t = 1 \text{ k}\Omega$$
$$V_n = \sqrt{4 \cdot K \cdot T \cdot \text{BW} \cdot R_t}$$
$$= \sqrt{4 \times 1.38 \times 10^{-23} \times 0.8 \times 10^6 \times 1 \times 10^3 \times 310}$$
$$= 1.1 \times 10^{-6}$$

Therefore, $V_n = 1.1 \ \mu\text{V}$.

Noise Temperature   It was mentioned earlier in this chapter that electronic noise is the result of thermally-agitated free-electron motion. This random motion generates a mean square voltage given by the expression $V_n = \sqrt{4 \cdot K \cdot T \cdot \text{BW} \cdot R}$ and noise power given by the expression $P_n = R \cdot T \cdot \text{BW}$. Since all electronic equipment incorporates circuits composed of active and passive components, the noise voltages described earlier are inherent to this equipment.

A close examination of the equations defining noise voltage and noise power reveals that the principal factor determining noise power is noise temperature (T), because K is a constant and BW and R are characteristics of the system. It is therefore possible to compute the **equivalent noise temperature** (T_e) from the average noise power at a specified operating **bandwidth.** It has been customary to express noise power in terms of equivalent noise temperature. Equivalent noise temperature is calculated by assuming that all internal noise is thermal. The equivalent noise temperature can then be calculated if the average noise power and bandwidth of the system are known.

$$P_n = K \cdot T_e \cdot \text{BW}$$

Solving for T_e we have:

$$T_e = \frac{P_n}{K \cdot \text{BW}} \tag{1.18}$$

EXAMPLE 1.3

Calculate the equivalent noise temperature (T_e) of a microwave amplifier with an average noise power of $1 \times 10^{-13} \text{ W}$ and an operating bandwidth of 27 MHz.

Solution

$$P_n = 1 \times 10^{-13} \text{ W}$$
$$\text{BW} = 27 \times 10^6 \text{ Hz}$$
$$T_e = \frac{P_n}{K \cdot \text{BW}}$$

$$= \frac{1 \times 10^{-13}}{1.38 \times 10^{-23} \times 27 \times 10^6}$$

$$= \frac{1 \times 10^{-13}}{37.26 \times 10^{-17}}$$

$$= \frac{1}{37.26} \times 10$$

$$\cong 268.4$$

Therefore, $T_e = 268.4$ K.

The equivalent noise temperature of 268.4 K generates an average noise power of 2×10^{-13} W under a limited bandwidth of 27 MHz. It is therefore evident that the equivalent noise temperature and system bandwidth are the two key components in determining the noise power of a system.

EXAMPLE 1.4

A microwave amplifier employed in a satellite receiving system generates an equivalent noise temperature of 140 K and has a gain of 50 dB and an operating bandwidth of 500 MHz. Determine the internal noise power RTI (return to input) and the output noise power.

Solution

$$T_e = 140 \text{ K}$$
$$G_p = 50 \text{ dB}$$
$$\text{BW} = 500 \text{ MHz}$$

Internal Noise Power

$$P_{n_i} = K \cdot T_e \cdot \text{BW}$$
$$= 1.38 \times 10^{-23} \times 140 \times 500 \times 10^6$$
$$= 9.7 \times 10^{-13}$$

Therefore, $P_{n_i} = 9.7 \times 10^{-13}$ W.

Output Noise Power

$$P_{n_o} = G_p P_{n_i}$$
$$= 1 \times 10^{-5} \times 9.7 \times 10^{-13}$$
$$= 9.7 \times 10^{-8}$$

Therefore, $P_{n_o} = 9.7 \times 10^{-13}$ W.

The knowledge of noise power is very important for establishing performance criteria in all microwave communications receivers. Instead of noise power, noise temperature is more often used in applications where the background temperature is different than 290 K (ambient temperature).

Quantization Noise

Quantization noise is directly associated with the error introduced during the sampling and coding process of analog signals (PCM). Extensive treatment of quantization noise will be dealt with in Chapter 2, which deals with the quantization process.

Phase Noise

Phase noise is the unwanted fluctuation of the phase of a carrier signal. The spectral purity of microwave signal sources is very important, especially in digital microwave and radar applications. Frequency stability is very important if such systems are to maintain strict operating parameters. By definition, frequency stability is referred to as the degree to which an oscillator oscillates at its predetermined frequency over a specified period of time. Although sophisticated techniques are used in the design and development of microwave oscillators, a certain degree of instability is unavoidable at the output, resulting in the degradation of a system's overall performance.

Digital and data communications systems use almost exclusively phase modulation techniques (modulating an analog carrier with a digital baseband). These phase modulation techniques are extremely sensitive to phase noise. Phase noise either degrades the system **bit error rate** (BER) or limits sensitivity and resolution in Doppler radar systems. Fluctuation of RF sources can be classified into two major categories: long term and short term.

Long-term fluctuations are frequency changes measured in parts per million per hour, day, or year, and are the result of the aging process of the frequency-determining component of the source. Short-term fluctuations involve elements that can cause frequency drifts for no longer than 5 s. The theoretical work performed by the **National Bureau of Standards (NBS)** relates phase noise to frequency stability as follows: Ideally, a sine wave is expressed as:

$$e(t) = E_0 \sin \omega t \qquad (1.19)$$

where
$e(t)$ = instantaneous amplitude (V)
E_0 = maximum amplitude (v)
$\omega = 2\pi f$ (**angular velocity**) (rad)
f_0 = frequency (Hz)

In practice, sine waves suffer from amplitude and frequency drifts. Therefore, the mathematical expression of Eq. (1.19) is modified in order to facilitate these unwanted drifts.

$$e(t) = (E_0 + E(t))\sin(2nf_0 t + \Delta\phi(t))$$

where
$E(t)$ = **amplitude fluctuation**
$\Delta\phi$ = **phase fluctuation**

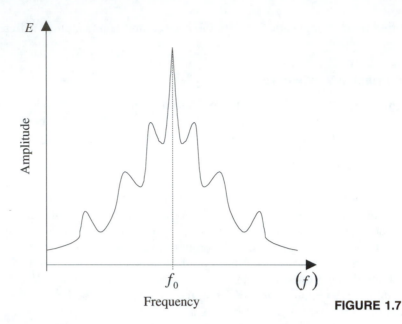

E

Amplitude

f_0

Frequency

(f)

FIGURE 1.7

Close examination of a nonideal sine wave in the frequency domain (Figure 1.7) reveals that the phase drift is composed of two distinct components: deterministic or spurious, and nondeterministic or random. Spurious components are the result of mixer intermodulation products and line-frequency drifts, whereas random components are related to flicker, shot, and thermal noises.

1.4 SIGNAL-TO-NOISE RATIO (SNR)

The **signal-to-noise ratio** (SNR) expresses in decibels the difference between baseband signal power and noise power at the input or output of a communications receiver. This ratio is perhaps the most important criteria of establishing performance for electronic equipment including communications receivers. The SNR is expressed in dB by

$$\text{SNR}_{\text{dB}} = 10 \log \frac{P_s}{P_n} \qquad (1.20)$$

where   SNR = signal-to-noise ratio (dB)
           P_s = signal power (W)
           P_n = noise power (W)

From Figure 1.8 we see that the signal power (P_s) is equal to 16 dB and the noise power (P_n) is equal to 2 dB; therefore,

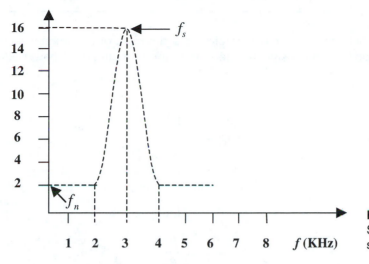

FIGURE 1.8 The SNR relationship of a signal at 3 KHz.

$$SNR_{dB} = P_{s_{dB}} - P_{n_{dB}}$$
$$= 16 - 2$$
$$= 14$$

Figure 1.8 indicates that there is a difference of 14 dB between the signal power and noise power.

EXAMPLE 1.5

Determine the SNR of communications equipment with a signal power of 55 μW and 5.5 nW of noise power at its input.

Solution

$$P_s = 55 \ \mu W$$
$$P_n = 5.5 \ nW$$
$$SNR = 10 \log \frac{P_s}{P_n}$$
$$= 10 \log \frac{55 \times 10^{-1}}{5.5 \times 10^{-9}}$$
$$= 10 \log(10 \times 10^3)$$
$$= 40$$

Therefore, SNR = 40 dB.

In this example, the signal power exceeds the noise power by 40 dB.

EXAMPLE 1.6

Compute the maximum noise power allowed at the input of a communications receiver in order to maintain a 40-dB signal-to-noise ratio for an input signal power equal to 20 pW.

Solution

$$P_n = ?$$

$$P_s = 20 \text{ pW}$$

$$\text{SNR} = 40 \text{ dB}$$

$$\text{SNR} = 10 \log \frac{P_s}{P_n}$$

$$40 = 10 \log \frac{P_s}{P_n}$$

$$4 = \log \frac{P_s}{P_n}$$

$$\frac{P_s}{P_n} = \text{antilog}(4)$$

$$\frac{P_s}{P_n} = 10^4$$

$$P_n = \frac{P_s}{1 \times 10^4}$$

$$= \frac{20 \times 10^{-12}}{1 \times 10^4}$$

$$= 20 \times 10^{-16} \text{ W}$$

$$\therefore P_n = 20 \times 10^{-16} \text{ W}$$

Therefore, the maximum noise power permitted at the input of the receiver is 20×10^{-16} W.

EXAMPLE 1.7

Determine the minimum signal power required at the input of a piece of communications equipment in order to maintain a minimum signal-to-noise ratio of 35 dB in the presence of 0.1 nW of noise power.

Solution

$$P_n = 0.1 \times 10^{-9} \text{ W}$$

$$\text{SNR} = 35 \text{ dB}$$

$$P_s = ?$$

$$\text{SNR} = 10 \log \frac{P_s}{P_n}$$

$$35 \text{ dB} = 10 \log\frac{P_s}{P_n}$$

$$3.5 = \log\frac{P_s}{0.1 \times 10^{-9}}$$

$$\frac{P_s}{P_n} = \text{antilog}(3.5) = 3162.3$$

$$P_s = (0.3162.3)(0.1 \times 10^{-6}) = 3.162 \times 10^{-7} \text{ W}$$

$$\therefore P_s = 0.3162 \times 10^{-6} \text{ W}$$

Therefore, the minimum signal power required to maintain a SNR of 35 dB in the presence of 0.1 nW of noise power is 0.3162 μW.

1.5 CARRIER-TO-NOISE RATIO (CNR)

Carrier-to-noise ratio is used instead of the signal-to-noise ratio only for modulated signals.

EXAMPLE 1.8

Compute the signal power required to maintain a 25-dB CNR in the presence of 20 nW of noise power at the input of communications equipment.

Solution

$$CNR = 25 \text{ dB}$$

$$P_n = 20 \text{ nW}$$

$$P_s = ?$$

$$SNR = 10 \log\frac{P_s}{P_n}$$

$$25 \text{ dB} = 10 \log\frac{P_s}{P_n}$$

$$2.5 = \log\frac{P_s}{20 \times 10^{-9} \text{ W}}$$

$$\frac{P_s}{20 \times 10^{-9} \text{ W}} = \text{antilog}(2.5) = 316.3$$

$$P_s = (316.3)(20 \times 10^{-9}) = 6.3 \times 10^{-6}$$

Therefore, $P_s = 6.3 \times 10\ \mu$W.

1.6 NOISE FIGURE (NF)

Since the early days of the development of electronic communications, thermal noise has been considered to be an inherent part of all electronic devices and circuits. This internally generated noise is largely responsible for the degradation of the overall system performance. It therefore became absolutely necessary to have thermal noise well defined and mathematically expressed, in order that it be meaningfully represented in the system specifications. Today, practically all manufacturers of communications equipment express internal noise in terms of noise figure (NF). This, of course, generated the need for devising an effective measurement technique, one capable of verifying noise figure theoretical computations. In 1945, H. T. Friis defined **noise figure** as the signal-to-noise ratio at the input of a network divided by the signal-to-noise ratio at the output of that network.

$$(\text{NF}) = \frac{(\text{SNR})_i}{(\text{SNR})_o} = \frac{S_i/N_i}{S_o/N_o} \tag{1.21}$$

where S_i = signal input power (W)
N_i = noise input power (W)
S_o = signal output power (W)
N_o = noise output power (W)

Since $N_i = P_{n_i} = KTBW$ and $N_o = P_{n_o} = KTBW$,

$$\text{NF} = \frac{P_{S_i}/KTBW}{\dfrac{P_{S_o}}{N_o}} = \frac{P_{S_i} \times N_o}{P_{S_o} \times KTBW} \tag{1.22}$$

The ratio P_{S_i}/P_{S_o} is the reverse of the system power, G_p, so

$$\text{NF} = \frac{I}{G_p} \times \frac{N_o}{KTBW}$$

Expressed in dB,

$$\text{NF}_{\text{dB}} = 10 \log \frac{I}{G_p} \times \frac{N_o}{KTBW} \tag{1.23}$$

where G_p = power gain of the system (no units)
K = Boltzmann's constant (1.38×10^{-23} J-K)
T = system noise temperature (K)
BW = system bandwidth (Hz)
N_o = output noise power (W)

An alternative method of establishing noise figure is as follows: The internally generated noise of any given device or equipment is compared to a reference noise power generated by a standard noise temperature of 290 K.

$$P_{n_o} = KT_0\text{BW} \qquad (1.24)$$

where $\quad P_{n_o}$ = **reference noise power** (W)
$\quad\quad\quad\quad K$ = Boltzmann's constant
$\quad\quad\quad\quad T_0$ = reference noise temperatre (290 K)
$\quad\quad\quad\text{BW}$ = operating bandwidth (Hz)

The internally generated noise power and the reference noise power are used to compute overall system noise power as follows:

$$F = \frac{P_n}{P_{n_o}} + 1 \qquad (1.25)$$

where $\quad F$ = **noise factor** (no units)
$\quad\quad\quad\quad P_n$ = **internal noise power** (W)
$\quad\quad\quad P_{n_0}$ = reference noise power (W)

The noise figure (NF) in dB is expressed by

$$\text{NF}_{\text{dB}} = 10 \log(F) \qquad (1.26)$$

or

$$\text{NF}_{\text{dB}} = 10 \log\left(\frac{P_n}{P_{n_o}} + 1\right) \qquad (1.27)$$

Since $P_n = KT\text{BW}$ and $P_{n_o} = KT_0\text{BW}$, the noise becomes

$$F = \frac{KT\text{BW}}{KT_0\text{BW}} + 1$$

$$F = \frac{T}{T_0} + 1$$

$$\frac{T}{T_0} = F$$

If T is the equivalent noise temperature (T_e), then

$$\frac{T_e}{T_0} = F - 1$$
$$T_e = T_0(F - 1)$$
$$F = \frac{T_e}{T_0} + 1 \tag{1.28}$$

The noise factor (F) is expressed in terms of the equivalent noise temperature generated by the equipment under test as compared to the standard noise temperature of 290 K. The noise figure of Eq. (1.28) can then be expressed by the equation.

$$\text{NF}_{\text{dB}} = 10 \log\left(\frac{T_e}{T_0} + 1\right) \text{dB} \tag{1.29}$$

Equation (1.29) indicates that there is an equivalent noise temperature for a specified value of the noise figure (NF).

EXAMPLE 1.9

Determine the noise figure of a microwave amplifier operating with a bandwidth of 27 MHz and internal noise power of 1×10^{-13} W (RTI).

Solution

$$\text{BW} = 27 \text{ MHz}$$
$$P_n = 1 \times 10^{-13} \text{ W}$$
$$P_{n_o} = K \cdot T_0 \cdot \text{BW}$$
$$= 1.38 \times 10^{-23} \times 290 \times 27 \times 10^{-6}$$
$$= 10.8 \times 10^3 \times 10^6 \times 10^{-23}$$
$$= 10.8 \times 10^{-14}$$
$$F = \frac{T_e}{T_0} + 1$$
$$= \frac{1 \times 10^{-13}}{10.8 \times 10^{-14}}$$
$$= 0.093 \times 10 + 1$$
$$= 0.93 + 1$$
$$\therefore F \cong 1.93$$
$$\text{NF}_{\text{dB}} = 10 \log(F)$$
$$= 10 \log(1.93)$$
$$= 2.855$$

Therefore: NF = 2.855 dB

Alternatively,

$$T_e = \frac{P_n}{K \cdot \text{BW}}$$

$$= \frac{1 \times 10^{-13}}{1.38 \times 10^{-14} \times 27 \times 10^6}$$

$$\therefore T_e = 268 \text{ K}$$

$$\text{NF}_{\text{dB}} = 10 \log\left(\frac{T_e}{T_0} + 1\right)$$

$$= 10 \log\left(\frac{268}{290} + 1\right)$$

$$\cong 2.85 \text{ dB}$$

Therefore, the noise figure is given by $\text{NF}_{\text{db}} \cong 2.85$ dB.

EXAMPLE 1.10

Determine the internal noise power (P_n) of a microwave amplifier operating with a bandwidth of 500 MHz and a specified noise figure of 2.5 dB.

Solution

$$\text{NF} = 2.5 \text{ dB}$$

$$\text{BW} = 500 \text{ MHz}$$

$$\text{NF} = 10 \log(F)$$

$$F = \text{antilog}(0.25)$$

$$\therefore F = 1.78$$

$$F = \frac{P_n}{P_{n_o}} + 1$$

$$P_n = P_{n_o}(F - 1)$$

$$= P_{n_o}(1.78 - 1)$$

$$= 0.78 \, P_{n_o}$$

$$P_{n_o} = K \cdot T \cdot \text{BW}$$

$$= 1.38 \times 10^{-23} \times 290 \times 500 \times 10^6$$

$$= 2 \times 10^5 \times 10^{-23} \times 10^6$$

$$= 2 \times 10^{-12}$$

$$\therefore P_{n_o} = 2 \times 10^{-12} \text{ W}$$

Therefore,

$$P_n = 0.78 \times P_{n_o}$$
$$= 0.78 \times 2 \times 10^{-12}$$
$$= 1.56 \times 10^{-12}$$
$$\therefore P_n = 1.56 \times 10^{-12}$$

Alternatively,

$$NF = 10 \log(F)$$
$$F = \text{inv.log}(0.25)$$
$$\therefore F = 1.78$$
$$F = \frac{T_e}{T_0} + 1$$
$$T_e = T_0(F - 1)$$
$$= T_0(1.78 - 1)$$
$$= 0.78\, T_0$$
$$T_e = 0.78\, T_0$$
$$= 0.78 \times 290$$
$$= 226.2 \text{ K}$$
$$\therefore T_e = 226.2 \text{ K}$$

Thus,

$$P_n = K \cdot T_e \cdot BW$$
$$= 1.38 \times 10^{-23} \times 226.2 \times 500 \times 10^6$$
$$= 1.56 \times 10^{-12}$$

Therefore, the internal noise power $P_n = 1.56 \times 10^{-12}$ W.

EXAMPLE 1.11

A low-noise amplifier operates with a bandwidth of 500 MHz and a noise figure of 1.78 dB. Compute the equivalent noise temperature and noise power.

Solution

$$NF = 1.78 \text{ dB}$$
$$BW = 500 \text{ MHz}$$
$$NF = 10 \log\left(\frac{T_e}{T_0} + 1\right)$$
$$1.78 = 10 \log\left(\frac{T_e}{T_0} + 1\right)$$

$$\frac{T_e}{T_0} + 1 = \text{antilog}(0.78)$$

$$= 0.506$$

$$T_e = 0.506 \times T_0$$

$$= 146 \text{ K}$$

$$\therefore T_e = 146 \text{ K}$$

$$P_n = K \cdot T_e \cdot \text{BW}$$

$$= 1.38 \times 10^{-23} \times 146.2 \times 500 \times 10^6$$

$$= 1 \times 10^5 \times 10^6 \times 10^{-23}$$

$$= 1 \times 10^{-12}$$

Therefore, $P_n = 1 \times 10^{-12}$ W.

1.6.1 Attenuator Noise Figure (NF)

Attenuators and signal attenuation are integral parts of all communication systems. Therefore, knowledge of the **attenuator noise figure** (NF) is very important for an overall system performance evaluation. Attenuators, like amplifiers, are able to operate at all frequency bands, from base bands to RF bands. By definition, an attenuator is a network whose input power to output power ratio is always larger than one.

$$L = \frac{P_i}{P_o} = \frac{S_i}{S_o} \tag{1.30}$$

where
$L = $ losses $L < 1$
$P_i = $ input power (W)
$P_o = $ output power (W)
$S_i = $ signal input (W)
$S_o = $ signal output (W)

In decibels

$$L_{\text{dB}} = 10 \log\left(\frac{P_i}{P_o}\right) \text{dB} \tag{1.31}$$

From the basic definition of noise figure, we have:

$$\text{NF} = \frac{S_i/N_i}{S_o/N_o} = \frac{S_i}{S_o}\frac{N_i}{N_o} = \frac{S_i}{S_o} \times \frac{N_i}{N_o}$$

where S_i = input signal (W)
N_i = input noise (W)
S_o = output signal (W)
N_o = output noise (W).

Since $\dfrac{S_i}{S_o} = L$, then

$$NF = L \cdot \frac{N_o}{N_i} \tag{1.32}$$

The operating principle of any attenuator is that the input and output impedance are equal $(Z_i = Z_o)$ and, assuming an equal internal impedance, the input noise power (N_i) is equal to the output noise power (N_o) (Figure 1.9). Therefore, Eq. (1.32) becomes

$$NF = L \tag{1.33}$$

Equation (1.33) indicates that the noise power (NF) of an attenuator is equal to the attenuation (L). In dB, Eq. (1.33) becomes

$$NF_{dB} = L_{dB} \tag{1.34}$$

1.6.2 Mixer Noise Figure

The third essential circuit in any communications receiver is that of the mixer; therefore, a knowledge of the **mixer noise figure** is also important. Two basic methods are employed

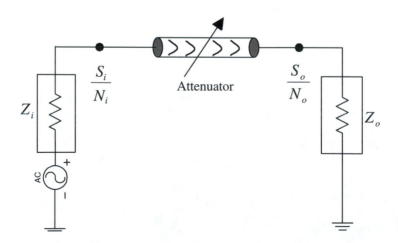

FIGURE 1.9   Attenuator.

in the design mixer circuits: single diode (**single-ended mixer**) and double diode (**balanced mixer**).

A mixer is a three-port network, consisting of two inputs and one output (Figure 1.10). By accepting a RF signal and a local oscillator signal from the respective input ports, an intermediate (IF) signal is generated at the output port. In essence, the mixer circuit has translated an RF signal into an IF signal. During this translation, or frequency conversion process, signal loss is observed. This signal loss is specified as conversion loss (L_c) and is an essential parameter of the mixer circuit. Another characteristic parameter of the mixer circuit is the mixer diode noise temperature factor (t_d).

Mixer Conversion Loss (L_c)

Mixer conversion loss is the power ratio of the RF input signal to IF output signal:

$$L_c = \frac{\text{power input (RF)}}{\text{power output (IF)}} \tag{1.35}$$

in dB,

$$L_{c_{dB}} = 10 \log\left(\frac{P_i}{P_o}\right) \tag{1.36}$$

Diode Noise Temperature Factor (t_d)

Diode noise temperature factor is defined as the ratio of diode noise temperature (T_d) to the ambient noise temperature T_0.

$$t_d = \frac{T_d}{T_0} \tag{1.37}$$

Therefore, mixer noise figure can be calculated on the basis of noise figure definition and mixer circuit parameters as follows:

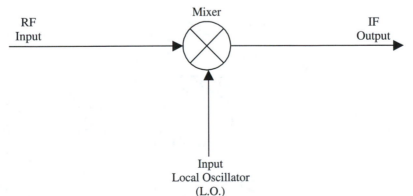

FIGURE 1.10
Single-ended mixer.

$$NF = \frac{S_i}{N_i} \times \frac{N_o}{S_o}$$

$$NF = \frac{S_i}{K \cdot T_0 \cdot BW} \times \frac{N_o}{S_o}$$

where $\quad N_o = K \cdot T_0 \cdot BW \cdot t_d \cdot \dfrac{L_c}{L_c}$ $\qquad\qquad$ **(1.38)**

$$S_o = \frac{S_i}{L_c}$$

Or

$$NF = \frac{S_i}{K \cdot T_0 \cdot BW} \times \frac{K \cdot T_0 \cdot BW \cdot t_d \cdot L_c / L_c}{(S_i/L_c)} = t_d \cdot L_c \qquad \textbf{(1.39)}$$

Therefore,

$$NF = t_d \cdot L_c \qquad\qquad\qquad \textbf{(1.40)}$$

In dB,

$$NF_{dB} = 10 \log(t_d \cdot L_c) \qquad\qquad \textbf{(1.41)}$$

The mixer noise figure is therefore equal to the diode noise temperature factor amplified by the mixer conversion loss.

EXAMPLE 1.12

Calculate the mixer noise figure having a conversion loss (L_c) of 3 and diode mixer temperature of 400 K.

Solution

$$L_c = 3$$
$$T_d = 400 \text{ K}$$
$$T_0 = 290 \text{ K}$$
$$t_d = \frac{T_d}{T_0} = \frac{400}{290} = 1.38$$
$$\therefore t_d = 1.38$$
$$NF_{dB} = 10 \log(t_d \times L_c)$$
$$= 10 \log(1.38 + 3)$$
$$= 10 \log(4.38)$$

Therefore mixer noise figure: NF = 6.4 dB.

1.6.3 Receiver Noise Figure (NF)

In any communication microwave link, the receiver section is a very important component. The performance of the receiver, as a critical element in that system, can be expressed in terms of its noise figure (NF). A closer look at the receiver function in reference to its signal-to-noise ratio reveals the following: First, the receiver performs the frequency conversion function. That is, the incoming (RF) signal after amplification performed by a separate unit, the low-noise amplifier, is down-converted to an IF signal by a single- or double-diode conversion process (mixer). This mixing process introduces two new elements: signal loss and noise.

Noise is expressed as the mixer diode noise temperature (T_d). For double frequency conversion where two mixers are involved, only the diode with the high noise temperature is taken into account. It was mentioned previously that since the incoming RF signal is very weak, it requires a fair amount of amplification along its linear section. This signal amplification is achieved by a number of amplifiers. These amplifiers not only provide signal amplification but also contribute significantly to the overall increase of internal system noise power. It is sometimes necessary to use attenuations in communication receivers for carrier-level settings. These passive components directly contribute to the overall receiver noise power incorporated into the receiver noise figure calculations. A complete study of the FM communications receiver and the relevant noise figure calculations will be given in Chapter 4, which deals with digital communication links.

1.7 RELATIONSHIPS BETWEEN THE CARRIER-TO-NOISE, CARRIER–TO–NOISE POWER DENSITY, AND BIT ENERGY–TO–NOISE POWER DENSITY RATIOS

In digital communication system design, bit energy (E_b) is required in order to establish the E_b/N_o ratio. **Bit energy** is given by:

$$E_b = \frac{C}{f_b} \qquad (1.42)$$

Since $f_b = 1/T_b$, then:

$$E_b = C \cdot T_b \qquad (1.43)$$

where E_b = energy of the bit in joules
f_b = **system bit rate**
T_b = **bit-time duration**

Dividing both parts of this equation by N_o (**noise power density**), we have,

$$\frac{E_b}{N_o} = \frac{C}{N_o} \times \frac{1}{f_b} \quad \text{in (W/J/s)} \cdot \text{(1/b/s)} \qquad (1.44)$$

In dB:

$$\frac{E_b}{N_o} = \frac{C}{N_o} \times \frac{1}{f_b}$$

If the receiver **noise bandwidth** in Hertz is BW and the total noise power measured at the receiver output is N, then

$$N_o = \frac{N(\text{W})}{\text{BW}(\text{Hz})}$$

where   BW = receiver noise bandwidth (Hz)

               N = receiver noise power at the output (W)

Since $N_o = \dfrac{N}{\text{BW}}$ and $N = KT_e\text{BW}$:

$$N_o = \frac{KT_e\text{BW}}{\text{BW}} = KT_e$$
$$N_o = KT_e \tag{1.45}$$

It is evident from this that the basic difference between noise power density (N_o) and noise power (N) is the fact that noise power is the product of noise density and system bandwidth. In dB, both expressions are given by:

$$N_{o_{\text{dB}}} = 10\log(K) + 10\log(T_e)$$

where $K = 1.38 \times 10^{-23}$ J-K

           T_e = equivalent noise temperature (K)

or

$$N_{o_{\text{dB}}} = -228.6_{\text{dBW}} + 10\log(T_e)$$
$$N = N_o \cdot \text{BW}$$

(in dB)

$$N_{\text{dB}} = N_{o_{\text{dB}}} + 10\log\text{BW} = -228.6 + 10\log(T_e) + 10\log\text{BW dB} \tag{1.46}$$

Substituting into E_b/N_o, we have

$$\frac{E_b}{N_o} = \frac{C}{N_o} \times \frac{1}{f_b} = \frac{C}{N/\text{BW}} \times \frac{1}{f_b} = \frac{C \cdot \text{BW}}{N \cdot f_b}$$

$$\frac{E_b}{N_o} = \frac{C \cdot \text{BW}}{N \cdot f_b} \tag{1.47}$$

In dB:

$$\frac{E_b}{N_{o_{dB}}} = \frac{C}{N_{dB}} + 10 \log(\text{BW}) - 10 \log(f_b)$$

If the noise bandwidth (BW) is equal to the bit rate (f_b) $(\text{BW} = f_b)$, then,

$$\frac{E_b}{N_o} = \frac{C}{N_o}$$

The ratio C/N_o relates the carrier power to the noise power per 1-Hz bandwidth. Convert C/N_o to C/N (carrier power to noise power in defined bandwidth)
Since

$$\frac{E_b}{N_o} = \frac{C \cdot \text{BW}}{N \cdot f_b} \quad \text{and} \quad \frac{E_b}{N_o} = \frac{C}{N} \times \frac{1}{f_b}$$

Then

$$\frac{C \cdot \text{BW}}{N \cdot f_b} = \frac{C}{N_o} \times \frac{1}{f_b} \quad \text{or} \quad \frac{C \cdot \text{BW}}{N} = \frac{C}{N_o}$$

Divide both sides by BW:

$$\frac{C \cdot \text{BW}}{N \cdot \text{BW}} = \frac{C}{N_o} \times \frac{1}{\text{BW}}$$

Therefore

$$\frac{C}{N} = \frac{C}{N_o} \times \frac{1}{\text{BW}}$$

In dB:

$$\frac{C}{N_{dB}} = \frac{C}{N_{o_{dB}}} - 10 \log(BW) \text{ dB} \tag{1.48}$$

In terms of receiver signal level, front-end noise figure, and system bit rate:

$$E_b = P_{r,min} - [10 \log(f_b) = 10 \log(KT) + NF]$$

1.8 FIGURE OF MERIT (G/T$_{sys}$ dB/K)

The figure of merit expresses the ratio of the receiver antenna gain (G) of a satellite communications receiver to the total system noise temperature (T_{syst}).

$$\frac{G}{T_{syst}} \text{ dB/K} \tag{1.49}$$

QUESTIONS

1. Define *electronic noise*.
2. Describe the difference between deterministic and nondeterministic signals.
3. List all the terrestrial and extraterrestrial noise sources.
4. What is the difference between thermal and nonthermal emission?
5. Name the sources of electronic noise that maintain a minimum 3 K.
6. List all the types of internal noise.
7. Define thermal noise.
8. Describe the relationship between equivalent noise temperature and noise power.
9. What is the relationship between the internal noise of electronic equipment and the noise figure?
10. Define mixer-conversion loss.

PROBLEMS

1. Compute the shot-noise power spectral density and noise voltage of a semiconductor device with a saturation current of 10 mA.

2. Determine the noise power (P_n) and noise voltage (V_n) of a circuit operating at a bandwidth of 5 MHz and temperature of 420 K with a load of 1 MΩ. Assume an ideal noise source.

3. A circuit generates a noise voltage of 3 μW. If the noise temperature is 300 K and the load resistance is 500 Ω, compute the operating bandwidth.

4. A microwave low-noise-amplifier (LNA) is operating with a bandwidth of 36 MHz. If the noise power (RTI) is 2.5×10^{-13} W, determine its equivalent noise temperature (T_e).

5. Compute the signal-to-noise ratio (SNR) in dB at the input of communications equipment with a signal strength of 20 nW and a noise level of 100 pW.

6. A communications receiver operates with a bandwidth of 36 MHz and equivalent noise temperature of 120 K. Compute (a) the noise power (RTI) and (b) the noise power (NF) in dB.

7. A microwave amplifier operating with a bandwidth of 500 MHz is listed with a noise figure of 5.5 dB. Determine (a) the noise power (RTI) and (b) the carrier signal required in order to maintain at its input a carrier-to-noise ratio of 15 dB.

8. Compute the signal power at the input of an attenuator with a noise figure of 3 dB and an output signal power of 10 nW.

9. Determine the conversion loss of a mixer circuit with a noise figure of 4.5 dB and a noise temperature of 320 K.

2

Voice Signal Digitization

Objectives

- Identify the need for voice signal digitization.
- Describe the various methods of voice signal digitization.
- Derive the signal–to–quantization noise ratio.
- Compare the advantages and disadvantages of PCM and DM.
- Identify the various digital hierarchies based on the CCITT recommendations.
- Define jitter and describe its impact on system performance.

Key Terms

Harmonic distortion
Digitizers
Pulse-amplitude
  modulation
Nyquist theorem
Sampling rate
Pulse-position modulation
Pulse-width modulation
Pulse-code modulation
Sample-and-hold circuit
Quantizer

Encoder
Decoder
Step size
Quantization error power
Linear quantization
Nonlinear quantization
Companding
The A-law
The μ-law
Slope overload
Granule noise

ADM
TDM
CCITT
Frame time
Bit rate
Jitter
ISI
Unipolar
Bipolar
Clock recovery
Equalizer

INTRODUCTION

Through the study of digital communications, it has become evident that the most important application of this fairly new technology is in the area of long-distance transmission of voice channels. This technique of digital voice transmission must satisfy voice-quality specifications at the receiver end. Voice signal quality can be expressed in terms of signal-to-noise ratio (SNR), **harmonic distortion,** and frequency response.

The international communications committee (**CCITT**) has classified these voice signal specifications as follows.

Signal-to-noise ratio better than 30 dB

Harmonic distortion better than 26 dB

Frequency response of 300 Hz–3.4 KHz

It was mentioned in the introduction of this text that analog transmission is very susceptible to noise and interference, whereas digital transmission is resistive to noise. During the last 30 years, this fact has generated considerable interest among communications experts from the industrialized nations.

Voice digitization is performed by voice **digitizers.** These circuits are classified in terms of their application as (a) storing, (b) transmitting, and (c) switching digitizers. Transmitting voice digitizers can further be classified as wideband and narrow band. For standard telephone applications, wideband digitizers are used; for low-data-rate transmission, narrow-band digitizers are utilized.

The most commonly used technique for voice signal digitization is **pulse-code modulation** (PCM).

Before discussing PCM, we must first discuss **pulse-amplitude modulation** (PAM).

2.1 PULSE-AMPLITUDE MODULATION (PAM)

2.2.1 The Sampling Theorem

The process by which an analog signal of specified frequency is uniformly sampled at discrete intervals of time is called the "sampling process," the mathematical interpretation of such a process is called the "sampling theorem." The result is the generation of discrete output signals with amplitudes proportional to the input signals. Figure 2.1 shows an analog input signal, sampled at specific intervals of time, and an output signal composed of uniform duration pulses related to clock pulses and varying amplitudes proportional to the input analog signal amplitude.

In 1933, Henry Nyquist developed the mathematical relationship between the analog input signal and the sampling frequency as follows:

$$f_s \geq 2\,\text{BW} \tag{2.1}$$

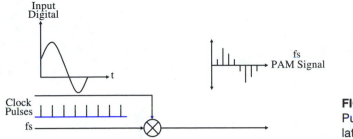

FIGURE 2.1.
Pulse-amplitude modulation (PAM).

where   BW = bandwidth of the sampled signal (Hz)
         f_s = clock frequency (Hz)

This relationship of Eq. (2.1), known as the **Nyquist Theorem,** determines the minimum required clock frequency necessary for sampling a continuous time-varying waveform without the possibility of distortion at the reproduction stage.

The relationship of the PAM signal to information (analog) and sampling signals is given by

$$f_s(t) = f(s) \cdot f(t) \tag{2.2}$$

Ideally, $f(s)$ is a pulse with an infinitesimal width. In practice, a non-ideal signal of constant amplitude and finite width is utilized. The mathematical derivation of the sampling theorem is rather involved and is beyond the scope of this text. Only the interpretation of such results will be examined.

1. If a band-limited information signal with bandwidth BW is sampled with a sampling frequency larger than 2 BW, the sample contains all the analog signal information and can be recovered through a filter without distortion. (Figure 2.2)

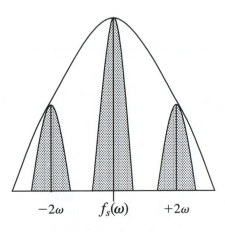

-2ω $f_s(\omega)$ $+2\omega$

FIGURE 2.2. PAM signal spectral representation: sampled at rate $f(s) > 2f(t)$.

Voice Signal Digitization

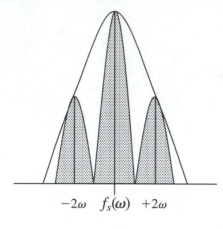

FIGURE 2.3. PAM signal spectral representation: sampled at rate $f(s) = 2f(t)$.

-2ω $f_s(\omega)$ $+2\omega$

2. If the sampling frequency is twice the bandwidth of the input signal, this input signal can also be recovered through a filter. This sampling frequency signal is absolutely band limited (Figure 2.3).
3. For a sampling frequency less than twice the analog input signal ($f_s < 2BW$), the signal's base band will overlap and recovery of the input signal is practically impossible. This phenomenon is referred to as "foldover distortion" or "aliasing" (Figure 2.4).

2.1.2 Sampling Circuits

Sampling analog signals is the first step toward voice signal digitization. Figure 2.5 illustrates the block diagram of a **sample-and-hold** (S/H) circuit. In Figure 2.5 the input sig-

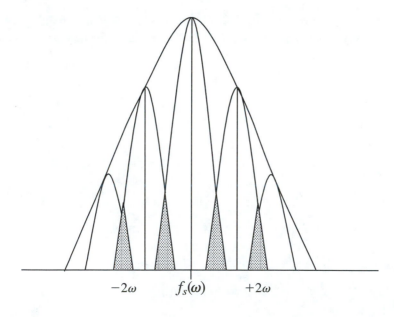

-2ω $f_s(\omega)$ $+2\omega$

FIGURE 2.4. Signal spectral representation: sampled at rate $f(s) < 2f(t)$.

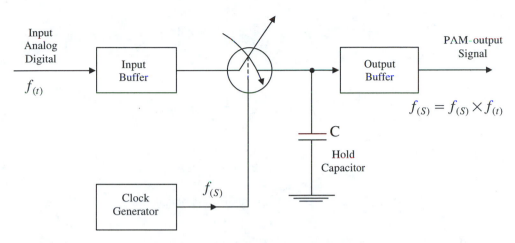

FIGURE 2.5. Sampling circuit block diagram.

nal is band limited by the input buffer and then periodically sampled by the electronic switch driven by the sampling clock. The generated output waveform is shown in Figure 2.6(c) and (d) with the capacitor disconnected; the PAM output waveform is shown in Figure 2.6(c). This type of signal is called a return-to-zero (RTZ) signal and is used mainly for multiplexing.

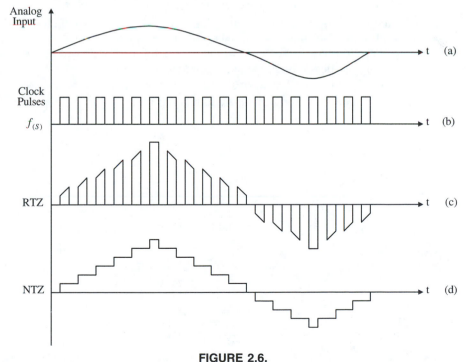

FIGURE 2.6.

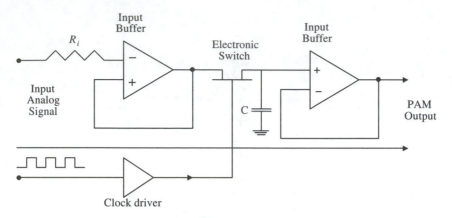

FIGURE 2.7. Open-loop sample-and-hold circuit.

With the capacitor (C) connected to the circuit, the sample is held across the capacitor until the next sample pulse arrives. The increase or decrease of the amplitude of the individual samples related to the corresponding increase or decrease of the input analog signal will adjust the charge across the capacitor proportionally. The output waveform is shown in Figure 2.6(d). This type of signal is referred to as a non-return-to-zero signal (NRZ) and is used primarily for coding.

There are several types of sample and hold (S/H) circuits, classified as closed-loop and open-loop configuration circuits. Figure 2.7 shows an open-loop sample-and-hold circuit. The input buffer op-amp has a very high input impedance and a very fast response time, while the field-effect-transistor switch exhibits a very low impedance in the ON state, and a very high impedance in the OFF state. The hold-capacitor connected at the input of the high-impedance op-amp contributes to the generation of the non-return-to-zero output waveform (NRZ). For better linearity and accuracy, a modified version of the circuit of Figure 2.7 is shown in Figure 2.8. In this circuit, the hold capacitor is used to form an integrator circuit with the output buffer operational amplifier.

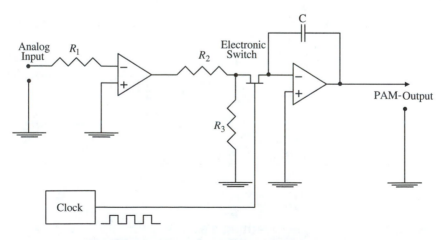

FIGURE 2.8. Closed-loop sample-and-hold circuit.

2.2 PULSE-POSITION MODULATION (PPM)

Another form of voice signal digitization is pulse-position modulation (PPM). In pulse-amplitude modulation, the amplitude of the sampled signal is made proportional to the analog input signal. In PPM, the amplitude of the modulated signal is maintained at a constant level while its pulse position is shifted at a rate proportional to the rate of change of the input signal Figure 2.9.

2.3 PULSE-WIDTH MODULATION (PMW)

The pulse-width modulation (PWM) process generates an output signal with constant amplitude and pulse widths proportional to the input signal amplitude rate of change (Figure 2.10). The circuit diagram of a pulse-position and pulse-width modulation scheme is illustrated in Figure 2.11. The analog signal applied at the voltage-controlled oscillator input disturbs the VCO's locking condition so that the output of the phase detector is shifted up and down in an attempt to establish equilibrium. This change in the phase detector output implies that there is a difference between the VCO signal and the input signal, or that the voltage-controlled oscillator output is phase shifted at a rate proportional to the amplitude of the input signal. If this PPM signal and the input analog signal are fed in to an exclusive-OR gate, a high output will be produced only when either input is at logic one. A comparison is always made between the PPM and the analog input signal, so the width of the signal at the output of the exclusive-OR gate will be proportional to the input analog signal.

2.4 PULSE-CODE MODULATION (PCM)

The most commonly used technique for voice-signal digitization is pulse code modulation. **Pulse-code modulation** is the process whereby an analog signal is converted to digital form in order to be transmitted by digital means. The block diagram of a simple PCM transmitter system is shown in Figure 2.12.

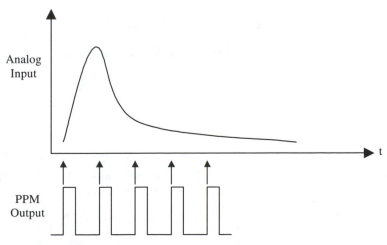

FIGURE 2.9. Pulse-position modulation input and output waveforms.

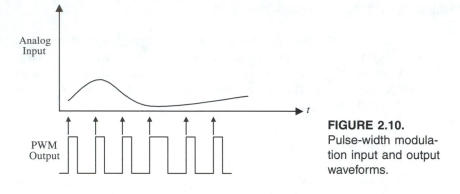

FIGURE 2.10.
Pulse-width modulation input and output waveforms.

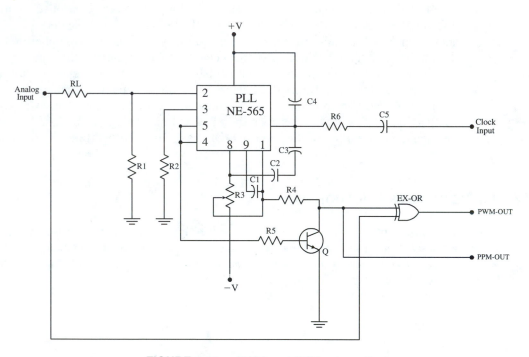

FIGURE 2.11. PWM and PPM schematic.

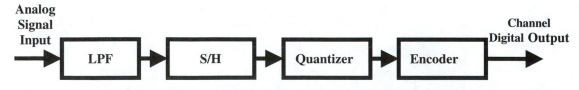

FIGURE 2.12. Pulse-code-modulation block diagram (transmitter).

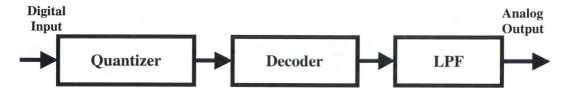

FIGURE 2.13. Pulse-code-modulation block diagram (Receiver).

In order to satisfy Nyquist sampling criteria, the analog signal at the input is band limited by the low-pass-filter, then sampled with a **sampling rate** equal to or larger than twice its bandwidth. After quantization, the samples are encoded through an encoder to a predetermined pattern before transmission. In essence, the quantizer and the encoder circuits perform the fundamental function of analog-to-digital conversion. At the receiver end, the reverse process is performed (Figure 2.13). The digitally coded analog signals are quantized, decoded and, through a low-pass filter, reproduced to their original analog form.

The basic design concern of PCM systems is to minimize the impact of channel noise at the final stage of the reconstruction process. This is achieved mainly by the **quantizer** circuit at the receiver end. The quantizer's function is to determine whether the incoming digital signal has a voltage level corresponding to binary zero, or binary one. If this recognition has been achieved satisfactorily, then the level of external noise at the input of the quantizer circuit will have no significance. The other two important parameters determining the overall accuracy of the regenerated analog signal, are quantization noise and bit error rate (BER).

2.4.1 Quantization

It was mentioned previously that an analog signal is quantized in order to be easily separated from the channel noise, and then reconstructed to its original form. This quantization process is shown in Figure 2.14. The analog signal (V) is confined between a minimum and maximum value, and then divided into a number of equal segments. The magnitude of these segments is called **step size** (S), and is calculated by dividing the peak-to-peak amplitude of the analog signal by the number of steps. Therefore,

$$\text{Step Size} \quad S = \frac{V_{\max} - V_{\min}}{M} \qquad \textbf{(2.3)}$$

where M = number of steps

From Figure 2.14, it is evident that the quantized signal closely resembles the original signal. The difference between the original and the quantized signal is referred to as "quantization noise." It is also evident that a decrease in the number of quantization steps will

Voice Signal Digitization

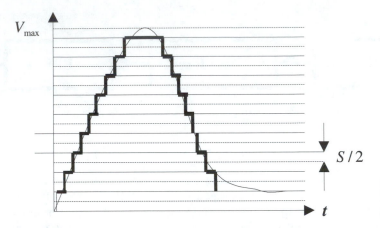

FIGURE 2.14.
Representation of the quantization process.

increase the quantization error, whereas an increase in the quantization steps will correspondingly decrease the quantization error.

At the receiver end, the received quantized levels will differ from the transmitted levels due to the transmission channel additive noise. If this noise has a magnitude less than half of the step size, the quantizer will generate an exact duplicate of the transmitted levels disregarding channel noise. On the other hand, if the noise level is equal to or higher than half the step size, then the probability of error is quite significant.

Quantization error power (P_{qn})

By definition, quantization error is the difference between the original and the quantized signal. This quantization error or quantization noise is calculated as follows: If the difference between the amplitude of the original analog signal and the quantized signal is e (error), then the mean square value of e is $(\bar{e})^2$ and is expressed by:

$$(\bar{e})^2 = \int_{-\infty}^{+\infty} e^2 \cdot P(e) \cdot de \qquad (2.4)$$

where $P(e)$ = probability density function

The error (e) must exist between $-(S/2) < e < +(S/2)$ while the probability-density function takes values between 0 and $1/S$. Therefore Eq. (2.4) becomes

$$(\bar{e})^2 = \int_{-\infty}^{+\infty} e^2 \cdot P(e) \cdot de = \int_{-1/S}^{+1/S} \frac{1}{S} e^2 \cdot de = \frac{1}{S} \int_{-S/2}^{+S/2} e^2 \, de = \frac{S^2}{12}$$

Therefore,

$$\text{Mean Square Value of Error} \quad (\bar{e})^2 = \frac{S^2}{12} \qquad (2.5)$$

Equation 2.5 represents the noise power of the quantized signal, and it can be represented as

$$P_n = \frac{S^2}{12} \tag{2.6}$$

Output Signal Power (P_{so})

A very important parameter determining the efficiency of a PCM system is the signal–to–quantization noise ratio. The quantization noise power was given by Eq. 2.6 as $S^2/12$. The output signal power (P_{so}) can be calculated as follows: Assuming an analog signal is sampled with a rate equal to or larger than twice its bandwidth, and if the number of quantization levels is (M), then the signal power output is expressed by:

$$P_{so} = \frac{1}{M}[0^2 + S^2 + 2S^2 + 2S^2 + \cdots + (M-1)S^2] = \frac{S^2}{M}\left(\frac{M(M-1)(2M+1)}{6}\right)$$

Assuming M is greater than one:

$$P_{so} = \frac{M^2 S^2}{3} \tag{2.7}$$

Equation (2.7) indicates that the power of the PCM output signal is proportional to the square of its quantization levels and step size.

Signal–to–Quantization Error Ratio

From Eqs. (2.6) and (2.7), the signal–to–quantization error ratio of a PCM system can be established as:

$$\frac{P_{so}}{P_{qn}} = \frac{M^2 \cdot S^2/3}{S^2/12} = \frac{12M^2 \cdot S^2}{3S^2}$$

$$\therefore \frac{P_{so}}{P_{qn}} = 4M^2 \tag{2.8}$$

Each level is quantized by a number of bits. Therefore M is equal to 2^N:

$$M = 2^N \tag{2.9}$$

where M = number of levels
N = number of bits per level

Thus,

$$\frac{P_{so}}{P_{qn}} = 4M^2 = 4(2^N)^2$$

In dB,

$$\frac{P_{so}}{P_{qn_{dB}}} = 10 \log[4(2^N)^2] = 10 \log 4 + 10 \log(2^{2N})$$
$$= 6 + 2N \cdot 10 \log 2 = 6 + 6N$$
$$= 6(N + 1)$$

Therefore,

$$\frac{P_{so}}{P_{qn_{dB}}} = 6(N + 1) \qquad (2.10)$$

Equation 2.10 indicates that, for an analog signal quantized and coded with N bits, the signal–to–quantization error ratio is equal to $6(N + 1)$ provided the analog signal does not exceed its quantization boundary.

EXAMPLE 2.1

Determine the signal–to–quantization error ratio of a PCM system using 8-bit words for quantization while not exceeding its predetermined voltage boundaries.

Solution
In dB,

$$\frac{P_{so}}{P_{qn_{dB}}} = 6(N + 1) = 6(8 + 1) = 6 \times 9 = 54$$

$$\therefore \frac{P_{so}}{P_{qn}} = 54 \text{ dB}$$

It is evident that for an 8-bit quantization, a theoretical 54-dB signal power–to–quantization noise ratio is maintained. It is also evident that for a smaller input analog signal, the P_{so}/P_{qn} is reduced, whereas for a larger input analog signal the P_{so}/P_{qn} is increased correspondingly. This variation of the P_{so}/P_{qn} does not comply with CCITT standards requiring that voice transmission must maintain a minimum P_{so}/P_{qn} equal to 30 dB regardless of the input signal amplitude variations. To rectify this problem, the concept of nonlinear quantization is utilized.

Nonlinear Quantization

During the quantization process, it was shown that an analog input signal confined within its predetermined boundaries reflects a quantization error of $\pm S/2$, where (S) is equal to step size. If the analog signal exceeds these established boundaries, the quantization error will increase significantly. Again, if the input analog signal is smaller and therefore does not swing fully between its boundaries, the error will be smaller. Referring back to the quantization error equation, $P_{qn} = S^2/12$, it is obvious that a decrease in the step size (S) will decrease the quantization error. But such a decrease in the step-size is the result of a corresponding increase in the number of sample levels (M), which ultimately leads to an increase in the PCM baseband bandwidth requirements.

Another scheme capable of achieving a better signal–to–quantization noise ratio without violating baseband bandwidth restrictions samples the smaller input signals with a higher sampling rate while employing smaller sampling rates for larger input analog signals.

The most effective method of achieving a better P_{so}/P_{qn} with a constant rate of sampling while maintaining PCM baseband bandwidth restrictions, is that of **nonlinear quantization.** In such a scheme, the input analog signal is compressed at the higher amplitude levels and expanded at the lower amplitude levels before quantization. Since the input analog signal is compressed and expanded, the scheme is called **companding,** and the circuits performing the companding functions are called "companders." There are three methods of companding: (a) instantaneous companding, (b) nonuniform companding, and (c) uniform distribution.

Instantaneous Companding

Instantaneous companding is the process whereby the multiplexed signal is processed through a nonlinear circuit prior to the encoding process, while at the receiver end the reverse process is applied before demultiplexing (Figure 2.15).

Nonuniform Companding

In nonuniform companding, the compression and expansion circuitry is incorporated into the encoding and decoding processes and can be implemented by either analog or digital means (Figure 2.16).

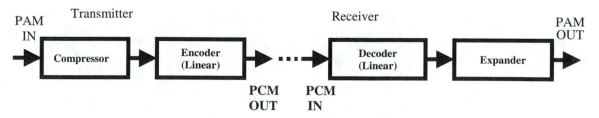

FIGURE 2.15. Instantaneous companding block diagram.

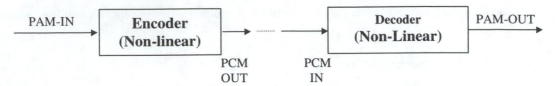

FIGURE 2.16. Nonuniform companding block diagram.

Uniform Distribution

The uniform distribution process incorporates a linear **encoder** and a digital compressor at the transmitter end, and a digital expander and linear **decoder** at the receiver end (Figure 2.17). The digital compressor reduces the high-resolution code into a lower-resolution uniform code with an acceptable degree of accuracy, while the opposite process takes place at the receiver end.

Companding Laws

The International Communications Committee (CCITT) has recommended two companding laws: (a) the European **A-law** and (b) the North American **μ−law.** These two companding laws reflect the scale of nonuniformity in reference to the amplitude of the input analog signal. In order to understand the two international companding laws, the following is necessary. The characteristic slope of a quantized signal is given by

$$\frac{V_0}{V_i} = \frac{dy}{dx} = \frac{1}{Kx} \tag{2.11}$$

Integrating Eq. (2.11),

$$y = \frac{1}{K}\int \frac{1}{x}\,dx = \frac{1}{K}\ln|x| + C$$

$$\therefore y = \frac{1}{K}\ln|x| + C \tag{2.12}$$

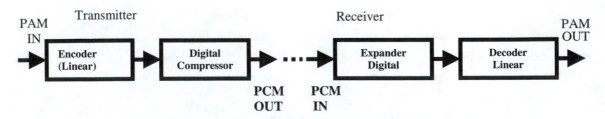

FIGURE 2.17. Uniform distribution block diagram.

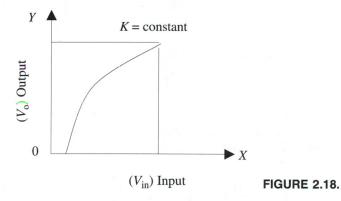

K = constant

Y

(V_o) Output

0

X

(V_{in}) Input

FIGURE 2.18.

Equation (2.12) is illustrated by the graph of Figure 2.18.

The A-Law   By interpolating the origin and the logarithmic law through a linear slope, it is possible to obtain logarithmic quantization for large signals and **linear quantization** for smaller signals. The mathematical representation of the A-law (Figure 2.19) is given by:

$$y = \sin(x)\frac{Ax}{1 + \ln(A)} \quad \text{for} \quad 0 \le X \le \frac{1}{A} \qquad \textbf{(2.13)}$$

where A = constant
x = normalized input amplitude with range (0, 1)
y = normalized output amplitude with range (0, 1)

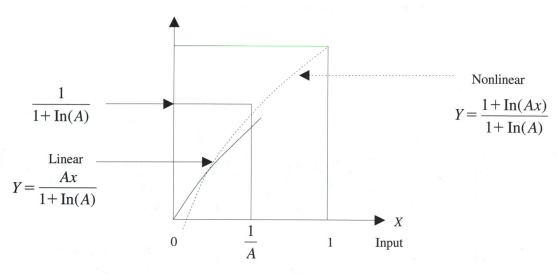

$\dfrac{1}{1 + \ln(A)}$

Nonlinear

$$Y = \frac{1 + \ln(Ax)}{1 + \ln(A)}$$

Linear

$$Y = \frac{Ax}{1 + \ln(A)}$$

0        $\dfrac{1}{A}$ 1   Input

X

FIGURE 2.19.   A-law.

Voice Signal Digitization

or:

$$y = \frac{1 + \ln(Ax)}{1 + \ln(A)}$$ (2.14)

Equation (2.13) is applicable to small-signal linear quantization; Eq. (2.14) is applicable to long-signal logarithmic quantization. Figure 2.20 shows A-law characteristic curves for different values of A. For linear P_0 (small signals),

$$y = \frac{Ax}{1 + \ln(A)} \qquad 0 \le X \le \frac{1}{A}$$

For logarithmic P_0 (large signals),

$$y = \frac{1 + \ln(Ax)}{1 + \ln(A)} \qquad \frac{1}{A} \le X \le +1$$

The μ-Law   The mathematical representation of the μ-law is given by:

$$y = \sin(x) = \frac{\ln[1 + \mu(x)]}{\ln(1 + \mu)}$$ (2.15)

where μ = constant
x = normalized input amplitude with range $(0, 1)$
y = normalized output with range $(0, 1)$

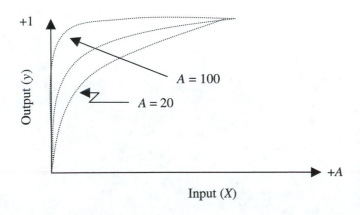

FIGURE 2.20.   A-law characteristic curves for different values of A.

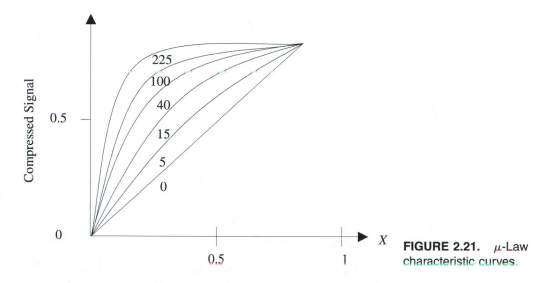

FIGURE 2.21. μ-Law characteristic curves.

The A-law and the μ-law express mathematically the mechanism through which a voice input signal is compressed before uniform encoding. Figure 2.21 shows the μ-law characteristic curves for different values of μ. At the receiver end and after decoding, the reverse process is implemented in order for the quantized signal to be restored to its original amplitude.

Uniform encoding and a companding constant (μ) equal to 100 were extensively used in early North American systems. However, today's systems have adapted the nonuniform encoding method for the generation of the compression characteristics. The segmentation of the companding characteristics for both μ-law and A-law systems contributed to the overall design simplification of nonuniform encoding systems.

2.5 DELTA MODULATION (DM)

The major disadvantage of transmitting analog signals though digital means is the large bandwidth requirements. This large bandwidth is directly related to the quantization process of the analog signals. If a quantization method is devised that is capable of reducing the overall number of quantization bits required while maintaining the same signal quality at the receiver end, then a smaller RF bandwidth will be required. The circuit capable of performing analog signal quantization with smaller bandwidth requirements is the delta-modulator (DM) circuit. This circuit operates on the principle that a binary output representing the most recent sampled amplitude will be determined on the basis of previous sampled amplitude levels (Figure 2.22). If the amplitude level to be quantized is smaller than the amplitude of the previous level, a zero bit will appear at the output of the circuit. If the amplitude level is larger than the previous level, then a one bit will appear at the output of the delta modulator.

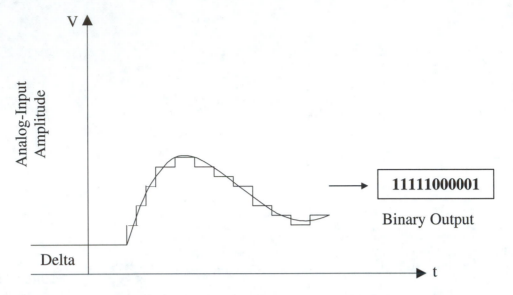

FIGURE 2.22. Delta-modulator wave representation.

Another advantage of delta-modulator circuits is the fact that they can easily perform the encoding and decoding function. Nevertheless, these encoded signals are very difficult to combine with other signals and therefore cannot be easily multiplexed. Figure 2.22 indicates the binary output 11111000001, which directly corresponds to the amplitude change of the input analog signal. The block diagram of a delta-modulator circuit is given in Figure 2.23. The circuit is composed of a low-pass filter at the input of the current to provide band-

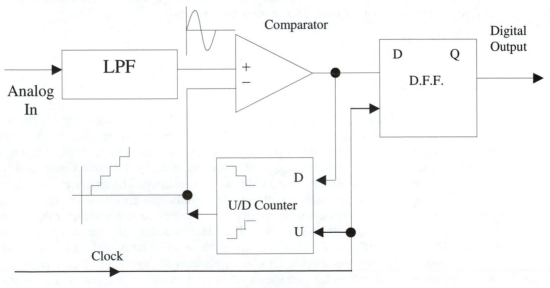

FIGURE 2.23.

width restriction for the input analog signal, a comparator, a staircase generator, and a clock generator.

The analog input at the comparator noninverting input is compared with the staircase output. If the analog input is larger than that of the staircase, the comparator output will go to (high) logic-one. If the input signal level is lower, then the output of the comparator will go to (low) logic-zero. Therefore, the digital output will follow the input analog signal variations.

This delta-modulator circuit suffers from two basic problems. One is slope overload, and the other is granule noise.

Slope Overload If the input analog signal amplitude changes faster than the speed of the modulator, then a **slope overload** will result. Figure 2.24 indicates that the amplitude of the input signal is higher than the rate of change of the delta modulator controlled by the feedback loop. For slope overload to be minimized or eliminated entirely, the product of the sampling step size and the sampling rate must be equal to or larger than the rate of change of the amplitude of the input analog signal.

Granule Noise **Granule noise** is defined as the difference between step size and sampled voltage (Figure 2.25). Granule noise can be minimized by an increase of the sampling rate, a decrease of the step size of the modulator, or both. These of course increase the overall circuit bandwidth requirements.

2.5.1 Adaptive Delta Modulator (ADM)

The overall performance of a delta modulator can be improved without a significant increase of the bandwidth requirements. This improvement of delta modulator performance

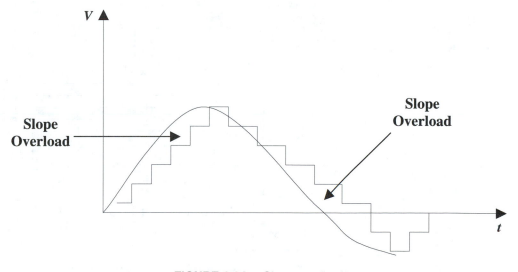

FIGURE 2.24. Slope overload.

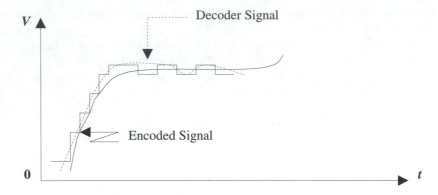

FIGURE 2.25. Granule noise.

can be achieved if the step size of the modulator does not remain constant, but rather changes (adapts) to the input signal amplitude variations. This method is known as continuously variable slope delta modulation (CVSDM). CVSDM modulators demonstrate a significant improvement in their dynamic range and noise levels over standard modulators (Figure 2.26).

Figure 2.26 basically illustrates a standard delta-modulator circuit with the addition of both a shift register capable of monitoring the last few bits of the modulator and a variable gain integrator. The monitoring of the few last bits through the shift register enables the circuit to determine whether these digits are all ones or all zeros. In either case, they are indicative of slope overload at the integrator output. The output of the logic control is fed into the input of the low-pass filter (LPF). The output of the filter is fed into the input

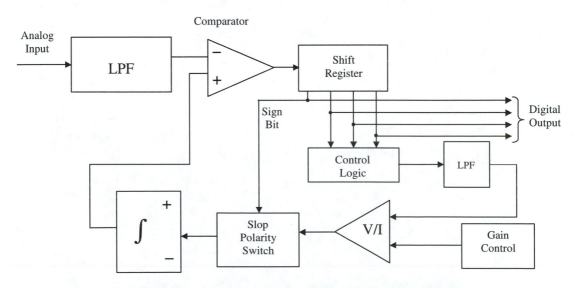

FIGURE 2.26. Adaptive-delta-modulator (ADM) block diagram.

of the voltage-to-current converter circuit, which in turn controls the integrator gain. The up/down control of the modulator is determined by the sign bit connected to the slope polarity switch. ADMs are capable of achieving dynamic ranges of 80 dB.

2.5.2 Quantization Noise in DM

The two fundamental parameters for a delta modulator are sampling rate, and step size. If the circuit is to follow the variations of faster input analog signals, either the sampling rate or the step size must be increased. An increase in the sampling rate will probably violate set bandwidth restrictions, while an increase in step size introduces quantization error (Figure 2.27). The delta-modulator quantization error is calculated as follows: Figure 2.27 shows an analog waveform and its quantized equivalent. The difference between the original and its approximate waveform is defined as the error waveform. Let $A(t)$ represent the analog input waveform, and $\overline{A}(t)$ represent the quantized approximate waveform. The difference between the original and its approximation is defined as the error waveform and is expressed by Eq. (2.16).

$$\delta(t) = A(t) - \overline{A}(t) \tag{2.16}$$

If $\delta(t)$ does not exceed the value of (S), a condition that eliminates the possibility of overload, then the probability density is given by

$$F_{(\delta)} = \frac{1}{2^S} \quad \text{where} \quad -S < \delta < +S \tag{2.17}$$

Therefore, the error power is given by

$$(\overline{\delta})^2(t) = \int_{-S}^{+S} \delta^2 F(\delta) \, d\delta = \frac{S^2}{3}$$

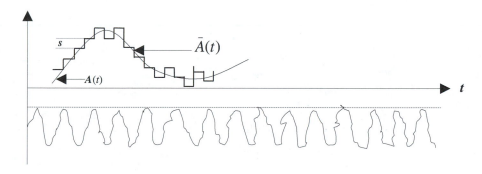

FIGURE 2.27.

$$(\bar{\delta})^2(t) = \frac{S^2}{3} \tag{2.18}$$

The utilization of a bandpass filter at the output of the delta modulator will eliminate a substantial amount of this noise power, provided the filter cutoff frequency (f_c) is equal to $1/\tau$, where τ is the duration of the bit.

$$f_c = \frac{1}{\tau} \tag{2.19}$$

Therefore, the quantization noise generated by a delta-modulator circuit at the output of the low-pass filter is given by the expression.

$$P_{qn} = \frac{S^2}{3} \times \frac{f_c}{f_b} \tag{2.20}$$

where $S^2/3$ = quantization noise power
            f_c = low-pass-filter cutoff frequency
            f_b = bit rate

If $f_c = f_b$, then

$$P_{qn} = \frac{S^2}{3} \quad \text{(quantization noise power)} \tag{2.21}$$

2.5.3 Delta Modulation Signal Power

Assuming the input analog signal of a delta modulator to be a sine wave $e(t)$,

$$e(t) = E \sin \omega_c t \tag{2.22}$$

where E = amplitude
            $\omega_c = 2f_c$

The signal power (P_{so}) is given by

$$P_{so} = e^{-2}(t) = \frac{E^2}{2} \tag{2.23}$$

Since the delta-modulation average slope approximation is S/f_b and the slope of the input signal is $E(\omega_c)$, then the amplitude of the input signal (E) must be

$$E(\omega c) = S \cdot f_b \tag{2.24}$$

$$E = \frac{S f_b}{\omega_c} \tag{2.25}$$

Substituting Eq. (2.25) into Eq. (2.23) we have

$$P_{so} = \frac{E^2}{2} = \frac{(S \cdot f_b)^2 / (\omega_c)^2}{2} = \frac{S^2 \cdot (f_b)^2}{2(\omega_c)^2}$$

$$P_{so} = \frac{S^2 \cdot (f_b)^2}{2(\omega_c)^2} \tag{2.26}$$

It is evident from Eq. 2.26 that delta-modulator signal power output is proportional to the square of the step size and the square of the bit rate, and inversely proportional to the square of the input signal frequency.

5.2.4 DM Signal–to–Quantization Noise Ratio

The DM signal–to–quantization noise ratio can be derived as follows. Since power signal output is expressed as

$$P_{so} = \frac{S^2 \cdot (f_b)^2}{2(\omega_c)^2}$$

and the quantization noise power is expressed as

$$P_{qn} = \frac{S^2 \cdot f_c}{3 \cdot f_b}$$

then

$$\frac{P_{so}}{P_{qn}} = \frac{S^2 \cdot (f_b)^2 / 2(\omega_c)^2}{\dfrac{S^2}{3} \times \dfrac{f_c}{f_b}} = \frac{3S^2 \cdot (f_b)^3}{2(\omega_c)^2 \cdot f_c} = \frac{3(f_b)^2}{2(\omega_c)^2 f_c} = \frac{3(f_b)^3}{2(2\pi) \cdot (f_c)^2 \cdot f_c}$$

$$\cong 0.04 \times \frac{(f_b)^3}{(f_c)^3}$$

or

$$\frac{P_{so}}{P_{qn}} \cong 0.04 \times \left(\frac{f_b}{f_c}\right)^3 \qquad\qquad (2.27)$$

EXAMPLE 2.2

Determine the signal–to–quantization noise ratio of a delta modulator with a bit rate of 64 Kb/s and an input signal bandwidth of 4 kHz.

Solution

$$f_c = 4 \text{ KHz}$$

$$f_b = 64 \text{ Kb/s}$$

$$\frac{P_{so}}{P_{qn}} = 0.04\left(\frac{f_b}{f_c}\right)^3 = 0.04 \times \frac{64 \times 10^3}{4 \times 10^3} = 163.84$$

In dB,

$$\frac{P_{so}}{P_{qn}} = 10 \log(163.84)$$

$$\therefore \frac{P_{so}}{P_{qn}} \cong 22 \text{ dB}$$

EXAMPLE 2.3

If a 25-dB signal–to–quantization noise ratio should be maintained in a delta-modulator circuit, determine the quantization bit rate for an analog input signal with a bandwidth of 3.4 kHz.

Solution

$$\frac{P_{so}}{P_{qn}} = 25 \text{ dB}$$

$$f_c = 3.4 \text{ kHZ}$$

$$P_{so} = 0.04\left(\frac{f_b}{f_c}\right)^3$$

or

$$25 = 10 \log\left[0.04\left(\frac{f_b}{f_c}\right)^3\right]$$

$$0.04\left(\frac{f_b}{f_c}\right)^3 = \text{anti} \log(2.5) = 316.2$$

$$\frac{0.04}{0.04}\left(\frac{f_b}{f_c}\right)^3 = \frac{316.2}{0.04} = 7905$$

$$\frac{f_b}{f_c} = 20$$

$$f_b = 20\,(f_c) = 20\,(3.4\text{ kHz})$$

$$\therefore f_b = 68\text{ Kb/s}$$

EXAMPLE 2.4

Determine the maximum input signal bandwidth in a delta-modulator circuit required to maintain a 20-dB signal–to–quantization noise ratio with a quantization bit rate of 50 kb/s.

Solution

$$\frac{P_{so}}{P_{qn}} = 20\text{ dB}$$

$$f_b = 50\text{ kb/S}$$

$$\frac{P_{so}}{P_{qn}} = 0.04\left(\frac{f_b}{f_c}\right)^3$$

$$20 = 10\log\left[0.04\left(\frac{f_b}{f_c}\right)^3\right]$$

$$0.04\left(\frac{f_b}{f_c}\right)^3 = \text{anti }\log(2)$$

$$\frac{0.04}{0.04}\left(\frac{f_b}{f_c}\right)^3 = \frac{100}{0.04}$$

$$\left(\frac{f_b}{f_c}\right)^3 = 2500$$

Solve for f_c

$$f_c = \frac{f_b}{13.57} = \frac{50\text{ Kb/s}}{13.57} = 3.68\text{ kHz}$$

$$\therefore f_c = 3.68\text{ kHz}$$

2.6 PCM AND DM VOICE SIGNAL COMPARISON

Both PCM and DM voice signal digitization methods exhibit certain advantages as well as disadvantages in their system implementation. To assess the advantages or disadvantages, the performance of each circuit must be evaluated separately in terms of the signal–to–quantization noise power ratio, and the circuits must then be compared with each other. For example, a linear PCM encoder operating with 8-bit quantization and a 4-kHz voice base-band bandwidth exhibits a signal–to–quantization noise power ratio given by

$$\frac{P_{so}}{P_{qn}} = 6(N + 1) = 6(8 + 1) = 54 \text{ dB}$$

$$\therefore \frac{P_{so}}{P_{qn}} = 54 \text{ dB}$$

At the same time, if the same voice signal is converted to a digital signal through the delta-modulation process, the signal-to-noise ratio is given by

$$\frac{P_{so}}{P_{qn}} = 0.04 \left(\frac{64}{4} \right)^3 = 22 \text{ dB}$$

$$\therefore \frac{P_{so}}{P_{qn}} = 22 \text{ dB}$$

It is evident that the delta-modulator method of voice signal digitization does not compare favorably with the PCM method. In practice, PCM systems using nonlinear quantization reflect a constant signal–to–quantization noise ratio of about 30 dB over an input signal dynamic range of approximately 40 dB. For DM systems to overcome their comparative disadvantage with PCM systems, the adaptive-delta-modulator scheme described earlier was developed. As a result of the comparative performance characteristics of PCM (dynamic range and signal–to–quantization noise power ratio), ADM schemes are preferable in some applications that are intended to further enhance system capacity.

2.7 TIME-DIVISION MULTIPLEXING OF PCM SIGNALS

So far we have examined the digitization process of a voice channel through pulse-code-modulation and delta-modulation methods. In a large communications system, several voice signals are combined and then transmitted through a single channel. The process of digital signal combining is referred to as digital multiplexing, or more specifically **time-division multiplexing (TDM).** Figure 2.28 illustrates an oversimplified process of time-division multiplexing. Here, three voice channels are sampled in a logical sequence and at a rate satisfying Nyquist's sampling relationship. The combined amplitude-modulation signal is shown at the output. A more detailed logic circuit implementing the time-division multiplexing process is shown in Figure 2.29.

Figure 2.29 illustrates a simplified diagram of a four-channel PCM–TDM logic circuit. In the circuit shown, each AND gate will be enabled only when all its inputs go to logic high. The control logic provides the required timing sequence enabling each AND gate in accordance to the timing sequence. The following truth table better illustrates the operation of this circuit:

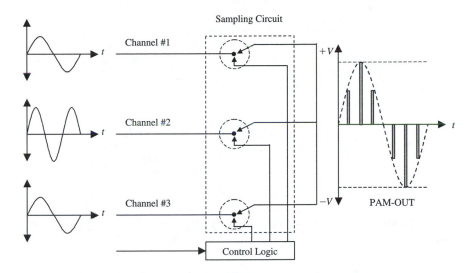

FIGURE 2.28. Time-division multiplexing (TDM).

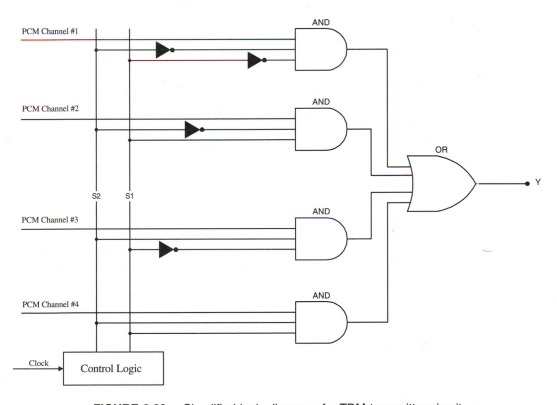

FIGURE 2.29. Simplified logic diagram of a TDM transmitter circuit.

TRUTH TABLE

S_2	S_1	I_0	I_1	I_2	I_3
0	0	1	0	0	0
0	1	0	1	0	0
1	0	0	0	1	0
1	1	0	0	0	1

$$Y = \overline{S_2}\,\overline{S_1}I_0 + \overline{S_2}S_1I_1 + S_2\overline{S_1}I_2 + S_2S_1I_3$$

2.8 THE INTERNATIONAL TELEGRAPH AND TELEPHONE CONSULTATIVE COMMITTEE (CCITT)

In 1957, the **CCITT** was established to set standards and make recommendations relating to matters of international communications to the International Telecommunications Union (ITU), the coordinating body of all member countries. The international consultative committees are divided into study groups. Each study group reports its findings to an assembly, which in turn generates the final recommendations. These recommendations are accepted by all member countries and have established the basis for the present operations and the future development of global communications. Recommendations G.732 and G.733 refer to primary PCM systems.

2.8.1 The North American (BELL) 24-Channel T_1 System (G.733)

This system accommodates 24 voice channels time-division multiplexed with a sampling rate of 8,000 samples per second per channel, and an 8-bit word for coding. The T_1 system generates a primary **bit rate** equal to 1.544 Mb/s as follows:

24-voice channels $\times$ 8 bits/word = 192 bits.
Add one bit for framing, 192 + 1 = 193

The frame, or synchronization bit is necessary for the receiver to recognize which bit corresponds to which of the 24 channels. Therefore, 193 bits $\times$ 8000 samples per second equal a primary bit rate of 1544 Mb/s.

Frame Time

The first bit for each of the 24 channels plus the synchronization bits constitute the time frame of the scheme, and the bit rate is calculated as follows: Each voice channel has a bandwidth of 3.3 KHz. This bandwidth is rounded off to 4 KHz. To satisfy Nyquist's requirement, the voice channel is sampled with a sampling rate $f(s)$ equal to 4 KHz $\times$ 2 = 8000 samples per second:

$$\text{Frame time} \quad t(s) = \frac{1}{8000} = 125 \ \mu s$$

$$t(s) = 125 \ \mu s$$

Therefore, the T_1 system is composed of a total of 8,000 frames, with each occupying a time slot of 125 μs and each time slot accommodating 193 bits of information. The pulse duration of each bit is 125 μs divided by 193, which equals 0.647668393 μs. From this figure, the system bit rate can now be established:

$$T_1 \ (\text{24-voice-channel}) \ \text{bit rate} = \frac{1}{\dfrac{\text{Frame time (s)}}{193 \ \text{bits}}} = \frac{1}{\dfrac{125 \times 10^{-6} \ s}{193 \ \text{bits}}} = 0.647668$$

$$\therefore \ \text{Bit rate} \ (T_1) = 1.544 \ \text{Mb/s}$$

T_1 systems are further combined to generate higher-order digital multiplexing hierarchies such as T_2, T_3, and T_4 schemes (Figure 2.30). The T_2 scheme accommodates 96 voice channels with a corresponding bit stream of 6.312 Mb/s, while the T_3 accommodates 672 voice channels with a corresponding bit rate of 44.736 Mb/s. Finally the T_4, with a total of 4032 voice channels, generates a bit stream of 274.176 Mb/s.

The second-level digital hierarchy has a total bit stream of 6.312 Mb/s. If this second level is divided by the four primary levels, a stream of 1.578 Mb/s instead of 1.544 Mb/s is indicated. The additional bits are required for framing, alignment, and control.

2.8.2 The Japanese Digital Hierarchy

Japanese and North American digital hierarchies are the same for T_1 and T_2 but differ for higher groups (Figure 2.31). The third-level Japanese hierarchy accommodates five (T_2) groups instead of the seven for North America, with a bit stream equal to 32.064 Mb/s. The fourth digital hierarchy accommodates three (T_1) instead of the six for North America, with a bit stream of 97.729 Mb/s. Furthermore, a fifth level is added, generating a total bit stream of 397.2 Mb/s, reflecting a capacity of 5760 voice channels compared with 4232 voice channels for the North American system.

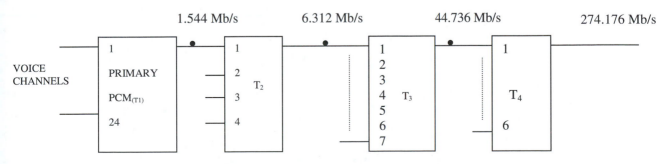

FIGURE 2.30. North American digital heirarchy.

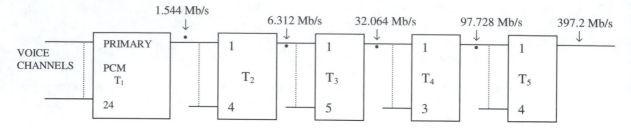

FIGURE 2.31.   The Japanese digital hierarchy.

2.8.3 The European 30-Channel System CEPT(G732)

The European system accommodates 30 voice channels through time-division multiplexing, using 8000 samples per second per channel (sampling rate), and an 8-bit word for coding (Figure 2.32). The European system generates a primary rate of 2.048 Mb/s as follows:

$$30 \text{ voice channels} \times 8\text{-bit quantization} = 240 \text{ bits}$$
$$\text{Add:} \quad 8 \text{ bits for signaling} \quad 8$$
$$\text{Add:} \quad 8 \text{ bits for framing} \quad \underline{\quad 8 \quad}$$
$$256 \text{ bits}$$
$$256 \text{ bits per frame} \times 8000 \text{ frames} = 2.048 \text{ million bits.}$$

The frame period is equal to 125 divided by 30 voice channels plus two for signaling and framing which gives 3.90625 μs (frame period) or 125 μs divided by 256 bits per frame, which generates a 488-ns bit period. Inverting, 1/488-ns is equal to 2.048 Mb/s.

The clock used in the multiplexing process is referred to as the "writing clock," while the clock used in the demultiplexing process is referred to as the "reading clock." Both reading and writing clocks must be absolutely synchronized in order to avoid the possibility of error. Under absolutely ideal synchronization conditions, the ratio of data-stream to synchronization bits is set as 45 to 1. In the event that the reading clock is slightly out of phase with the writing clock, an imbalance will occur between the writing and the reading rates. This rate of imbalance is rectified by the justification.

Justification is the process whereby a time slot is added to or subtracted from the frame in order to maintain the balance between writing and reading bits. Since two clocks

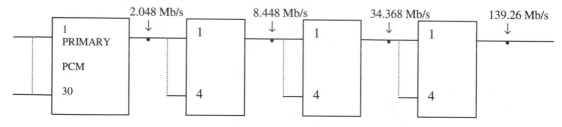

FIGURE 2.32.   The European digital hierarchy.

are used in the multiplexing/demultiplexing process, three possibilities will exist. When the writing clock is slightly faster than the reading clock, the reading clock is unable to completely follow the writing clock, and one bit must be subtracted from the frame. Because of the reduction of a bit from the writing frame, the process is called "negative justification." On the other hand, if the reading clock is faster than the writing clock, a bit must be added to the writing clock. Because the addition of a bit to the writing frame is required, the process is referred to as "positive justification." If no difference exists between writing and reading clocks, no justification bits are required. The writing and reading clock drifts described here are referred to as "multiplex jitter."

2.8.4 Jitter

Practically all long-distance digital communications systems are coherent. That is, a mechanism for carrier, clock, and framing recovery is utilized for demodulation, sampling, and demultiplexing of digital signals. In such a system, at the receiver end the carrier frequency is exactly reproduced and maintained through a phase-lock loop, thus providing relative stability and maintaining system performance within the predetermined limits for the entire operational life of the system.

For shorter terms, certain unpredictable variations in the transmission path characteristics result in phase drift of the recovered carrier in relation to the transmitter carrier. This drift is referred to as "jitter." The CCITT has defined **jitter** as "short-term variations of the significant instants of a digital signal from their ideal positions in time." By "significant instant," the committee referred to any variation of the pulse from its original position in time (Figure 2.33). Jitter can be considered as a form of unwanted phase modulation which ultimately contributes to system performance degradation.

Figure 2.33 shows the observation of the digital signal over a period of time. It is evident that a displacement of the signal has occurred at different intervals over that period of observation time. The accumulated displacement illustrated in Figure 2.33 can be considered as a sine wave having a specific frequency and amplitude. The most common measure of the amplitude of the jitter is in unit intervals (UI).

A unit interval is defined as the time required for the transmission of one bit of information. For example, in the T_1 system with a bit rate equal to 1.544 Mb/s, the unit interval is 1 divided by 1544 Mb/s, or 0.648 μs. If a jitter has peak-to-peak deviation equal to 2 μs, then the jitter amplitude is 2 μs divided by the unit interval of 0.64 μs, or 3-clock pulses.

For quantitative analysis, jitter can also be considered as frequency modulation whereby the amplitude is given in unit intervals, and the bit rate and frequency deviation, are related by

$$J_{p=p} = \frac{m}{\pi} \tag{2.28}$$

where $J_{p=p}$ = jitter amplitude (peak to peak)
   m = index of modulation

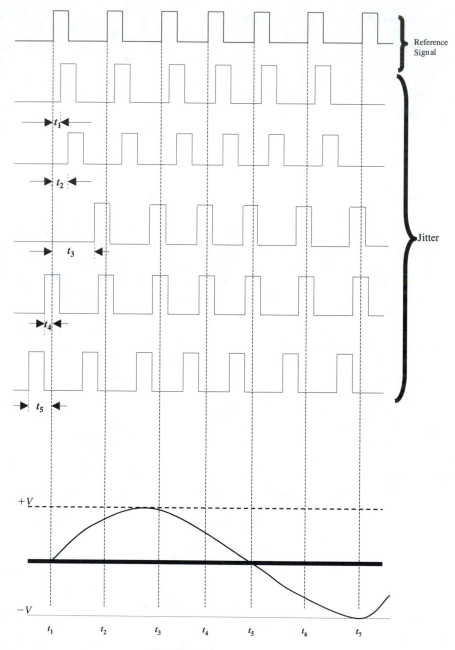

FIGURE 2.33. Jitter function.

EXAMPLE 2.5

Determine the range of variation for a T_1 PCM–TDM system with jitter amplitude of 3 UI and jitter frequency equal to 200 Hz.

Solution

1. Index of Modulation (m)

$$J_{\text{p=p}} = \frac{m}{\pi}$$

$$m = J\pi = 3\pi = 9.42$$

$$\therefore m = 9.42$$

2. Bit-stream variations:

$$\Delta f_b = m \times f_j$$

where
Δf_b = bit stream variations
m = index of modulation
f_j = jitter frequency
$\Delta f_b = 9.42 \times 200 = 1884$ b/s

Range of Variation

T_1 system = 1.544 Mb/s free of jitter,
with jitter range 1.544 Mb/s $\pm$ 1884 b/s

System Degradation

If the amplitude of the jitter is not controlled to acceptable levels, it will accumulate over a period of time, resulting in system degradation. This degradation will take the form of amplitude distortion of the reconstructed analog information at the digital-to-analog converter end, increasing the error probability or introducing slips into the digital information.

Sources of Jitter

Major jitter sources contributing significantly to the degradation of the overall system performance are multiplexers and demultiplexers, digital degradators, and electronic components. Multiplexers/demultiplexers as a primary source of jitter, can be attributed to the process of justification.

By examining the process of multiplexing/demultiplexing, it was shown that the two major digital hierarchies are the Bell North American and the European (CEPT) systems. In accordance with CCITT recommendations, the primary North American digital hierarchy accommodates 24 voice channels with a bit stream equal to 1.544 Mb/s and the European system accommodates 30 voice channels with a bit stream equal to 2.048 Mb/s. For each voice channel in both systems, the total bit stream is equal to $4000 \times 2 \times 8$, or 64 Kb/s.

The second-level digital hierarchy has a total bit stream equal to 6.312 Mb/s. If this second level of 6.312 Mb/s is divided by four primary levels, a bit rate of 1.578 Mb/s instead of 1.544 Mb/s is indicated. The additional bits beyond the 1.544 Mb/s are framing, alignment, and control bits.

A more detailed explanation of the justification process is given in PCM–TDM systems. The regenerators are another major source of jitter in digital transmission systems. The basic function of a regenerator in a digital transmission link is to receive, recondition, and retransmit the incoming signal to the next regenerator. The accumulated jitter effect from all regenerators through a transmission link is detrimental to the overall system performance and must be confined within acceptable limits. The block diagram of a regenerator unit is given in Figure 2.34.

From Figure 2.34, the incoming signal, altered and somewhat distorted through the transmission path, is applied to the input of the **equalizer.** The equalizer circuit amplifies and restores the incoming signal to its original level. The **bipolar-to-unipolar** converter is required to convert the signal from bipolar to unipolar for **clock recovery,** and the decision circuit will determine whether the 1 bit or the 0 bit is present. The timing circuit will reconstruct the binary signal, and the unipolar-to-bipolar circuit will generate the final signal for transmission to the next regenerator.

The major sources of jitter within the regenerators are pattern-dependent sources, and random sources. Pattern dependent jitter sources induce signal impairments such as **intersymbol interference (ISI),** amplitude/phase impairments, and finite pulse-width effects. Intersymbol interference is the phenomenon whereby electrical characteristics of the transmission line such as nonlinearities and frequency responses alter the shape of the traveling pulse through the transmission line or force the pulse to shift at the receiver beyond the acceptable limits, and by doing so, introduce error through the receiver sampling process (Figure 2.35). The electrical characteristic of the clock recovery circuit prevents it from responding to the pulse shifts, thus making it incapable of distinguishing between pulses and intersymbol interference.

Another source of jitter is the threshold detector of the clock recovery circuit. The main function of the threshold detector is to produce a square-wave output signal, triggered by the resonated bit rate output of a tuned circuit. If the amplifier of the threshold detector induces impairments, which are attributed to the component and to component thermal effects, these effects will be translated into jitter (Figure 2.36). From this figure, it is evident that amplitude/phase conversion jitter induces pattern generator jitter.

If the width of the pulse at the input of the tuned circuit of the clock recovering is wider than the input at the clipper circuit, a displacement will occur. Figure 2.37 indicates that if displacement is to be avoided, both the input and output signal at the tuned circuit of the clock recovering module must simultaneously pass through the zero point. If the input pulse is slightly wider, it will shift the zero crossing point of the recovered clock, with the possibility of error at the decision circuit output.

Other jitter sources do not depend on the digital regenerator. These are cross-talk, tuned circuit mistuning, and differential pulse delays. Cross-talk occurs when more than one digital transmission utilizes the same transmission medium. Cross-talk can be divided into two major categories: near end, and far end. Near-end cross-talk is measured at the input of the transmission medium (i.e., wire pair), and must maintain a level approximately 30 dB below the signal level, for a corresponding bit error rate of 10^{-8}. The far-end cross-talk is measured at the output of the medium and must maintain a level 15 dB below the signal level.

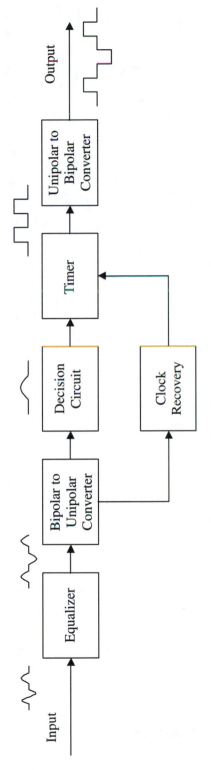

FIGURE 2.34. Block diagram of a regenerator circuit.

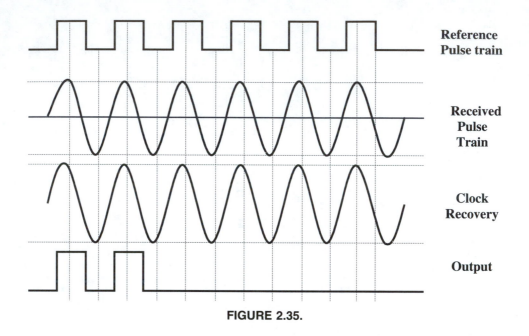

Reference
Pulse train

Received
Pulse
Train

Clock
Recovery

Output

FIGURE 2.35.

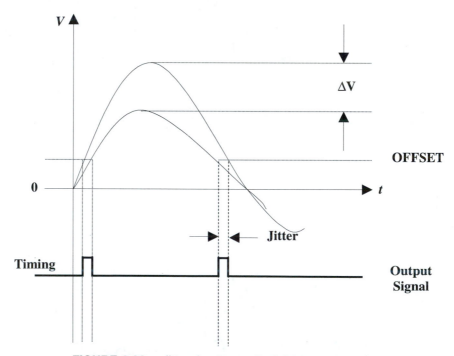

FIGURE 2.36. Jitter due to amplitude/phase conversion.

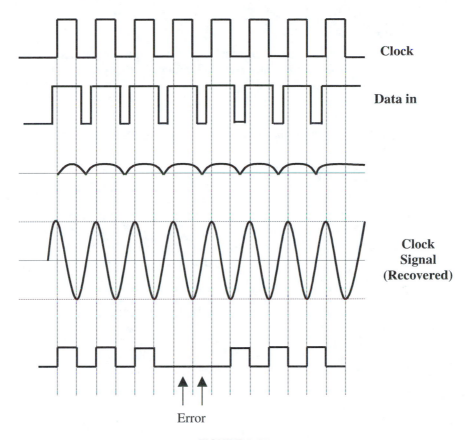

Clock

Data in

Clock
Signal
(Recovered)

Error

FIGURE 2.37.

The cross-talk-induced jitter is the result of (LC) component mistuning within the clock recovery circuit. This mistuning may cause a static phase shift, introducing a transmission delay in the regenerator, or it may cause a dynamic phase shift proportional to the data bit stream. Most of the time the overall effect of the tuned-circuit mistuning is not all that significant, especially when several regenerators are used. In this case, there is a high probability that half of the tuned circuits will be mistuned in one direction and the other half in the other. Under these circumstances the possibility of a partial or complete elimination of this type of jitter is quite high.

The conversion of a unipolar to a bipolar signal at the output of the regenerator circuit is achieved with implementation of semiconductor devices operating at saturation or cutoff modes. When in saturation mode, a slight variation of the base-to-emitter junction capacitance (C_b) will result in a shift in time of the output pulse. This type of jitter is noticeable in high frequencies, and it can be removed by filtering.

QUESTIONS

1. Name the three fundamental voice-signal quality parameters specified by CCITT.
2. Define the Nyquist Theorem and comment on its significance as applied to voice-signal quantization.
3. Define the terms PAM, PPM, and PWM, and briefly describe their differences and similarities.
4. Briefly describe the process of pulse-code modulation (PCM) and list its advantages.
5. What is quantization noise?
6. Define the term *nonlinear quantization*. Why is it required?
7. Describe the difference between instantaneous, nonuniform, and uniform companding.
8. What are the companding laws? Explain their differences and similarities.
9. Briefly describe the delta-modulation quantization process.
10. List the advantages and disadvantages of delta modulation and pulse-code modulation.
11. What is the difference between a delta modulator and an adaptive delta modulator?
12. List the differences between the Bell North American, Japanese, and European digital hierarchies.
13. What is jitter? List the main sources of jitter.
14. Describe the effect of jitter on system performance.

PROBLEMS

1. Compute the signal–to–quantization noise ratio of a pulse-code-modulation system employing a 10-bit quantization word.
2. In a PCM system employing 7-bit quantization, the signal–to–quantization noise ratio is 35 dB. If the signal power is 25 mW, calculate the maximum permissible noise power.
3. An analog signal with a bandwidth of 8 kHz is to be digitized through a delta-modulator circuit with a bit rate of 64 kb/s. Determine the signal–to–quantization noise ratio.
4. Determine the signal power required at the input of a delta modulator with an 8-bit quantization word in order to maintain a noise power level equal to 50 nW. The input analog signal bandwidth is 3.7 kHz, and Nyquist minimum criteria must be maintained.
5. A delta-modulator circuit with a 64 kb/s quantization rate is to maintain a signal–to–quantization noise ratio equal to 30 dB. Compute the maximum input signal bandwidth, satisfying minimum Nyquist criteria.
6. A PCM–TDM modulation scheme is operating with a jitter amplitude of 4 UI (unit intervals) and a jitter frequency equal to 280 Hz. Determine the range of variations for the T_1 Bell North American hierarchy.
7. In a PCM–TDM T_1 system, the bit rate is measured to vary between 1.546 Mb/s and 1.542 Mb/s. Determine the jitter amplitude for a jitter frequency of 500 Hz.

3

Digital Radio

Objectives

- Identify the various digital modulation schemes.
- Describe in detail each digital modulation scheme.
- Define coherent and noncoherent detection.
- Calculate the error performance and error probability for each modulation scheme.
- Compare the digital modulation schemes in terms of their performance.

Key Terms

Digital radio
Digital modulation
FSK
PSK
QAM
I/Q modulator
Data combiner
MSK
Balanced modulator
Phasor diagram

Series-to-parallel converter
Constellation display
  diagram
DPSK
Spectral efficiency
Differential encoder
Bit energy
Noise spectral density
Bit energy–to–noise
  density ratio

Clock recovery
Costas loop
I channel
Q channel
Error performance
Error probability
Error distance
CNR
Symbol rate

INTRODUCTION

Digital radio is a radio system that transmits digitally coded information (base band) by altering the fundamental properties (amplitude, frequency, or phase) of an RF carrier to a discrete number of states and then, through a transmission medium, receives the digitally coded and modulated signal, recovers the base band, converts it to a binary stream, and descrambles and records it to CCITT standards. The combination of the transmitter, the receiver, and the transmission medium constitutes the digital radio link.

Since digital radio links are designed to carry information from one point to another, usually separated by a long distance, the volume of information that can be carried at any point in time is referred to as "radio link capacity." A small-capacity digital radio system can carry information at a rate of up to 10 Mb/s; a medium-capacity system can carry up to 100 Mb/s, and a large-capacity system can carry information well beyond the 100 Mb/s mark.

The volume of information or capacity of a digital communication system can only be limited by that system's sophistication and complexity. The volume of data carried by a RF bandwidth is determined by the modulation scheme employed. Several modulation schemes are available to the system designer including frequency-shift keying (**FSK**), phase-shift keying (**PSK**), and quadrature amplitude modulation (**QAM**). High-level QAM systems are very complex systems, and although they are capable of carrying more data per RF bandwidth, they are very difficult to implement because of their high degree of susceptibility to noise and interference.

The simpler **digital modulation** techniques such as FSK and low-level PSK are normally used in satellite communications and military applications, while the more complex modulation schemes are used almost exclusively in line-of-sight (LOS) and optical fiber systems.

3.1 DIGITAL RADIO BLOCK DIAGRAM

The simplified versions of digital transmitter and receiver systems are illustrated in Figures 3.1 and 3.2 respectively. Although certain similarities between digital and analog systems do exist, their modulation/demodulation schemes are very different.

The Transmitter   The main building blocks of a digital radio transmitter are the coder, the filter, the modulator, the up-converter, the microwave power amplifier, the output filter, and the antenna assembly.

The Receiver   At the receiver end the intercepted RF signal is amplified by the low-noise amplifier (LNA) then down-converted (D/C) to a predetermined intermediate frequency (70 MHz or 140 MHz). The automatic-control circuit (AGC) provides the control signal level required at the input of the demodulator circuit (Figure 3.2).

The Coder   The coder provides the standard CCITT interface for the digital network and also facilitates the service channels and other framing bits which are combined to a main

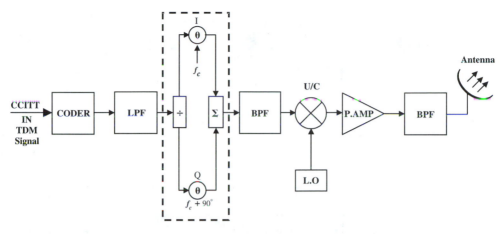

FIGURE 3.1 Digital radio transmitter block diagram.

bit stream through a digital multiplexing process. The block diagram of the coder is illustrated in Figure 3.3

The scrambler is designed to remove the long runs of zeros and ones in order to maintain spectral restrictions. The differential encoder divides the data stream into two equal and parallel paths. The series stream of each path is then converted to an *N*-bit word driving a digital-to-analog converter (D/A). The filter removes the transition edges from the analog pulse stream generated by the A/D converter before it is fed into the input of the digital modulator.

The Modulator Figure 3.4 illustrates the block diagram of a digital modulator circuit. In this modulator, the carrier signal is divided into sine and cosine components known as in phase (I) and quadrature phase (Q). The **I/Q modulator** circuit generates two IF subcarrier signals with a phase difference of 90°. The output of each low-pass filter amplitude

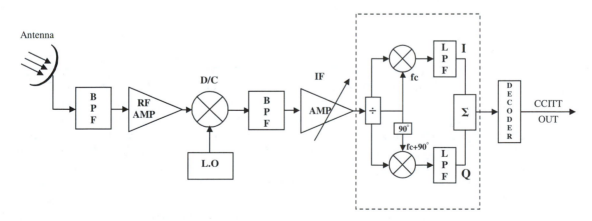

FIGURE 3.2 Digital radio receiver block diagram.

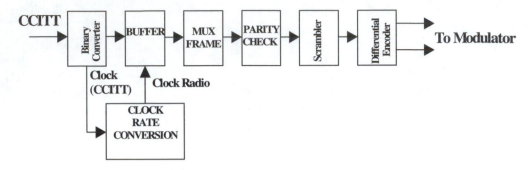

FIGURE 3.3 Digital radio coder block diagram.

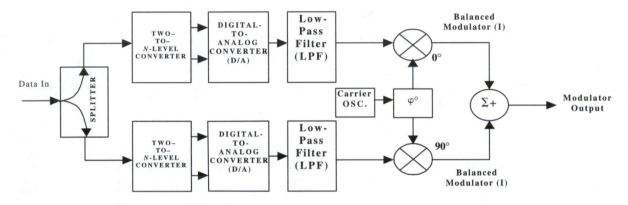

FIGURE 3.4 Digital modulation block diagram.

modulates the IF subcarriers. The two subcarrier signals are combined to form one amplitude- and phase-modulated signal. The output of the summing circuit is fed into the input of the up-converter in order for the RF signal to be generated. The RF signal is then amplified by the traveling-wave tube (TWT) or by a microwave solid-state amplifier, then fed into the antenna assembly for final transmission.

The Demodulator Figure 3.5 illustrates the block diagram of a digital radio demodulator. A brief description of such a circuit is as follows: The local-oscillator circuit is phase locked with the incoming signal, generating the exact carrier frequency. A phase shift circuit generates the 0° and 90° phase differences required by the IQ demodulator. The two baseband signals are fed into the inputs of the corresponding LPFs. The function of the filter is to remove the high-frequency components and maintain bandwidth restrictions. The analog signals are then sampled with the exact transmitter carrier frequency provided by the timing recovery circuit. The data slicer converts the voltage levels to N-bit binary words, and the analog-to-digital converter generates the binary bit stream from the multilevel input signals. The two parallel binary streams of data are de-scrambled, then fed into the combiner circuit to form the standard CCITT signal. The CCITT signal is then demultiplexed, decoded, and finally processed to the appropriate transponder. A detailed description of digital radio subsystems is given in the following pages.

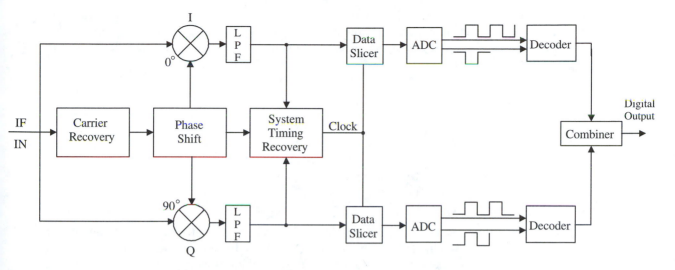

FIGURE 3.5 Digital radio demodulator block diagram.

3.2 DIGITAL MODULATION

In traditional communications systems, the three modulation techniques are amplitude modulation (AM), frequency modulation (FM), and phase modulation (PM). Common to all modulation schemes is the fact that the modulating and carrier signals are both analog. In digital modulation, the information signals, whether audio, video, or data, are all digital. Therefore, digital information modulates an analog carrier, thus the name "digital modulation." To differentiate between analog modulation and digital modulation, the terminology has been changed. Amplitude-shift keying (ASK) has been adopted instead of amplitude modulation; frequency-shift keying (FSK) instead of frequency modulation; and phase-shift keying (PSK) instead of phase modulation.

Phase-shift keying modulation is also subdivided into two-level and multilevel (M-array) systems. The two-level system can transmit only two bits of information for each level of carrier phase change, whereas the M-array modulation schemes can transmit more than two information bits per carrier phase-level change. For example, a 16-level (M-16) system can theoretically transmit four bits of binary information. It is obvious that the utilization of higher-level modulation schemes requiring eight or more bits results in more complex systems with a higher degree of difficulty in their implementation. At the receiver end, the digitally modulated signal is detected and, through the detection process, digital information is retrieved from the carrier and channeled out for further processing.

There are two basic detection processes: coherent and noncoherent. In coherent detection, the receiver must generate the exact carrier in terms of frequency and phase. This is an absolutely necessary requirement, because digital information is recovered from the phase of the carrier; even the smallest phase difference between the transmitter carrier and the receiver-generated carrier will be interpreted as error. This error is referred to as "phase noise." In noncoherent detection, a reference carrier signal is not required. A more detailed discussion of coherent and noncoherent detection will follow later in this chapter.

By definition, digital modulation is the process whereby the fundamental properties of the carrier signal (amplitude, frequency, or phase) are altered in accordance with the digital base band information signal. The mathematical expression of a pure sine wave (carrier) is given by

$$e(t) = E_0 \sin(\omega t + \phi) \tag{3.1}$$

where $e(t)$ = instantaneous amplitude (V)
E_0 = maximum amplitude (V)
f = frequency (Hz)
ϕ = phase (0°)

If the amplitude (E) of the carrier is altered by the digital information, the modulation scheme is called amplitude-shift keying (ASK) or ON–OFF keying (OOK). If the frequency (f) of the carrier is altered in accordance with the digital information, the modulation scheme is called frequency-shift keying (FSK). If the phase of the carrier is altered, then the modulation scheme is called phase-shift keying (PSK). A fourth and very important modulation method is rapidly gaining recognition as the most effective modulation method, that is, quadrature-amplitude modulation, (QAM). Through this modulation process, both the amplitude and the phase of the carrier are altered by the digital information, thus enhancing spectral efficiency.

3.2.1 Amplitude-shift Keying (ASK)

Amplitude-shift keying (ASK) refers to the digital modulation technique whereby the digital information alternates the amplitude of the carrier between two distinct levels (Figure 3.6). This digital modulation method is also referred to as ON–OFF keying, (OOK). The ON state represents binary 1; the OFF state represents binary 0. The modulated output signal is the product of the two-level unipolar information signal and the carrier signal. Amplitude-shift keying is perhaps the simplest of all digital modulation schemes. At the receiver end, the noncoherent method of detection is employed. This gives rise to a higher

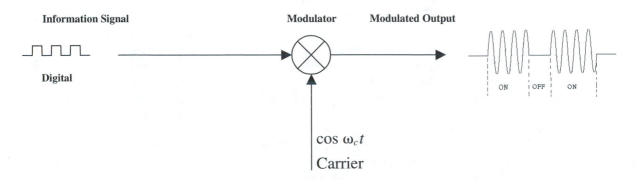

FIGURE 3.6 Amplitude-shift keying digital modulation scheme.

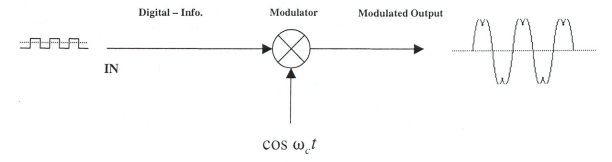

FIGURE 3.7 Phase-reverse keying.

degree of bit error probability than exists in coherent methods employed by other modulation schemes. To achieve a better error probability with ASK modulation, the information signal is converted into a two-level symmetrical signal. The resultant digitally modulated carrier reverses its phase by 180° (Figure 3.7). This modified ASK modulation method is also referred to as phase-reverse keying (PRK). With phase-reverse keying digital modulation, the digital information is not carried in the amplitude of the carrier but rather in the phase. Phase-reverse keying (PRK) requires coherent detection at the receiver end, resulting in a significant bit-error-rate (BER) improvement. The output waveform of Figure 3.6 shows the amplitude of the ASK signal to be constant for a period (T) corresponding to the presence of a bit and suppressed when a bit is absent. The output voltage is expressed by

$$u(t) = E \cos(\omega_c t + \phi) \qquad \text{for binary 1}$$
$$u(t) = 0 \qquad\qquad\qquad \text{for binary 0} \qquad\qquad\qquad\qquad (3.2)$$

3.2.2 Frequency-Shift Keying (FSK)

Frequency-shift keying modulation incorporates two carrier frequencies for the transmission of two bits of binary data. Carrier frequency f_1 corresponds to bit 1, and carrier frequency f_2 corresponds to bit 0. The FSK waveform is represented by

$$u(t) = E \cos \omega_1 t \qquad \text{for binary 1}$$
$$u(t) = E \cos \omega_2 t \qquad \text{for binary 0} \qquad\qquad\qquad\qquad (3.3)$$

The simplified block diagram of the FSK modulation scheme is represented in Figure 3.8. The composite FSK signal at the output of the modulator circuit is expressed by

$$u(t) = E \cos[2\pi(f_c \pm \Delta f)t] \qquad\qquad\qquad\qquad (3.4)$$

The FSK signal spectrum shows the carrier frequency (f) deviating in accordance with the input binary data. At input binary 0, the modulator is switching to carrier (f_1), and at

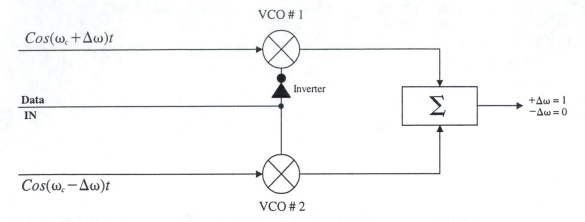

FIGURE 3.8 Block diagram of a frequency-shift-keying modulation scheme.

binary 1, the modulator is switching to carrier (f_2). Therefore,

$$f_1 = f_c + \Delta f \qquad f_2 = f_c - \Delta f$$

The total spectral width of the FSK modulation scheme is determined as follows:

$$f_1 = f_c + \Delta f$$
$$f_2 = f_c - \Delta f$$

Solve for Δf:

$$(1) \quad \Delta f = f_1 - f_c$$
$$-\Delta f = f_2 - f_c \quad \text{or}$$
$$(2) \quad \Delta f = f_c - f_2$$

Add (1) and (2):

$$2\Delta f = f_1 - f_c + f_c - f_2$$
$$2\Delta f = f_1 - f_2$$
$$\Delta f = \frac{f_1 - f_2}{2} \qquad\qquad (3.5)$$

The minimum frequency deviation in the FSK modulation scheme is based on the pulse duration of the input binary bit. For a bit duration of τ s, the corresponding time is 2τ s:

$$2\tau = \frac{1}{f_1 - f_2}$$

$$f_1 - f_2 = \frac{1}{2\tau} \tag{3.6}$$

Substituting Eq. (3.6) into Eq. (3.5), we have

$$\Delta f = \frac{1/2\tau}{2} = \frac{1}{2(2\tau)} = \frac{1}{4\tau}$$

$$\Delta f = \frac{1}{4\tau} \tag{3.7}$$

Substituting Eq. (3.7) into Eq. (3.4), we have

$$u(t) = [E \cos(2\pi f_c)t \cos(\pm\pi/2\tau)t] - [E \sin(2\pi f_c)t \cdot \sin(\pm\pi/2\tau)t] \tag{3.8}$$

Equation (3.8) represents the output voltage of the frequency-shift keying modulation scheme.

FSK Bandwidth

The bandwidth of an FSK signal is defined as the maximum occupied frequency range of the signal spectrum and is determined by the frequencies f_1 and f_2, and the pulse duration of the input binary data (Fig. 3.9).

$$BW = f_A - f_B \tag{3.9}$$

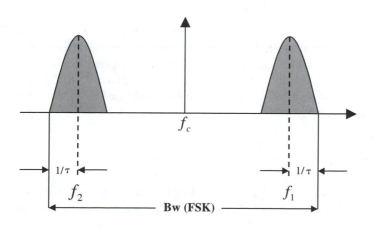

FIGURE 3.9   FSK Bandwidth.

where
$$f_A = f_1 + \frac{1}{\tau}$$

$$f_B = f_2 - \frac{1}{\tau}$$

τ = pulse duration

$$\text{BW} = f_1 + \frac{1}{\tau} - \left(f_2 - \frac{1}{\tau}\right)$$

$$= f_1 + \frac{1}{\tau} - f_2 + \frac{1}{\tau}$$

$$= f_1 - f_2 + \frac{2}{\tau}$$

but
$$f_1 - f_2 = 2\Delta f$$

Therefore,
$$\text{BW} = 2\Delta f + 2\left(\frac{1}{\tau}\right)$$

$$= 2\left(\Delta f + \frac{1}{\tau}\right)$$

$$\text{BW} = 2\left(\Delta f + \frac{1}{\tau}\right) \tag{3.10}$$

From Eq. (3.10), it is evident that the FSK bandwidth is twice the sum of Δf and the binary input pulse duration (τ).

Minimum-Shift Keying (MSK)

A modified version of FSK modulation is minimum-shift keying (**MSK**). In an MSK modulation, the two carrier frequencies are at an exact 180° phase difference. This way, a minimum signaling difference generates a maximum phase difference—a distinct advantage at the receiver end. MSK is also referred to as continuous-phase FSK (CPFSK). The most important advantages of an MSK modulator are the condensed spectrum and the excellent **bit energy–to–noise density ratio**.

3.2.3 Phase-Shift Keying (PSK)

Phase-shift keying is another digital modulation technique, perhaps the one most commonly used in today's digital communications systems. In a PSK modulation scheme, the phase of the carrier is altered in accordance with the input binary-coded information. The mathematical representation of a PSK signal is given by

Allowable phase state (levels), or

$$u(t) = E \sin\left(\omega_c + \frac{2\pi(n-2)}{M}\right)t \qquad (3.11)$$

where $\quad M = 2^n$
$\qquad n = 1, 2, 3, 4$

In reference to the number of carrier phase levels used, PSK modulation schemes are further subdivided into BPSK, where the number of levels (M) are 2, quadrature-shift keying (QPSK), where M is equal to 4, 8-PSK, where M is equal to 8, 16-PSK, where M is equal to 16 and so on. The progressive increase of the number of levels in a PSK digital modulation scheme reflects a proportional increase in that system's information capacity. Since maximum attainable system capacity in digital communications systems is the main objective of the system designer, more sophisticated and consequently more complex digital modulation schemes are employed above the BPSK level. Regardless of the modulation technique used, the system must maintain a fixed bit error rate (BER). Bit error rate is a very important element in system performance evaluation. An increase in system capacity corresponds to a proportional increase in the BER, with a constant signal-to-noise ratio (SNR). To maintain a constant BER (characteristic of system performance), a proportional increase of the signal-to-noise-ratio is required. In summary, higher-order PSK modulation schemes are characterized by their higher bandwidth efficiency. That is, they can transmit more digital information per given channel bandwidth at the expense of a higher SNR.

Binary Phase-Shift Keying (BPSK)

In the BPSK modulation method, the output carrier switches between two phases in accordance with the input binary information signal. If the input binary information is one, the output-modulated carrier is in phase with the oscillator frequency. If the input binary information is 0, the output modulated carrier is out of phase with the carrier oscillator by 180° (Figure 3.10). More specifically, the unipolar binary input signal is converted to a bipolar signal through a comparator circuit. The two dc levels generated by the binary comparator alter the flow of current (in or out) of the **balanced modulator,** resulting in a phase shift of the local oscillator frequency between 0 and 180°. The bandpass filter (BPF) is required in order to maintain channel bandwidth restrictions, and to reduce intersymbol interference (ISI). The **phasor diagram** of a BPSK modulator is illustrated in Figure 3.11.

The **constellation display diagram** (Figure 3.12) also represents digital modulation schemes. Constellation diagrams, in contrast with the phasor diagrams, show only the position of the phasor relative to local oscillator reference frequency. The waveform illustrating the relationship between the input binary data and the carrier frequency is shown in Figure 3.13.

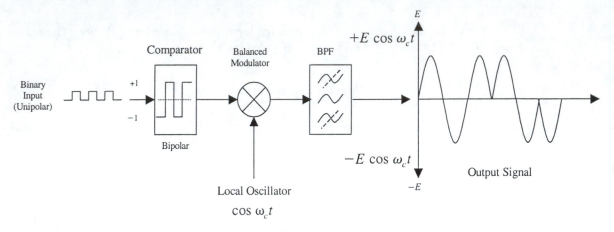

FIGURE 3.10 Binary phase-shift keying (BPSK) modulation.

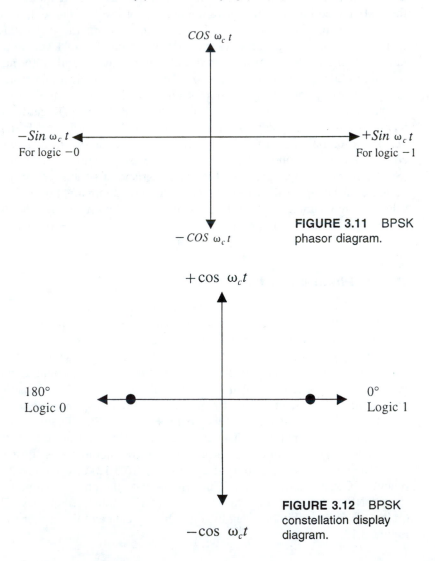

FIGURE 3.11 BPSK phasor diagram.

FIGURE 3.12 BPSK constellation display diagram.

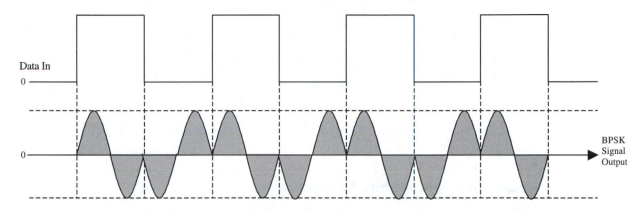

FIGURE 3.13 BPSK output waveform.

BPSK Signal Bandwidth (BW) Since the binary input signal of a BPSK modulation method is converted to a bipolar signal, the modulator output is the product of the two equal but opposite alternating voltage levels and the local oscillator reference frequency. More specifically, the BPSK signal output is the product of the local oscillator frequency and the binary input fundamental frequency. The BPSK bandwidth is then established as follows: Let $\sin \omega_c t$ be the carrier frequency and $\sin \omega_m t$ be the modulating fundamental frequency; then

$$u(t)_{\text{BPSK}} = (\sin \omega_c t) \times (\sin \omega_m t) \tag{3.11}$$

Applying trigonometric relationships, we have

$$u(t) = (\sin \omega_c t) \times \sin \omega_m t = \frac{1}{2} \cos(\omega_c - \omega_m)t - \frac{1}{2}\cos(\omega_c + \omega_m)t$$

Since $\omega_c = 2\pi f_c$ and $\omega_m = 2\pi f_m$, the total Nyquist spectral bandwidth (BW) is

$$f_c - f_m - (f_c + f_m) = f_c - f_m - f_c - f_m = 2f_m$$

Therefore

$$\text{BW} = 2f_m \tag{3.12}$$

The binary data fundamental frequency is half the data bit rate:

$$f_m = \frac{f_b}{2} \tag{3.13}$$

Substituting Eq. (3.13) into Eq. (3.12), we have:

$$BW = 2\left(\frac{f_b}{2}\right) = f_b$$

$$\therefore BW = f_b$$

Therefore, with the BPSK digital modulation method, the required minimum channel bandwidth must be equal to the binary input data rate.

Quadrature Phase-Shift Keying (QPSK)

QPSK digital modulation is one level above the BPSK scheme in terms of circuit complexity and performance outcome. With QPSK, the modulated output signal is shifted by four phases in accordance with the input binary data. Since there are four phase shifts at the output of the modulated signal, the QPSK method requires two input bits for each phase shift. Quadrature phase-shift keying modulation exhibits better spectral efficiency but at the expense of more complex circuitry and more critical performance requirements. The block diagram of a QPSK modulation method is illustrated in Figure 3.14.

In the QPSK modulator of Figure 3.14, the series binary input data is converted into two parallel data streams—even and odd. The even stream is fed into the input of binary comparator 1, and the odd is fed into the input of binary comparator 2. The multilevel signal output from both comparators is fed into the low-pass filters, where the high-frequency components are removed. Both outputs are now fed into the inputs of the corresponding balanced modulators. The output signal of balanced modulator 1 is a modulated signal in phase with the carrier signal, modulated by the even binary data stream (I). The output sig-

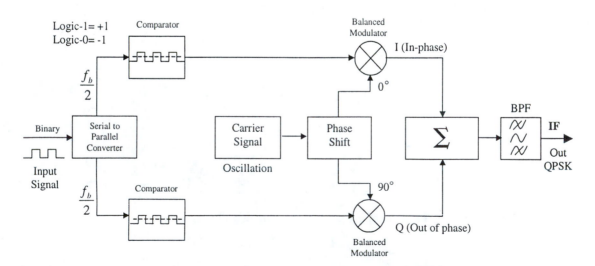

FIGURE 3.14   Q-PSK modulator block diagram.

nal of balanced modulator 2 is also a modulated signal, out of phase by 90° with the carrier, modulated by the odd binary data stream (Q). Both I and Q signals are now combined through the combiner circuit to form the (IF) QPSK signal output. The mathematical representation of a QPSK output signal is determined as follows: If the carrier signal is represented by the expression $\sin(\omega t)$ for the In-phase channel (**I channel**), and $\cos(\omega t)$ for the quadrature-phase channel (**Q channel**), then the outputs from each balanced modulator are given by

$$I_0 = E_I \sin \omega_c t \qquad (3.14)$$
$$Q_0 = E_Q \cos \omega_c t \qquad (3.15)$$

Assuming a balanced QPSK modulator with $E_I = E_Q$.

Now, both I and Q signals are fed into the input of the combiner circuit. At the output of the combiner, the combination of both input signals will be present. That is,

$\sin \omega_c t + \cos \omega_c t$
$-\sin \omega_c t + \cos \omega_c t$
$\sin \omega_c t - \cos \omega_c t$
$-\sin \omega_c t - \cos \omega_c t$

From this mathematical expression, the QPSK phasor diagram can be produced (Figure 3.15).

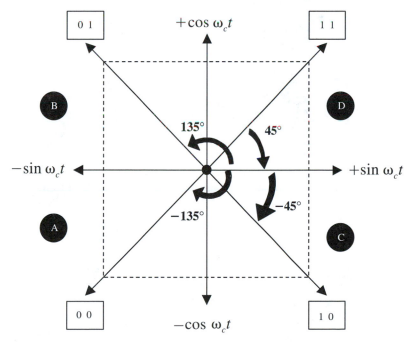

FIGURE 3.15 QPSK phasor diagram.

Since two binary input bits are required for each phase change in a QPSK system, then assume:

Binary Input	Output (V)
1	+1
0	−1

TRUTH TABLE

	I	Q
1	0	0
2	0	1
3	1	0
4	1	1

From the truth table: Binary   Voltage
                        input   level

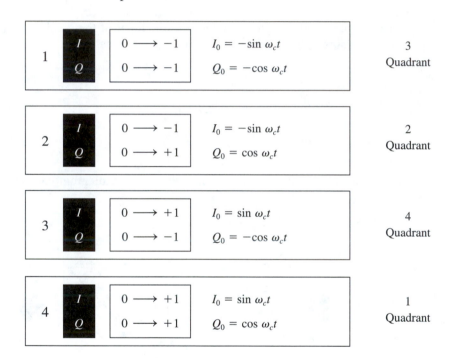

		Binary input	Voltage level	Equations	Quadrant
1	I / Q	$0 \longrightarrow -1$ / $0 \longrightarrow -1$		$I_0 = -\sin \omega_c t$ / $Q_0 = -\cos \omega_c t$	3 Quadrant
2	I / Q	$0 \longrightarrow -1$ / $0 \longrightarrow +1$		$I_0 = -\sin \omega_c t$ / $Q_0 = \cos \omega_c t$	2 Quadrant
3	I / Q	$0 \longrightarrow +1$ / $0 \longrightarrow -1$		$I_0 = \sin \omega_c t$ / $Q_0 = -\cos \omega_c t$	4 Quadrant
4	I / Q	$0 \longrightarrow +1$ / $0 \longrightarrow +1$		$I_0 = \sin \omega_c t$ / $Q_0 = \cos \omega_c t$	1 Quadrant

Transferring this to Figure 3.15, the QPSK phasor diagram illustrates the phase shift of the carrier with a combination of two binary input bits as follows:

Input Bits		Phase Shift
0	0	−135°
0	1	+135°
1	0	−45°
1	1	+45°

It is evident that a 90° phase shift is always maintained with each two-bit binary input combination. As mentioned before, an extension of the phasor diagram is the constellation display diagram, which relates the amplitude and phase of the QPSK output signal for each

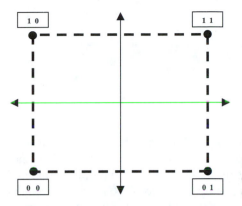

FIGURE 3.16 QPSK constellation display diagram.

combination of two-input binary bits (Figure 3.16). The constellation display diagram will be used to identify and analyze a number of digital radio impairments. This type of problem identification will be thoroughly examined in the appropriate section dealing with digital radio measurements.

QPSK Signal Bandwidth Bandwidth is a very important parameter used to compare various digital modulation methods and also to establish system spectral efficiency. Since the input binary data in a QPSK system is divided into two parallel streams I and Q, the fundamental frequency in each stream is one-fourth of the input binary stream, or,

$$\omega_m = \frac{2\pi f_b}{4}t \qquad (3.16)$$

where f_b = binary input stream (b/s)

The convolution of the two waveforms, that of the reference carrier, $\sin \omega_c t$, and that of the fundamental of the I-channel parallel stream, $\sin \omega_m t$, is expressed as follows:

$$u(t) = \sin \omega_c t \times \sin \omega_m t = \frac{1}{2}\cos(\omega_c - \omega_m)t - \frac{1}{2}\cos(\omega_c + \omega_m)t$$

or

$$BW = f_c - f_m - (f_c + f_m) = f_c - f_m - f_c - f_m = -2f_m$$

Since

$$f_m = \frac{f_b}{4}$$

Digital Radio

89

Then

$$BW = 2\left(\frac{f_b}{4}\right) = \frac{f_b}{2}$$

Therefore, $\qquad\qquad\qquad$ $BW = f_b/2.$

In summary, the QPSK digital modulation scheme is, in essence, a double BPSK circuit. Each non-return-to-zero data stream in the I and Q channels is modulated by the reference carrier (*I*, in phase; *Q*, out of phase by 90°) in the time domain. The resultant QPSK output signal is then fed into the up-converter (U/C) circuit through a bandpass filter (BPF) with an ideal bandwidth equal to $f_b/2$. This leads to the assumption that an ideal QPSK digital modulation method exhibits bandwidth efficiency equal to 2 b/s/Hz. Bandwidth efficiency is defined as the number of binary input data bits a modulation method can effectively transmit per hertz of channel bandwidth.

Differential Phase-Shift Keying (DPSK)

A critical element in a coherent communications system is the clock recovery at the receiver end. Clock recovery requires complex and sophisticated circuit design in order to activate absolute synchronization between the transmitter carrier frequency and the receiver recovery carrier frequency. Even the smallest phase difference between the receiver and transmitter carrier frequencies will be interpreted as additional phase noise, inducing as a consequence, additional system bit error.

An alternative method to coherent transmission is the implementation of differential phase-shift keying at both the transmitter and the receiver ends. In a **DPSK** scheme, the state of the output data will be determined by comparing the bit applied at the input of the modulator with the next bit. If the phase of the two bits is zero, then the detected bit is a zero. Likewise, if the phase of the two bits is not zero, then the detected bit is a one.

DPSK Modulator The basic block diagram of a DPSK modulator is illustrated in Figure 3.17. The Ex-Nor gate will produce a logic 1 at its output when both inputs are either at 0-0 or 1–1 logic levels. Assume the binary data 01 is applied at input *a* of the DPSK modulator. Also assume the reference 0 bit is applied at input *b*. The output of the Ex-Nor gate is shown in Table 3.1.

The unipolar signal at the output of the Ex-Nor gate is converted to a bipolar signal so that the two dc levels generated by the modulator circuit will shift the carrier frequency between 0° and 180° for logic 1 and logic 0 respectively.

TABLE 3.1 Input data: 1101

a	b	y
1	0	0
1	0	0
0	0	1
1	1	1

Output data: *y* = 0011

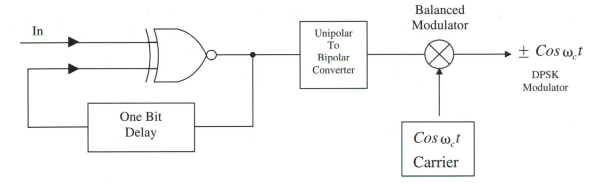

FIGURE 3.17 DPSK modulator block.

3.3 QUADRATURE AMPLITUDE MODULATION (QAM)

So far we have examined digital modulation schemes that carry the binary information in the phase of the carrier signal, but in our quest for higher bandwidth efficiency and ultimately a higher bit energy–to–noise density ratio (E_b/N_o), other various digital modulation schemes have been developed. Among them is the very efficient quadrature amplitude modulation (QAM). In a QAM system, the digital information is carried in both the phase and the amplitude of the carrier signal. In other words, with QAM, the amplitude and the phase of the carrier are simultaneously varied in accordance with the input digital information. This modulation technique exhibits certain definite advantages over the PSK scheme. For example, with a 16-PSK modulator, the theoretical maximum spectral efficiency is 4 b/s/Hz ($2^x = 2^4$; therefore, $x = 4$), reflecting an error probability equal to 10^{-4}. With a 16-QAM modulator, the same theoretical maximum spectral efficiency can be achieved while reflecting an error probability of 10^{-8}. It is therefore evident that with QAM modulation systems, an error probability improvement equal to 10^{-4} can be obtained.

The block diagram of an 8-QAM modulation scheme is given in Figure 3.18. From Figure 3.18, the input data f_b is divided into three parallel streams, the I channel, the Q channel, and the control channel—each carrying an equal $f_b/3$ bit rate. This block diagram is very similar to the 8-PSK modulation diagram, the only difference being that here, the control channel is simultaneously fed into the inputs of both digital-to-analog (D/A) converters without inversion. The resultant output signals from both D/A converters are equal in magnitude with polarities determined by the I- and Q-channel bits.

The mathematical expression of an 8-level QAM modulator is very similar to that of the 8-PSK modulator. The only exception is that here there is no inversion of the control bits applied to the Q channel, that is, only one voltage level is generated from both I and Q channels. Therefore,

$$I_0 = E\sin \omega_c t \tag{3.17}$$

$$Q_0 = E\cos \omega_c t \tag{3.18}$$

Here again, the polarity of the D/A converter #1 signal output is determined by the I channel (two bits) and the magnitude is determined by the control channel (one bit). Similarly,

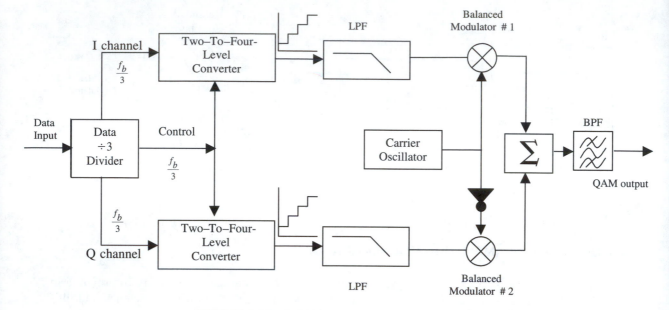

FIGURE 3.18 8-QAM modulator block diagram.

the polarity of the D/A converter #2 signal output is determined by the Q channel (two bits) and the magnitude is determined by the same control bit (no inversion). Therefore, two signal magnitudes and two polarities are fed into the summing circuit. The summing circuit output signal is composed of the following products:

$$+E \sin \omega_c t + E \cos \omega_c t$$
$$-E \sin \omega_c t + E \cos \omega_c t$$
$$+E \sin \omega_c t - E \cos \omega_c t$$
$$-E \sin \omega_c t - E \cos \omega_c t$$

3.3.1 Phasor Diagram

The phasor diagram of an 8-level QAM modulator is structured as follows:

	I	Q	C
A	0	0	0
B	0	0	1
C	0	1	0
D	0	1	1
E	1	0	0
F	1	0	1
G	1	1	0
H	1	1	1

where:

$$I = 0 \rightarrow -\sin \omega_c t \left.\right\} \text{ polarities}$$
$$I = 1 \rightarrow +\sin \omega_c t$$

$$Q = 0 \rightarrow -\cos \omega_c t \left.\right\} \text{ polarities}$$
$$Q = 1 \rightarrow +\cos \omega_c t$$

$$C = 0 \rightarrow 0.4142 \text{ V} \left.\right\} \text{ magnitude}$$
$$C = 1 \rightarrow 1 \text{ V}$$

For A

I	$0 \rightarrow -\sin \omega_c t$
Q	$0 \rightarrow -\cos \omega_c t$
C	$0 \rightarrow 0.4142$ V

Therefore, the summing output is given by $A = -0.4142 \sin \omega_c t - 0.4142 \cos \omega_c t$, or

Magnitude: $\quad V_A = \sqrt{-0.4142^2 + (-0.4142)^2} = 0.586$ V

$\qquad\qquad V_A = 0.586$ V

Phase: $\quad \theta = \tan^{-1}\left(\dfrac{-0.4142}{-0.4142}\right) = 45° \quad$ or $\quad \theta = -180 + 45 = -135°$

$\qquad\qquad \therefore \theta = -135°$

For B

I	$0 \rightarrow -\sin \omega_c t$
Q	$0 \rightarrow -\cos \omega_c t$
C	$1 \rightarrow 1$ V

Therefore, the summing output is expressed by $B = -\sin \omega_c t - \cos \omega_c t$, or

Magnitude: $\quad V_B = \sqrt{1^2 + 1^2} = 1.414$ V

$\qquad\qquad V_B = 1.414$ V

Phase: $\quad \theta = \tan^{-1}\left(\dfrac{-1}{-1}\right) = 45° \quad$ or $\quad \theta = -180° + 45° = -135°$

$\qquad\qquad \therefore \theta = -135°$

For C

I	$0 \rightarrow -\sin \omega_c t$
Q	$1 \rightarrow \cos \omega_c t$
C	$0 \rightarrow 0.4142$ V

Digital Radio

The summing output is expressed by $C = -0.4142 \sin \omega_c t + 0.4142 \cos \omega_c t$, or

$$\text{Magnitude:} \quad V_c = \sqrt{0.4142^2 + 0.4142^2} = 0.586 \text{ V}$$
$$V_c = 0.586 \text{ V}$$
$$\text{Phase:} \quad \theta = \tan^{-1}(-1) = -45°$$
$$\therefore \theta = +135°$$

For D

I	$0 \rightarrow -\sin \omega_c t$
Q	$1 \rightarrow \cos \omega_c t$
C	$1 \rightarrow 1 \text{ V}$

The summing output for D is expressed by $D = -\sin \omega_c t + \cos \omega_c t$, or

$$\text{Magnitude:} \quad V_D = \sqrt{1^2 + 1^2} = 1.41 \text{ V}$$
$$V_D = 1.414 \text{ V}$$
$$\text{Phase:} \quad \theta = \tan^{-1}\left(\frac{1}{-1}\right) = -45°$$
$$\therefore \theta = +135°$$

For E

I	$1 \rightarrow \sin \omega_c t$
Q	$0 \rightarrow -\cos \omega_c t$
C	$0 \rightarrow 0.4142$

The summing output for E is expressed by $E = 0.04142 \sin \omega_c t - 0.4142 \cos \omega_c t$, or

$$\text{Magnitude:} \quad V_E = \sqrt{0.4142^2 + 0.4142^2} = 0.586 \text{ V}$$
$$V_E = 0.586 \text{ V}$$
$$\text{Phase:} \quad \theta = \tan^{-1}\left(\frac{-0.4142}{0.4142}\right) = -45°$$
$$\therefore \theta = -45°$$

For F

I	$1 \rightarrow \sin \omega_c t$
Q	$0 \rightarrow -\cos \omega_c t$
C	$1 \rightarrow 1 \text{ V}$

The summing output for F is expressed by $F = \sin \omega_c t - \cos \omega_c t$, or

$$\text{Magnitude:} \quad V_F = \sqrt{1^2 + 1^2} = 1.414 \text{ V}$$
$$V_F = 1.414 \text{ V}$$

$$\text{Phase:} \qquad \theta = \tan^{-1}\left(\frac{1}{-1}\right) = -45°$$

$$\therefore \theta = -45°$$

For G

I	$1 \rightarrow \sin \omega_c t$
Q	$1 \rightarrow \cos \omega_c t$
C	$0 \rightarrow 0.4142$

The summing output for G is expressed by $G = 0.4142 \sin \omega_c t + 0.4142 \cos \omega_c t$, or

$$\text{Magnitude:} \qquad V_G = \sqrt{0.4142^2 + 0.4142^2} = 0.586 \text{ V}$$
$$V_G = 0.586 \text{ V}$$
$$\text{Phase:} \qquad \therefore \theta = 45°$$

For H

I	$1 \rightarrow \sin \omega_c t$
Q	$1 \rightarrow \cos \omega_c t$
C	$1 \rightarrow 1 \text{ V}$

The summing output for H is expressed by $H = \sin \omega_c t + \cos \omega_c t$, or

$$\text{Magnitude:} \qquad V_H = 1.414 \text{ V}$$
$$\text{Phase:} \qquad \theta = 45°$$

These values for magnitude and phase for each I, Q, and C combined will produce the phasor diagram of Figure 3.19.

3.3.2 Constellation Display

The constellation display of an 8-level QAM modulation scheme is shown in Figure 3.20.

3.3.3 Eight-Level QAM Signal Bandwidth

The signal bandwidth of an 8-level QAM modulation scheme is identical to that of the 8-PSK because here, as with 8-PSK, the binary input data is also divided into three streams; the I channel, the Q channel, and the control channel, each with a channel capacity equal to $f_b/3$. The mathematical analysis of the 8-QAM bandwidth is identical to that of the 8-PSK. Therefore, 8-QAM bandwidth

$$\text{BW} = \frac{f_b}{3}$$

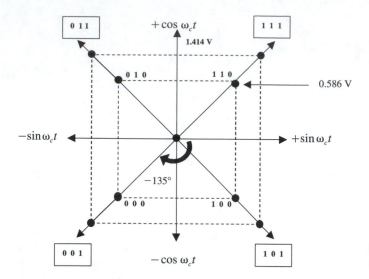

FIGURE 3.19   Phasor diagram.

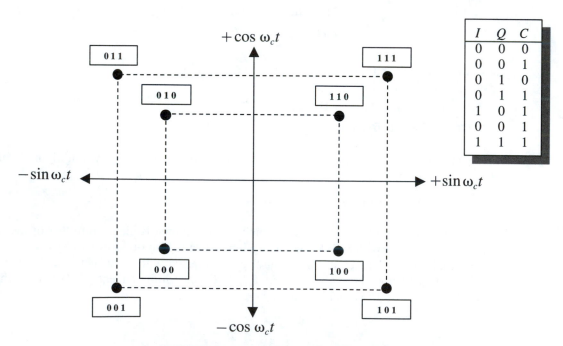

FIGURE 3.20   8-QAM constellation display.

EXAMPLE 3.1

Calculate the theoretical and practical spectral efficiency of the preceding 8-QAM digital modulation scheme required to process 20 Mb/s of input data. Assume LPF filter $\alpha = 0$, for theoretical and 0.5 for practical considerations.

Solution

The 20 Mb/s serial input binary data is divided into two parallel streams of 10 Mb/s each. Each 10 Mb/s stream is converted to three-level symbols by the D/A converter.

Number of Symbols

$$\text{Symbol rate} = \frac{10 \text{ Mb/s}}{\log_2(L)} \quad \text{where} \quad L = 3 \text{ (for 8-QAM, } 2^3 = 8)$$

$$= \frac{10 \text{ Mb/s}}{\log_2(3)} = \frac{10 \text{ Mb/s}}{1.585} = 6.31 \text{ million symbols/s}$$

Therefore, the **symbol rate** = 6.3 million symbols/s.

The minimum low-pass-filter Nyquist bandwidth is half the symbol rate or 6.31 Mb/s ÷ 2 = 3.155 MHz. The double side bandwidth at the output of the summing circuit is 2 × 3.155 MHz = 6.31 MHz. Therefore, it will require an absolute minimum theoretical IF bandwidth of 6.31 MHz to process a 20 Mb/s binary input data with an 8-QAM modulation scheme.

Spectral efficiency of 8-QAM
Spectral efficiency is given by

$$\frac{\text{Data input}}{\text{Bandwidth}} = \frac{20 \text{ Mb/s}}{6.31} \cong 3\tfrac{\text{b}}{\text{s}}/\text{Hz}$$

$\therefore$ Spectral efficiency $= 3\tfrac{\text{b}}{\text{s}}/\text{Hz}$ (Theoretical maximum with a filter $\alpha = 0$)

For practical considerations, filter requirements are relaxed to a reasonable $\alpha = 0.5$. This filter relaxation requires at least a 50% increase in IF bandwidth. Therefore,

$$\text{Spectral efficiency} = \frac{20 \text{ Mb/s}}{(1.5)(6.31)} = 2.1$$

$\therefore$ Spectral efficiency $\cong 2\tfrac{\text{b}}{\text{s}}/\text{Hz}$ (practical)

QAM digital modulation schemes have been evolved from relatively simple to very complex multiphase schemes such as 16-QAM, 64-QAM, 128-QAM, 512-QAM, 1024-QAM, and 2048-QAM. Although advanced QAM schemes above 64-QAM exhibit a higher spectral efficiency, they are very difficult to implement because of the higher carrier-to-noise requirements. A detailed table comparing different digital modulation techniques in reference to their spectral efficiency (b/s/Hz), **error probability** (P_e), and carrier-to-noise ratio (*C/N*) is given later in this chapter.

3.4 DIGITAL DEMODULATION

At the receiver end, the binary data is recovered from the carrier and further processed through demultiplexing and decoding circuits. The detector circuits implementing the digital signal detection are envelope detectors in the case of noncoherent detection and product detectors for coherent detection.

3.4.1 Coherent Detection

For long-distance digital communications links, simpler modulation techniques are used. One of these modulation techniques is the BPSK. This digital modulation scheme is the equivalent of suppressed-carrier orthogonal wave modulation. At the receiver end, the suppressed carrier is regenerated with a frequency and phase identical to that of the transmitter carrier frequency. This of course is an ideal situation. In actual practice, some frequency and phase errors must be expected. The regeneration of the carrier at the receiver end during the process of demodulation is referred to as "coherent demodulation." The block diagram of a coherent demodulator is shown in Figure 3.21.

Figure 3.21 shows that the digitally modulated RF signal at the input of the demodulator circuit undergoes a predetector filtering to eliminate all the unnecessary frequency components. The output of the filter is still an RF signal with Gaussian noise added to it. The product detector multiplies the complex RF signal with the carrier recovery signal, cos ωt. The LPF rejects the high-frequency component, and the low-frequency component is processed through.

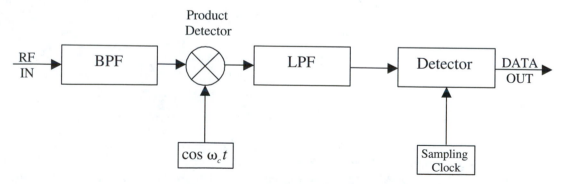

FIGURE 3.21   Block diagram of coherent demodulator.

3.4.2 FSK Demodulator

Basically there exist two methods of FSK demodulation: noncoherent, and coherent.

Noncoherent Demodulation

The FSK modulation scheme is presented here from a historical perspective. In actual practice, this method of digital modulation is very inefficient in terms of channel capacity and bit-error-rate performance. Because digital microwave systems are designed to maximize their available baseband bandwidth while maintaining a preset level of bit error rate, FSK schemes are incapable of satisfying the two fundamental system requirements. Therefore, they are only used for low-density systems.

A noncoherent FSK demodulator block diagram is illustrated in Figure 3.22. The FSK modulated signal is applied to both baseband filters BPF 1 and BPF 2. Baseband filter 1 is designed with bandpass bandwidth equal to transmitter frequency $f_c + \Delta f$; baseband filter 2 is designed with a bandpass bandwidth equal to $f_c - \Delta f$. The detected voltage at the output of each detector circuit is fed into the inverting and noninverting inputs of the comparator circuit.

Assume a positive voltage at the output of the comparator. If the output voltage of detector circuit 2 is higher than the output voltage of detector 1 by the comparator threshold voltage, then the output of the comparator will go to a negative voltage. It is therefore evident that the output of the detector circuit will shift between positive and negative voltage levels in accordance with the input FSK signal (bipolar signal). Since the two carrier signals composing the FSK transmitter signal are not regenerated at the demodulator circuit and are unable to provide phase or frequency synchronization, this demodulation process is classified as noncoherent.

Coherent FSK Demodulation

The most commonly used method for FSK coherent detection is the method utilizing a phase-lock loop (PLL) (Figure 3.23). A PLL circuit is composed of a phase detector, a loop filter and a voltage-controlled oscillator (VCO). The phase detector's function is to measure the phase difference between the VCO output and the FSK input signal, and to gener-

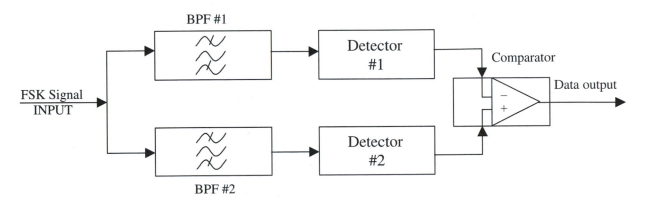

FIGURE 3.22 FSK demodulator (noncoherent)

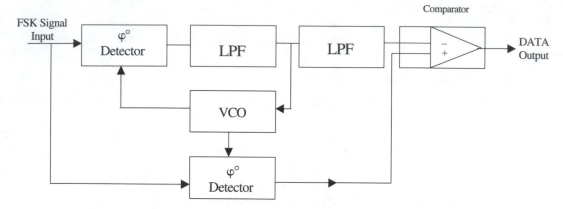

FIGURE 3.23 FSK demodulation employment PLL circuit (coherent detection).

ate either a positive or negative voltage at its output. If the phase of the FSK signal is leading the VCO phase, a positive voltage will be detected at the output of the phase detector. Likewise, if the phase of the FSK input signal is lagging the phase of the VCO signal, a negative voltage will be detected at the output of the phase detector.

From the circuit of Figure 3.23, it is evident that the PLL circuit is locked in, with the incoming FSK signal at two distinct frequencies $(f_c + \Delta f)$ and $(f_c - \Delta f)$ while its output is shifting between two voltage levels, one positive and one negative. This bipolar output signal is further processed and converted into the original modulation binary data stream.

FSK Demodulation Performance

The two fundamental parameters determining the performance of any digital modulation scheme are the ratio of **bit energy** to **noise spectral density**, E_b/N_o, and error power, $P(e)$. For noncoherent FSK detection, the relationship between these two parameters is given by

$$P(e) = \frac{1}{2}e^{-E_b/N_o} \tag{3.19}$$

For different values of E_b/N_o, the FSK modulation noncoherent performance graph of Figure 3.24 is constructed from the values of Table 3.2.

TABLE 3.2

E_b/N_o (dB)	$P(e)$
7	4.1×10^{-2}
8	2.1×10^{-2}
9	9.4×10^{-3}
10	3.4×10^{-3}
11	9.2×10^{-4}
12	1.8×10^{-4}
13	2.3×10^{-5}
14	1.7×10^{-6}

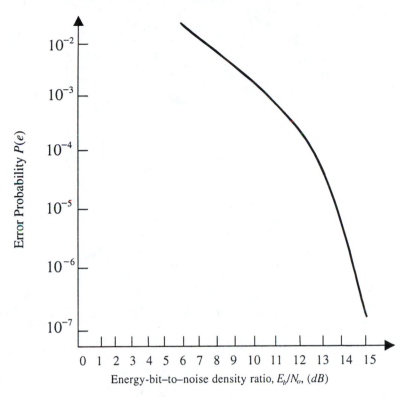

FIGURE 3.24 FSK noncoherent demodulation performance curve.

For

$$\frac{E_b}{N_o} = 7 \text{ dB}$$

$$7 = 10 \log\left(\frac{E_b}{N_o}\right)$$

$$\frac{E_b}{N_o} = \text{antilog}(0.7) = 5$$

$$\frac{E_b}{2N_o} = 2.5$$

$$P(e) = \frac{1}{2}e^{-2.5} = 4.1 \times 10^{-2}$$

$$\therefore P(e) = 4.1 \times 10^{-2}$$

Repeating this for $E_b/N_o = (8, 9, 10, 11, 12,$ and $13)$ dB, $P(e)$ is shown in Table 3.2. For coherent FSK detection, the performance curve is derived from

$$P(e) = \text{erf } C\left(\frac{E_b}{N_o}\right)^{1/2} \tag{3.20}$$

TABLE 3.3

E_b/N_o (dB)	$P(e)$
6	5.3×10^{-3}
7	1.7×10^{-2}
8	4.1×10^{-4}
9	7.1×10^{-5}
10	8.1×10^{-6}
11	5.4×10^{-7}

where

$$\text{erf}(x) \doteq \frac{e^{-x^2}}{x\sqrt{\pi}} \tag{3.21}$$

By solving Eq. (3.20) for different values of E_b/N_o, the corresponding values for $P(e)$ are obtained (Table 3.3).

For example, for $E_b/N_o = 6$ dB,

$$\frac{E_b}{N_o} = \text{antilog}(0.6) = 3.98$$

$$\text{erf}(2.0) = \frac{e^{-3.98^2}}{2\sqrt{\pi}} = 5.3 \times 10^{-3}$$

$$\left(\frac{E_b}{N_o}\right)^{1/2} \cong 2.0$$

$$\therefore P(e) = 5.3 \times 10^{-3}$$

The performance curve of a coherent FSK modulation scheme is illustrated in Figure 3.25. Comparing Figure 3.24 with Figure 3.25, it is evident that the coherent FSK modulation scheme is more efficient than the noncoherent method. Of course the coherent method requires more complex circuitry in its implementation.

3.4.3 BPSK Demodulator with Clock Recovery

It was mentioned previously that an accurate demodulation process requires the recovery of the carrier signal at the receiver end. The accurate regeneration of the carrier is required so that it can be compared with the incoming RF signal. This allows for the precise determination of the phase difference, a necessary requirement for the binary data detection.

Under normal circumstances, that is, in systems where the carrier signal is not suppressed at the transmitter, a phase-lock loop (PLL) at the receiver end can easily be employed to provide the much needed carrier synchronization between the transmitter and re-

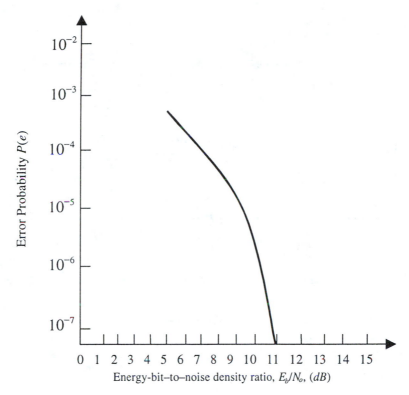

FIGURE 3.25 FSK coherent detection performance curve.

ceiver. Unfortunately, this method of synchronization cannot be employed in this case because the carrier is suppressed at the transmitter end. Therefore, another method must be implemented for carrier recovery. The most commonly used method is a "Costas-loop" or I-Q demodulator.

Costas Loop

The block diagram of the carrier recovery method employing the **Costas loop** is shown in Figure 3.26. The digitally modulated RF signal at the input of the circuit is given by

$$V(t) = V_{rms}d(t).\cos(\omega_c t + \theta) \tag{3.22}$$

where
$V(t)$ = instantaneous value (V)
V_{rms} = rms value (V)
$d(t)$ = data
θ = phase shift (°)

The power splitter divides the incoming signal into the I channel and Q channel (I $\rightarrow$ in phase, Q $\rightarrow$ quadrature phase). The low-pass filters eliminate the high-frequency

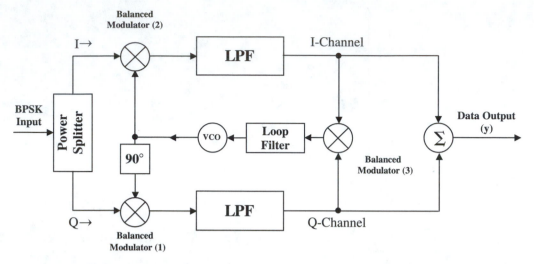

FIGURE 3.26 Block diagram of Costas-loop, carrier-recovery BPSK demodulator.

components, and both filter outputs are expressed by

$$V(t)_{\text{LPF I}} = \left(\frac{V_1}{2}\right) d(t) \sin(\theta_1 - \theta_2) \tag{3.23}$$

$$V(t)_{\text{LPF II}} = \left(\frac{V_2}{2}\right) d(t) \cos(\theta_1 - \theta_2) \tag{3.24}$$

where $(\theta_1 - \theta_2)$ = phase error

The two filter outputs are fed into the inputs of balanced modulator 3, which has its output expressed by

$$V(t)_{\text{BPF II}} = u(t)_{\text{LPF I}} u(t)_{\text{LPF II}} = \frac{(u_1 u_2)^2}{4} d^2(t) \sin 2(\theta_1 - \theta_2) \tag{3.25}$$

For $u_1 = u_2$ and $d(t) = \pm 1$, Eq. (3.25) becomes

$$u(t) = \frac{V^2}{4} \sin 2(\theta_1 - \theta_2) \tag{3.26}$$

The loop filter eliminates the high-frequency components, thus allowing the VCO to adjust its frequency in accordance with the loop-filter output voltage. When the VCO frequency has reached the suppressed carrier frequency, an error voltage will be detected at the loop-filter output. This voltage is proportional to the VCO phase error. Therefore, this error voltage will maintain the VCO frequency locked at the suppressed carrier frequency. The signal output of the circuit of Figure 3.26 is either

$$y = d(t) \cos(\theta_1 - \theta_2) \qquad \text{for } \theta_1 - \theta_2 = 0$$
$$y = d(t) \sin(\theta_1 - \theta_2) \qquad d(t) = 1$$

Substituting for $\theta_1 - \theta_2 = 0$ or $\theta_1 - \theta_2 = 180°$ and $d(t) = 1$ gives

$$y = d(t) \cos(0°) = 1 \quad \text{or} \quad y = d(t) \cos(180°) = 1$$

and

$$y = d(t) \sin(0°) = 0 \quad \text{or} \quad y = d(t) \sin(180°) = 0$$

By locking into two distinct phases of the carrier (0° or 180°), the BPSK demodulator will generate at its output the original data stream in addition to performing the clock recovery function.

BPSK Error Performance

The primary task or design objective of a demodulator circuit in a digital receiver is to differentiate the various phase states of the incoming signal from noise. For example, in a BPSK modulation scheme, the noise vector must generate a maximum phase shift of less than 180°. If the phase shift generated by the noise vector is equal to or larger than 180°, a bit error will occur. It is obvious that the criterion for evaluating performance of various PSK modulation schemes is the **error distance** (d) related to the spacing between adjacent phase states. This error distance is expressed by

$$d = 2 \sin\left(\frac{\pi}{N}\right) \tag{3.27}$$

where d = error distance (dB)
N = number of phase states

EXAMPLE 3.2

Determine the error distance in dB for a BPSK modulation scheme.

Solution:

$$d = 2 \sin\left(\frac{\pi}{N}\right) = 2 \sin\left(\frac{180}{2}\right) = 2$$
$$d_{dc} - 20 \log(2) = 6.02$$
$$\therefore d = 6.02 \text{ dB}$$

BPSK Error Probability

The error probability of a BPSK modulation scheme in reference to E_b/N_o is given by

$$P(e) = \frac{1}{2}\text{erfc}(x) \qquad (3.28)$$

where $\quad x = \sqrt{\dfrac{E_b}{N_o}}$

$$\text{erfc}(x) = \frac{e^{-x^2}}{x\sqrt{\pi}} \qquad (3.29)$$

It is evident from Eq. (3.29) that the error probability of a two-level digital modulation scheme is a function of the bit energy–to–spectral noise density ratio (E_b/N_o). This relationship can better be demonstrated by plotting the graph of $P(e)$ versus E_b/N_o (Figure 3.27). For different values of E_b/N_o, the corresponding values of $P(e)$ are obtained (Table 3.4).

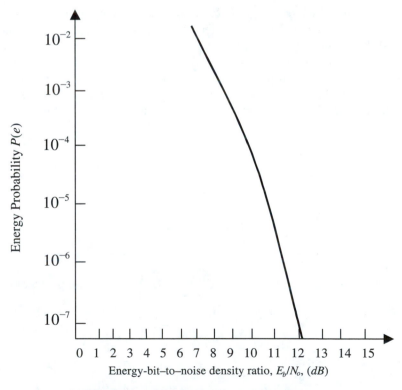

FIGURE 3.27 BPSK modulation performance curve.

TABLE 3.4

E_b/N_o (dB)	$P(e)$
7	8.4×10^{-4}
8	2.04×10^{-4}
9	3.6×10^{-5}
10	4×10^{-6}
11	2.7×10^{-7}
12	9.3×10^{-9}
13	1.3×10^{-10}

For

$$E_b/N_o = 7 \text{ dB},$$

$$\frac{E_b}{N_o} = \text{antilog}(0.7) = 5$$

$$\left(\frac{E_b}{N_o}\right)^{1/2} = 2.23$$

$$\text{erfc} = \frac{e^{-2.23^2}}{2.23\sqrt{\pi}} = 1.68 \times 10^{-4}$$

$$P(e) = \frac{1}{2}\text{erfc} = \frac{1}{2}(1.68 \times 10^{-4}) = 8.4 \times 10^{-5}$$

$$\therefore P(e) = 8.4 \times 10^{-5}$$

Repeating the process for all the E_b/N_o values of Table 3.4, the corresponding values of $P(e)$ have been obtained.

3.4.4 QPSK Demodulation

The block diagram of a QPSK demodulator is shown in Figure 3.28. The primary task of the QPSK demodulator is to recover the carrier signal (coherent detection). The recovered carrier signal, $\cos \omega_c t$, is mixed with the incoming QPSK signal by balanced modulator 1, while at balanced modulator 2, the recovered carrier is shifted by 90° (quadrature phase), then mixed with the incoming QPSK signal. The complex signal at the output of balanced modulator 1 is given by Eq. (3.30).

I channel

    Carrier signal $= \cos \omega_c t$.
    QPSK signal $= (\cos \omega_c t - \sin \omega_c t)$.
    Balanced modulator output

$$I = \cos\omega_c t(\cos \omega_c t - \sin \omega_c t) \tag{3.30}$$

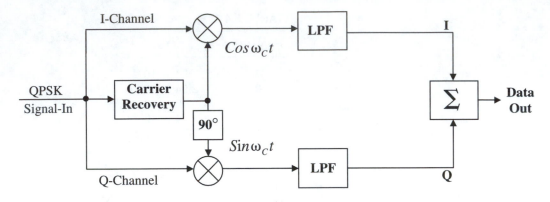

FIGURE 3.28   QPSK demodulator block diagram.

Applying trigonometric identities, we have

$$I = (\cos \omega_c t)(\cos \omega_c t - \sin \omega_c t) = \cos^2 \omega_c t - (\cos \omega_c t)(\sin \omega_c t)$$

$$= \frac{1}{2}(1 + \cos 2\omega_c t) - \frac{1}{2}\sin \omega_c t - \frac{1}{2}\sin(\omega_c t - \omega_c t)$$

$$= \frac{1}{2}(\cos 2\omega_c t) - \frac{1}{2}\sin \omega_c t$$

Therefore:

$$I = \frac{1}{2}(1 + \cos 2\omega_c t) - \frac{1}{2}\sin \omega_c t \qquad (3.31)$$

Q channel

Carrier signal $= \sin \omega_c t$.
QPSK signal $= \cos \omega_c t - \sin \omega_c t$.
Balanced modulator output: $Q = \cos \omega_c t(\cos \omega_c t - \sin \omega_c t)$.

Applying the same trigonometric identities as with the I channel, we have

$$I = \sin \omega_c t(\cos \omega_c t - \sin \omega_c t) = (\sin \omega_c t)(\cos \omega_c t) - \sin^2 \omega_c t$$

$$= \frac{1}{2}\sin \omega_c t + \frac{1}{2}\sin(\omega_c t - \omega_c t) - \frac{1}{2}(1 - \cos 2\omega_c t)$$

$$= \frac{1}{2}\sin \omega_c t - \frac{1}{2}(1 - \cos 2\omega_c t)$$

$$Q = \frac{1}{2}\sin \omega_c t - \frac{1}{2}(1 - \cos 2\omega_c t) \qquad (3.32)$$

The high-frequency components of Eqs. (3.31) and (3.32) are removed by the demodulator low-pass filters. The $\pm$ dc levels from the I and Q channels, representing the two parallel bit streams, are converted into a series bit stream and then further processed to appropriate circuitry for demultiplexing and decoding.

QPSK Error Performance

For a QPSK digital modulation scheme, the error performance will be examined in terms of error distance d, and error probability $P(e)$.

Error Distance (d) Error distance is determined by the length of the vector between two adjacent points in a PSK phasor diagram. If the vector is equal to or longer than the vector between two adjacent points, then a bit error will be induced. The error distance for any PSK modulation scheme is expressed by

$$d = 2\,\sin\!\left(\frac{\pi}{N}\right) \tag{3.33}$$

where $\quad d$ = error distance (dB)
N = number of different phases

EXAMPLE 3.3

$$d = 2\,\sin(\pi/N)$$

where $\quad N = 4(4 - PSK)$
$$d = 2\,\sin\!\left(\frac{180}{4}\right) = 1.41$$

In dB,

$$d = 20\,\log(1.41)$$
$$\therefore d = 3\text{ dB}$$

Applying the same relationship, the error distances for various levels of PSK modulation are shown in Table 3.5. From Table 3.5 it is evident that the error distance is progressively decreasing, with a corresponding increase of the $(N\phi)$ of the PSK modulation scheme.

Error Probability $P(e)$ The error probability of a PSK system is given by

$$P(e) = (\log_2 N_\phi)\,\text{erfc}(Z) \tag{3.34}$$

TABLE 3.5

Modulation Schemes	N_ϕ	Error Distance (dB)
BPSK	2	6.02
QPSK	4	3.01
8-PSK	8	−2.3
16-PSK	16	−8.17
32-PSK	32	−14.15

where N_ϕ = number of phases of the modulation scheme used

$$Z = \frac{E \sin(\pi/N_\phi)}{\sigma\sqrt{2}} \quad (3.35)$$

where E = amplitude of the signal at the detector input (V_{rms})
 σ = noise voltage (V_{rms}).

The signal amplitude (E) is given by

$$E = \sqrt{E_b(\log_2 N_\phi)(1/T)} \quad (3.36)$$

where E_b = energy of the bit (joules)
 N_ϕ = number of phases
 T = bit duration (s)

The noise voltage (σ) is given by

$$\sigma = \sqrt{N_\phi\left(\frac{1}{2T}\right)} \quad (3.37)$$

To compute the value of (Z), substitute Eqs. (3.36) and (3.37) into (3.35):

$$Z = \frac{E \sin(\pi/N_\phi)}{\delta\sqrt{2}} = \frac{\sqrt{E_b(\log_2 N_\phi)(1/T)}\ \sin(\pi/N_\kappa)}{\sqrt{N_o(1/2T)}\sqrt{2}}$$

$$= \frac{\sin(\pi/N_\phi)}{\sqrt{2}}\sqrt{\frac{E_b(\log_2 N_\phi)(1/T)}{N_o(1/2T)}} = \frac{\sin(\pi/N_\phi)}{\sqrt{2}}\sqrt{\frac{2E_b\log_2 N_\phi}{N_o}}$$

$$= \frac{\sin(\pi/N_\phi)}{\sqrt{2}}\sqrt{2}\sqrt{\frac{E_b}{N_o}\log_2(N_\phi)}$$

Therefore,

$$Z = \sin\left(\frac{\pi}{N_\phi}\right)\sqrt{\frac{E_b}{N_o}\log_2(N_\phi)} \tag{3.38}$$

From Eqs. (3.38) and (3.34), it is evident that the error probability is relevant to the modulation scheme used, (N_ϕ), and the bit energy–to–noise density ratio (E_b/N_o). A simplified formula relating error probability $P(e)$ and the carrier-to-noise ratio (**CNR**, *C/N*) for different PSK phase levels is given by

$$P(e) = e^{-(C/N)\sin^2(\pi/N_\phi)} \tag{3.39}$$

where C/N = carrier-to-noise ratio
                N_ϕ = number of phases

Solving Eq. 3.39 for various PSK levels, we get the following data (Table 3.6).

For BPSK

$$P(e) = e^{(-C/N)\sin^2(\pi/2)} = e^{-C/N}$$

For $P(e)10^{-7}$ where $\sin^2(180/2) = 1$,

$$-C/N = \ln(10^{-7})$$
$$-C/N = -13.8155$$

In dB,

$$\therefore C/N = 11/4 \text{ dB}$$
$$\frac{E_b}{N_o} = C/N\left(\frac{\text{BW}}{f_b}\right)$$

TABLE 3.6 For $P(e) = 10^{-7}$

Modulation Schemes	Theoretical Efficiency	Practical Efficiency	E_b/N_o (dB)	C/N dB
BPSK	1	0.8	11.4	11.4
QPSK	2	1.5	11.4	14.4
8-PSK	3	2.6	14.98	19.76
16-PSK	4	3.0	19.58	25.6
32-PSK	5	3.8	24.58	31.57
64-PSK	6	4.5	30	32.58

For a theoretical bandwidth efficiency BW $= f_b$,

$$\frac{f_b}{N_o} = C/N = 11.4 \text{ dB}$$

$$\therefore \frac{E_b}{N_o} = 11.4 \text{ dB}$$

For QPSK

$$P(e) = e^{(-C/N)\sin^2(\pi/N_\phi)}$$

$$10^{-7} = e^{-\frac{1}{2}C/N}$$

$$-0.5 \, C/N = \ln(10^{-7})$$

$$\therefore \, C/N = 14.4 \text{ dB}$$

$$\frac{E_b}{N_o} = C/N\left(\frac{\text{BW}}{f_f}\right)$$

The bandwidth efficiency (theoretical) for QPSK $= 2$, so

$$\frac{E_b}{N_o} = C/N\left(\frac{\text{BW}}{2\text{BW}}\right)$$

$$\frac{E_b}{N_o} = \frac{1}{2}C/N$$

$$\therefore \frac{E_b}{N_o} = 11.4 \text{ dB}$$

For 8-PSK

$$N_\phi = 8,$$

$$\sin^2(180/8) = 0.1464$$

$$P(e) = e^{-0.14C/N}$$

$$\therefore \, C/N = 19.76 \text{ dB}$$

$$\frac{E_b}{N_o} = C/N\left(\frac{\text{BW}}{3\text{BW}}\right)$$

The bandwidth efficiency (theoretical) for 8-PSK $= 3$, so

$$\frac{E_b}{N_o} = \frac{1}{3}C/N$$

$$\frac{E_b}{N_o} = 14.98 \text{ dB}$$

For 16-PSK

$$N_\phi = 16$$
$$\sin^2(180/16) = 0.038$$
$$P(e) = e^{-0.038C/N}$$

Similarly,

$$C/N = 25.6 \text{ dB}$$

The theoretical bandwidth efficiency = 4 b/s/Hz), so

$$E_b/N_o = C/N(\text{BW}/4\text{BW})$$
$$E_b/N_o = 19.58 \text{ dB}$$

For 32-PSK

$$N_\phi = 32$$
$$C/N = 31.57 \text{ dB}$$

The theoretical bandwidth efficiency = 5 b/s/Hz, so

$$E_b/N_o = 24.58 \ dB$$

For 64-PSK

$$N_\phi = 64$$
$$C/N = 37.58 \text{ dB}$$

The theoretical bandwidth efficiency = 6 b/s/Hz, so

$$E_b/N_o = 30 \ dB$$

PSK Modulation Graphs

Graphs relating various levels of E_b/N_o and error probability $P(e)$ for BPSK, QPSK, 8-PSK, 16-PSK, and 32-PSK are derived as follows:

BPSK

$$P(e) = e^{(-C/N)\sin^2(\pi/N_\phi)}$$

For

$$P(e) = 10^{-4}$$
$$\ln(10^{-4}) = -C/N$$
$$\frac{E_b}{N_o} = 8.4 \text{ dB}$$

where
$$N_\phi = 2 \text{ (number of phases)}$$
$$\sin^2(180/2) = 1$$

$$\frac{E_b}{N_o} = C/N\left(\frac{\text{BW}}{f_f}\right)$$
$$\text{BW} = f_b$$
$$\therefore \frac{E_b}{N_o} = C/N$$

Applying this relationship for various levels of $P(e)$, the corresponding values of E_b/N_o are obtained (Table 3.7).

QPSK
Here again, $N_\phi = 4$, $\sin^2(180/4) = 0.5$.

For
$$P(e) = e^{(-C/N)\sin^2(\pi/N_\phi)}$$
$$P(e) = 10^{-7},$$
$$C/N = 14.4 \text{ dB}$$

TABLE 3.7 BPSK

$P(e)$	E_b/N_o (dB)
10^{-4}	8.4
10^{-5}	9.6
10^{-6}	10.6
10^{-7}	11.4
10^{-8}	12

Theoretical bandwidth efficiency for QPSK = 2 b/s

$$\frac{E_b}{N_o} = C/N \left(\frac{BW}{f_b}\right) = \frac{1}{2} C/N$$

$$\frac{E_b}{N_o} = 11.4 \text{ dB}$$

For different levels of $P(e)$, the corresponding values of E_b/N_o are obtained (see Table 3.8).

8-PSK

Similarly,

$$P(e) = e^{(-C/N)\sin^2(\pi/N_\phi)}$$
$$N_\phi = 8 \qquad \sin^2(180/80) = 0.1464$$
$$\therefore C/N = 19.76 \text{ dB}$$

bandwidth efficiency = 3 b/s/Hz, so

$$\frac{E_b}{N_o} = C/N \frac{BW}{3BW}$$

$$\frac{E_b}{N_o} = \frac{1}{3} C/N$$

$$\therefore \frac{E_b}{N_o} = 14.98 \text{ dB}$$

Repeating this process for various values of $P(e)$, the corresponding values of E_b/N_o are obtained (see Table 3.9).

TABLE 3.8 QPSK

$P(e)$	E_b/N_o (dB)
10^{-4}	8.4
10^{-5}	9.6
10^{-6}	10.6
10^{-7}	11.4
10^{-8}	12

TABLE 3.9 8-PSK

$P(e)$	E_b/N_o (dB)
10^{-2}	7.2
10^{-3}	10.2
10^{-4}	11.96
10^{-5}	13.2
10^{-6}	14.18
10^{-7}	14.97

TABLE 3.10   16-PSK

$P(e)$	E_b/N_o (dB)
10^{-2}	11.8
10^{-3}	14.8
10^{-4}	16.5
10^{-5}	17.8
10^{-6}	18.8

16-PSK

For 16-PSK, $\sin^2(\pi/16) = 0.038$ and Maximum bandwith efficiency = 4. Following the same format as with 8-PSK, Table 3.10 is obtained. The graph representing all the PSK modulation schemes derived from Eq. (3.39) is shown in Figure 3.29. This expression for the derivation of the PSK modulation curves relating E_b/N_o and $P(e)$ is accurate for low phase levels up to QPSK. Beyond this level, an error is observed to a maximum of ± 1 dB.

3.5 QAM DEMODULATION

The block diagram of a QAM demodulator is shown in Figure 3.30. The M-ary QAM demodulator of Figure 3.30 is a coherent detector similar to a M-ary PSK demodulator circuit. This circuit also incorporates a carrier recovery circuit, as well as a symbol-timing recovery circuit for maximum error performance.

The IF input of the QAM signal is band-limited for the removal of excessive noise, then divided into the I channel and Q channel through a power divider circuit. The recovered carrier is multiplied with the incoming IF signal by both balanced modulators. The 90° phase shift generates the orthogonal signal $\sin \omega_c t$ required for the Q channel.

The low-pass filters smooth the multilevel signals of both the I channel and the Q channel of their high-frequency components while the threshold comparators convert the multilevel channel signals into two-level signals. The complexity of the comparator circuits is related to the number of levels at their input. That is, a high-level M-ary QAM demodulator requires a more complex and sophisticated comparator circuit design.

The times two data combiner converts the parallel input binary data from both I and Q channels, $(f_b/2) + (f_b/2)$, into a series data stream (f_b).

3.5.1 QAM Error-Performance Evaluation

The error performance of a multilevel QAM modulation scheme is determined by the error distance (d) and error probability $P(e)$ versus E_b/N_o.

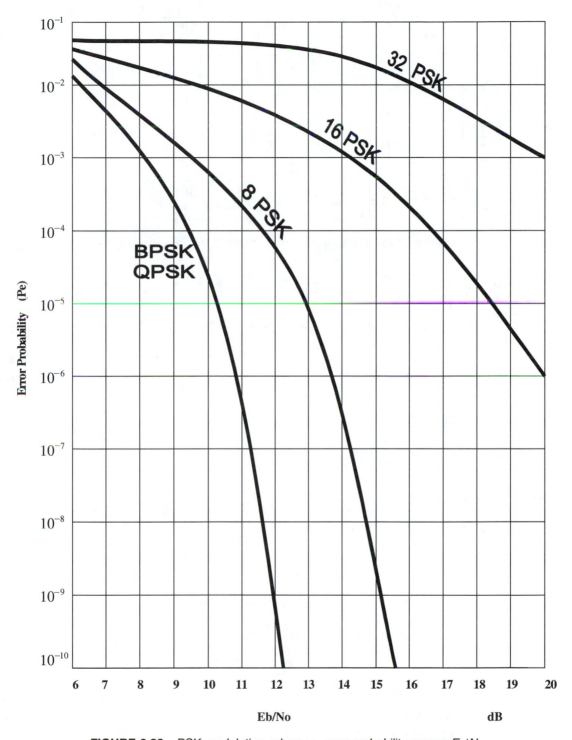

FIGURE 3.29 PSK modulation schemes: error probability versus E_b/N_o.

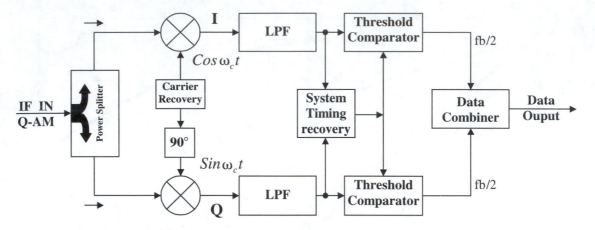

FIGURE 3.30 Block diagram of an *M*-ary QAM demodulator.

Error-Distance (d)

The error distance between two adjacent parts in a QAM modulation scheme is given in dB by

$$d = 20 \log \left[\sqrt{2}(L - 1)^{-1} \right] \tag{3.40}$$

Applying Eq. (3.40) for various QAM schemes results in the values of error shown in Table 3.11.

EXAMPLE 3.4

Determine the error distance of the 8-QAM modulation scheme, assuming that the peak power levels of adjacent points are equal.

Solution

$$d = 20 \log \left[\sqrt{2}(L - 1)^{-1} \right]$$

For 8-QAM: $L = 2$, (two levels for each quadrant), so

$$d = 20 \log \left[\sqrt{2}(2 - 1)^{-1} \right] = 1.5 \text{ dB}$$

Repeating this process for different modulation schemes, we get Table 3.11. From Table 3.11 it is evident that the error distance progressively decreases with increases of QAM levels.

TABLE 3.11

Modulation Scheme	Number of Levels (L) per Quadrant	Error Distance (dB)
8-QAM	2	1.5
16-QAM	4	−3.26
32-QAM	8	−6.95
64-QAM	16	−10.25

Error Probability (P_e)

The error probability of various levels of QAM modulation schemes in reference to E_b/N_o is given by

$$P(e) = (\log_2 L)^{-1}\left(\frac{L-1}{L}\right)\text{erfc}(x) \qquad (3.41)$$

where $\qquad L$ = number of levels per quadrant

$$\text{erfc}(x) = 1 - \text{erf}(x)$$

$$x = \frac{(\log_2 L)^{1/2}}{L-1}\left(\frac{E_b}{N_o}\right)^{1/2} \qquad (3.42)$$

EXAMPLE 3.5: 8-QAM

Compute the error probability $P(e)$ of an 8-QAM modulation scheme for various levels of E_b/N_o.

Solution
For $E_b/N_o = 7$ dB, and $L = 2$ (2 levels for quadrant),

$$\frac{E_b}{N_o} = \text{antilog}(0.7) = 5$$

$$\therefore \frac{E_b}{N_o} = 5 \quad \text{or} \quad \left(\frac{E_b}{N_o}\right)^{1/2} = 2.24$$

$$x = \frac{\log_2(2)}{2-1}(2.24)$$

$$\therefore x = 2.24$$

From tables,

$$\text{erf}(x) = 0.998140$$
$$\text{erfc}(x) = 1 - \text{erf}(x) = 1.86 \times 10^{-3}$$
$$P(e) = \frac{1}{2}\text{erfc}(x) \cong 9 \times 10^{-4}$$

Applying the same formula for various levels of E_b/N_o results in the corresponding values of error probability shown in Table 3.12. The complete table incorporating all necessary steps for the derivation of error probability in reference to E_b/N_o is shown in Table 3.13. The error probability values shown in Table 3.13 are approximate to within 1 dB of E_b/N_o.

TABLE 3.12

E_b/N_o (dB)	$P(e)$ 10^{-x}
7	9×10^{-4}
8	2×10^{-4}
9	3×10^{-5}
10	5×10^{-6}
11	1×10^{-7}
12	1×10^{-8}

TABLE 3.13   8-QAM ($L = 2$)

E_b/N_o (dB)	$\sqrt{E_b/N_o}$	x	erf(x)	erfc(x)	$P(e)$
7	2.2	2.2	0.998137	1.8×10^{-3}	9×10^{-4}
8	2.5	2.5	0.9995375	4.6×10^{-4}	2×10^{-4}
9	2.8	2.8	0.999925	7.5×10^{-5}	3×10^{-5}
10	3.16	3.16	0.999959	4.1×10^{-5}	5×10^{-6}
11	3.54	3.54	0.999998	2×10^{-6}	1×10^{-7}

EXAMPLE 3.6: 16-QAM

Calculate the error probability $P(e)$ of a 16-QAM modulation scheme for various levels of E_b/N_o.

Solution

For $E_b/N_o = 7$ dB, and $L = 4$ (4 levels per quadrant),

$$\frac{E_b}{N_o} = 5$$
$$\left(\frac{E_b}{N_o}\right)^{1/2} = 2.24$$

$$x = \frac{\sqrt{2}(\log_2 4)^{1/2}}{4-1}\left(\frac{E_b}{N_o}\right)^{1/2} = \left(\frac{2}{3}\right)(2.24) = 1.49$$

$$\therefore x = 1.49$$

From tables,

$$\text{erf}(x) \cong 0.966105$$
$$\text{erfc}(x) \cong 0.033895$$

$$P(e) = (\log_2 4)^{-1}\left(\frac{4-1}{4}\right)(3.3 \times 10^{-2})$$

$$= \left(\frac{1}{2}\right)\left(\frac{3}{4}\right)(3.3 \times 10^{-2})$$

$$= 1.3 \times 10^{-2}$$

$$\therefore P(e) \cong 1.3 \times 10^{-2}$$

Repeating the same process for different values of E_b/N_o gives the values of error probability that are listed in Table 3.14. Repeating the same process for different values of E_b/N_o, we get the corresponding values of error probability listed in Table 3.15a.

TABLE 3.14 16-QAM ($L = 4$)

E_b/N_o (dB)	$\sqrt{E_b/N_o}$	x	erf(x)	erfc(x)	$P(e)$
7	2.24	1.12	0.88100	0.11900	4×10^{-2}
8	2.51	1.23	0.922161	0.077839	2×10^{-2}
9	2.81	1.4	0.952285	0.047715	1×10^{-2}

TABLE 3.15a

E_b/N_o (dB)	$P(e)$
7	1.3×10^{-2}
8	8×10^{-3}
9	3.44×10^{-3}
10	8.2×10^{-4}
11	2.7×10^{-4}
12	5×10^{-5}
13	8.7×10^{-6}
14	1×10^{-6}

EXAMPLE 3.7: 32-QAM

Calculate the error probability $P(e)$ of a 32-QAM modulation scheme for various levels of energy bit–to–noise density ratio.

Solution

Repeating the preceding process for various levels of E_b/N_o, gives the values of $P(e)$ shown in (Table 3.16a).

TABLE 3.16a

E_b/N_o	$P(e)$
7	8×10^{-2}
8	7×10^{-2}
9	5×10^{-2}
10	4×10^{-2}
11	3×10^{-2}
12	2×10^{-2}
13	1×10^{-2}

EXAMPLE 3.8: 64-QAM

Compute the error probability $P(e)$ for a 64-QAM modulator scheme for various levels of energy bit–to–noise density ratio.

Solution

Repeating the same process as with 16-QAM, gives the data listed in Table 3.15b.

TABLE 3.15b

E_b/N_o	$P(e)$
7	1.6×10^{-1}
8	8×10^{-2}
9	7×10^{-2}
10	6×10^{-2}
11	5×10^{-2}
12	4×10^{-2}
13	3×10^{-2}

3.5.2 Comparison of Digital Modulation Schemes

The modulation schemes that have been examined are all available to the system designer for proper implementation in digital microwave links. In such a system, the three most important parameters taken into consideration during the design process are: (a) error probability or bit error rate (BER), (b) the system's capacity or maximum bit rate (f_b) based on the selected modulation scheme spectral efficiency, and (c) the required carrier-to-noise ratio (C/N). The selection of the appropriate modulation scheme is critical in maximizing the effectiveness of these parameters for optimum system performance.

Maximizing system capacity is the ultimate design goal. This maximum operating system bit rate is compared with the theoretical maximum given by the Shannon Theorem, expressed by

$$C = \text{BW} \log_2(1 + C/N) \qquad (3.43)$$

where
C = maximum theoretical channel capacity (b/s)
BW = channel bandwidth (Hz)
C/N = carrier-to-noise ratio (dB)

The theoretical maximum spectral efficiency for different carrier-to-noise ratios is calculated as follows:

EXAMPLE 3.9

Let $C/N = 5$ dB, or $C/N = $ antilog $(0.5) = 3.16$, $\log_2(1 + 3.16) = \log_2(4.16) = 2$. The theoretical spectral efficiencies for various levels of C/N are illustrated in Table 3.16b. The theoretical maximum spectral efficiency for Table 3.16 is shown in Figure 3.31. In practice, spectral efficiency is smaller than the theoretical maximum. The interrelationship between spectral efficiency, bit error rate, and carrier-to-noise ratio is illustrated in the following example.

EXAMPLE 3.10

A digital microwave link is to transmit 90 Mb/s with a system available bandwidth of 36 MHz while maintaining $E_b/N_o = 10^{-8}$.
Determine (a) spectral efficiency (b/s/Hz), (b) modulation scheme, (c) E_b/N_o, and (d) Filter (α = system BPF rolloff: $0 \le \alpha \le 1$).

Solution
$f_b = 90$ Mb/s
BW = 36 MHz
$E_b/N_o = 10^{-8}$

TABLE 3.16b

C/N (dB)	$\log_2\left(1 + \dfrac{C}{N}\right)$ (b/s/Hz)
5	2
10	3.5
15	5
20	6.5
25	8.3
30	9.9
35	11

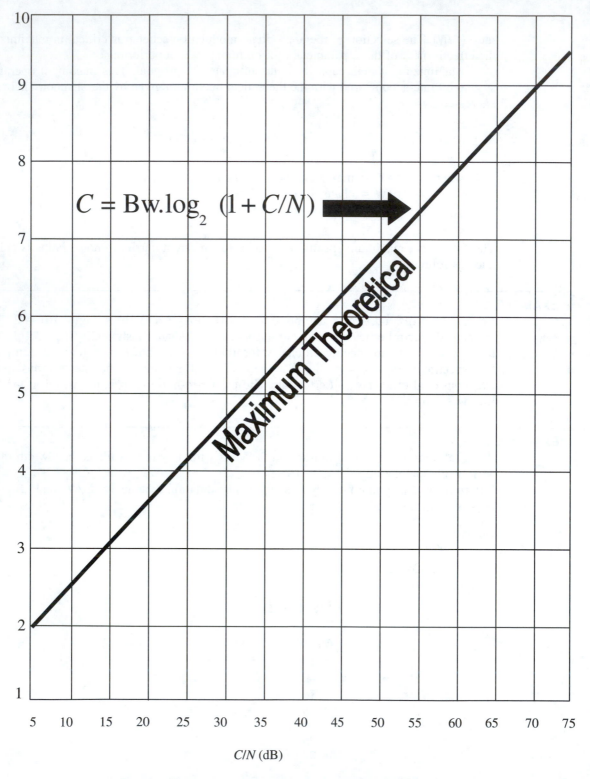

$$C = \mathrm{Bw}.\log_2 (1 + C/N)$$

Maximum Theoretical

Spectral Efficiency (b/s/Hz)

C/N (dB)

FIGURE 3.31 Digital modulation schemes spectral efficiency b/s/Hz.

Spectral Efficiency

$$\text{Spectral efficiency} = \frac{90 \text{ Mb/s}}{36 \text{ MHz}} = 2.5 \text{ b/s/Hz}$$

Round off to 3 b/s/Hz.

$$\therefore \text{ Spectral efficiency} = 3 \text{ b/s/Hz}$$

Modulation Scheme

The selection of the appropriate modulation scheme will be based on the comparison between modulation schemes with the same spectral efficiency and smaller E_b/N_o requirements. From graphs, the two modulation scheme candidates for this particular application are the 8-PSK and 8-QAM. Both 8-PSK and 8-QAM have a maximum spectral efficiency of 3 b/s/Hz, but 8-QAM requires approximately 18 dB of C/N, while 8-PSK requires 21 dB. Therefore, the more efficient 8-QAM is selected for this application.

$$\therefore \text{ Modulator scheme selected is 8-QAM}$$

E_b/N_o

The required E_b/N_o is determined as follows:

$$\frac{E_b}{N_o} = \frac{C}{N}\left(\frac{\text{BW}}{f_b}\right)$$

Since

$$\text{BW} = 14 \text{ dB}$$
$$f_b = 90 \text{ Mb/s}$$
$$\frac{C}{N} = 18 \text{ dB} \quad (\text{required for a } P(e) = 10^{-8})$$

Then

$$\frac{E_b}{N_o} = 63\left(\frac{36}{90}\right) = 25.2$$

In dB,

$$\frac{E_b}{N_o} = 14 \text{ dB}$$

Filter α

The filter α is determined as follows:

System available bandwidth BW = 36 MHz.

The theoretical minimum required is equal to system bit rate divided by spectral efficiency.

$$\frac{90 \text{ Mb/s}}{3 \text{ b/s}} = 30 \text{ MHz}$$

The ratio of theoretical minimum and available bandwidth is equal to

$$\frac{30 \text{ MHz}}{36 \text{ MHz}} = 0.83$$

$$\alpha = 1 - 0.83 = 0.17$$

Rounding off gives

$$\alpha = 0.2$$

EXAMPLE 3.11

Calculate the error probability $P(e)$ of a 16-QAM modulation scheme for $E_b/N_o = 7$ dB.

Solution

$$\frac{E_b}{N_o} = 7 \text{ dB}$$

$$\frac{E_b}{N_o} = \text{antilog}(0.7) = 5$$

$$P(e) = (\log_2 L)^{-1}\left(\frac{L-1}{L}\right)\text{erfc}(x)$$

If $L = 4$ (16-QAM),

$$P(e) = (\log_2 4)^{-1}\left(\frac{4-1}{4}\right)\text{erfc}(x)$$

$$= 0.37 \text{ erfc}(x)$$

$$x = \frac{(\log_2 L)}{L-1}\left(\frac{E_b}{N_o}\right)^{1/2} = \frac{(\log_2 4)^{1/2}}{4-1}(2.24) = 1.12$$

Therefore, $x = 1.12$ and $\text{erf}(x) = 0.881$, so

$$\text{erfc}(x) = 1 - \text{erf}(x) = 1 - 0.881 = 0.119$$

Therefore, $P(e) = (0.37)(0.119) = 4.4 \times 10^{-2}$.
This high $P(e)$ is the result of a very low of E_b/N_o of 7 dB.

QUESTIONS

1. List the main components of a digital radio transmitter and briefly describe their functions.
2. With the assistance of a block diagram, briefly describe the operation of a digital radio receiver.
3. Sketch the block diagram of a digital modulator and describe its function.
4. What is the main function of a coder in a digital radio? With the assistance of a block diagram, briefly explain its operation.
5. Describe the operation of a digital demodulator.
6. Derive the output voltage relationship of an FSK modulator.
7. Derive the bandwidth relationship of an FSK modulation scheme.
8. Describe the difference between FSK and MSK modulation schemes.
9. Describe the principal characteristics of a phase-shift-keying (PSK) modulation scheme.
10. Derive the signal bandwidth of a BPSK modulator.
11. Sketch the phase-state diagram of a QPSK modulator.
12. Define quadrature amplitude modulation (QAM). With the assistance of a block diagram, briefly describe its function.
13. Sketch the block diagram of a digital demodulator and describe its operation.
14. State the difference between coherent and noncoherent detection.
15. What are the fundamental components determining FSK demodulation performance and how are they related?
16. Why is clock recovery required in a BPSK demodulator circuit?
17. What are the fundamental components determining BPSK error performance and how are they related?
18. With the assistance of a block diagram, describe the function of a 16-QAM modulation scheme.
19. List the differences and similarities of the PSK and QAM modulation schemes.
20. List the fundamental components involved in the determination of the error performance of a QAM modulator and show their interrelationship.

Line-of-Sight Microwave Links

Objectives

- Identify the advantages of LOS microwave links.
- Describe in detail the various components of an analog LOS microwave link system.
- Describe in detail the important parameters involved in a LOS digital microwave system.
- Calculate the antenna height of a LOS microwave system.
- Design a complete digital LOS microwave system.

Key Terms

Transmission path
Repeater
Transmission path
  redundancy
Fade margin
Transmission path
  degradation
Pre-emphasis
De-emphasis
Traveling-wave tube (TWT)
Frequency-division
  multiplexing (FDM)

Time-division multiplexing
  (TDM)
Standard group
Standard super group
Standard super master
  group
FM Modulation index
Side bands of FM signals
Bessel function coefficients
FM bandwidth
FM improvement
  threshold

System gain
Signal fading
Multipath fading
Direct path
Reflective path
Path losses
Receiver sensitivity
Bit error rate
Receiver noise figure
First Fresnel zone
    clearance

INTRODUCTION

The main objective of a line-of-sight (LOS) microwave link is to transmit modulated carrier signals of high frequency (GHz) from one point and to successfully receive, demodulate and demultiplex the baseband signal at another point according to pre-established specifications and at a desired distance (Figure 4.1). Line-of-sight microwave links are classified as digital and analog. Digital microwave links are designed to process baseband signals in digital form; analog microwave links are intended to process baseband signals in analog form.

For LOS microwave links, the carrier frequencies are usually above 200 MHz. For digital transmission, the frequency range between 1.8 GHz and 7 GHz is utilized. This high frequency has two main advantages: first, it provides the large bandwidth necessary for high-bit-rate transmission; second, high carrier frequencies are less susceptible to atmospheric effects induced by the **transmission path.**

Perhaps the most important requirement of any communications system is its reliability factor. This is the percentage of time within a year that the system is fully operational while strictly maintaining pre-established system specifications. This percentage of reliability is usually set at 99.98% of the time; therefore, the maximum time that the link is allowed to be out of service is 1.7 hours for 1 year of operation.

The pre-established performance requirement for analog systems is the receiver signal-to-noise ratio (SNR); for digital systems it is the **bit error rate** (BER).

The two main components in a line-of-sight microwave link are the transmission path and the available terminal equipment. For the link to maintain the required high degree of reliability, both the transmission path and the terminal equipment must be operational at all times. This of course, requires that equipment as well as transmission path redundancy be incorporated into the system design.

Transmission path redundancy is required because the transmission path is susceptible to atmospheric changes (i.e., rain, snow, hail, etc.), and equipment redundancy is also required to provide a backup alternative in case of equipment failure. To remedy the **transmission path degradation** due to atmospheric impairment, the transmitter power is increased beyond the maximum required level under normal operating conditions. This increase of transmitter power is called **fade margin.** Systems operating at different carrier frequencies require different levels of fade margin. Since path fading due to atmospheric

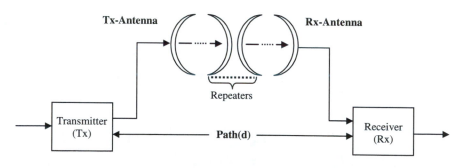

FIGURE 4.1   LOS microwave link block diagram.

impairment is proportional to the carrier frequency, the higher the operating frequency, the higher the transmission path degradation. Fade margins between 35 to 45 dB are normal for line-of-sight digital microwave links.

The wide variations of fade margin are mainly due to microwave link geographic location. Drier geographic locations require less fade margin than locations with heavy and long rainy seasons. An excessive fade margin for wet geographic locations can be used to increase the path length in dry areas. On the other hand, a strong fade margin under clear atmospheric conditions becomes problematic for the receiver demodulator.

This excessive transmitter power may saturate the demodulator circuit and increase the system performance degradation. For digital systems, an increase of the BER is possible. Normally, the receivers are equipped with sophisticated automatic-gain-control (AGC) circuits to counteract the excessive fade margin.

To cover the desired distance (say 500 km) several **repeaters** are required. Such repeaters consist of a transmitter, a receiver, and associated equipment and supporting facilities such as an antenna tower, buildings, roads, water, and electrical supplies. It is therefore obvious that the smaller the number of repeaters, the more economical the microwave link. It must also be noted that the maximum hop length (with no repeaters) of a microwave link is approximately 62 km.

4.1 ANALOG LINE-OF-SIGHT MICROWAVE LINKS

The basic block diagram of an analog line-of-sight microwave link is shown in Figure 4.2. It consists of the transmitter section, the transmission path, and the receiver section.

4.1.1 The Transmitter

The transmitter section of a line-of-sight microwave link is composed of a **pre-emphasis** circuit, the FM modulator, an up-converter, the power amplifier (usually a **traveling-wave tube (TWT)**) and the transmitter antenna (Figure 4.2a).

The Baseband Signal

The main objective of a LOS microwave link is to transmit and successfully receive voice or video signals from one point to another and at a predetermined distance. The voice signals must be multiplexed before transmission. There are two types of multiplexing: FDM and TDM. **Frequency-division multiplexing (FDM)** is the process whereby a number of analog signals are directly multiplexed with an analog subcarrier to form a combined analog signal. **Time-division multiplexing (TDM),** on the other hand, is the process whereby a number of analog signals are first converted to digital signals, then combined by sharing a preset frame time to form a binary data stream. When the baseband is an FDM signal modulating a carrier signal within a LOS microwave link, the link is classified as analog; when a TDM signal modulates the carrier signal, the link is classified as digital.

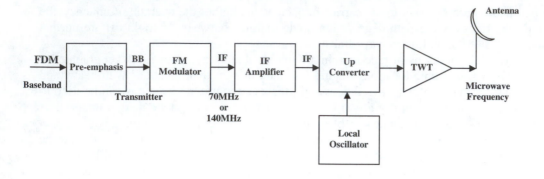

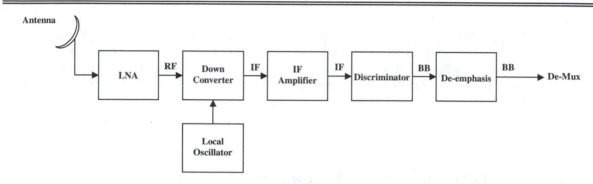

FIGURE 4.2 Analog microwave link transmitter–receiver

Frequency-Division Multiplexing (FDM)

An FDM system incorporates several voice channels. Each voice channel is assumed to have a bandwidth of 4 kHz. These voice channels are combined to form a single FDM signal at the output of the multiplexer. A low-pass filter (LPF) of 4-kHz bandwidth is required to confine the voice signal bandwidth to 4 kHz, thus eliminating any frequency components outside the 4-kHz limit.

The subcarrier oscillator generates the subcarrier and pilot frequencies required for modulation. Each modulator modulates the appropriate subcarrier with the corresponding voice channel. The output bandpass filter makes sure that each channel occupies the pre-assigned frequency band. This bandpass filter contributes greatly to an efficient and accurate demultiplexing process.

The pilot or reference frequency is required to ensure that the demultiplexing subcarrier oscillators generate the exact frequencies with the multiplexing subcarrier oscillators. Any frequency difference will result in a signal distortion. A 4-kHz frequency guard band is assigned for each voice channel to avoid channel overlap.

CCITT Groups

The Standard Group In its G.232 recommendation, CCITT has defined a **standard group** as 12 voice channels occupying the frequency range between 60 and 108 kHz with

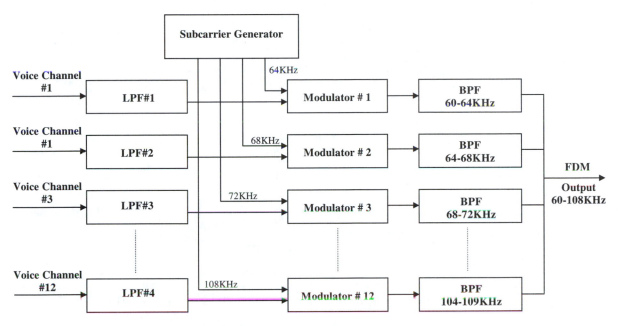

FIGURE 4.3 Voice-channel FDM block diagram (CCITT standard group).

each voice channel having a bandwidth of 4 kHz. The usual bandwidth of a voice channel is between 300 and 3400 Hz (rounded off to 4 kHz) (Figure 4.3).

The Standard Super Group The **standard super group** consists of five standard groups, totaling 60 voice channels and occupying the frequency range between 312 kHz and 552 kHz. The five subcarrier frequencies required for the formation of a standard super group are 420 kHz, 468 kHz, 516 kHz, 564 kHz, and 612 kHz (Figure 4.4).

The Standard Master Group The standard master group FDM scheme incorporates five super groups, totaling 300 voice channels, and occupying the frequency range between 812 kHz and 2.044 MHz (Figure 4.5).

Standard Super Master Group The **standard super master group** is composed of three super groups occupying the frequency range between 8.516 MHz and 12.388 MHz (Figure 4.6). The three subcarrier frequencies for the standard super master group are 10.56 MHz, 11.88 MHz, and 13.2 MHz. The generation of these frequencies is accomplished through the utilization of the harmonic components of the standard 4-kHz, 12-kHz, and 124-kHz oscillator frequencies.

The harmonics of a 4-kHz oscillator cover the standard group range between 64 kHz and 108 kHz; the harmonics of a 12-kHz oscillator are utilized to cover the frequency range

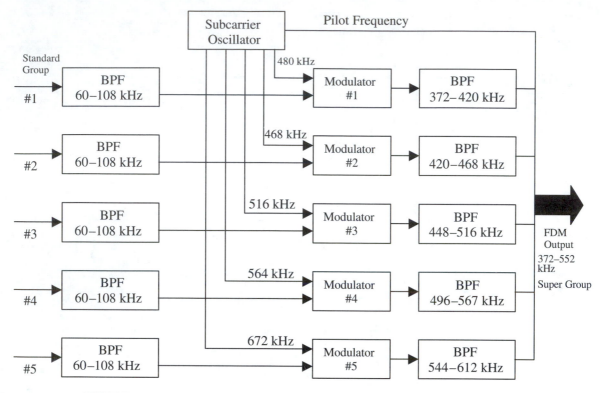

FIGURE 4.4   Voice-channel FDM block diagram (CCITT standard super group).

between 420 kHz and 612 kHz (super group); and the harmonics of a 124-kHz oscillator are used to cover the range between 1052 kHz and 2044 kHz (master group).

In an FDM system, the presence of pilot frequencies is absolutely necessary for two basic reasons: one is to regulate the system level and maintain level fluctuations within the ± 0.5-dB range, and the other is to maintain a constant frequency generation for the entire spectrum of frequencies required by all standard FDM groups. None of these frequencies must deviate by more than ± 2 Hz from the preassigned range. In all FDM systems, a master frequency source is used to generate all the required FDM frequency components. The receiver section of the system must be in absolute synchronization with the transmitter. This is accomplished through the pilot frequency. For digital transmissions (TDM), synchronization is achieved through coherent detection.

Pre-emphasis

In a line-of-sight microwave system, the signal-to-noise ratio at the output of the FM receiver must be maintained constant across the entire baseband frequency range. This constant SNR required by the LOS FDM system is contrary to the fundamental operation of the FM system in which the receiver output noise level increases proportionally to the increase of the modulating frequency. As a result, at higher modulation frequencies a higher

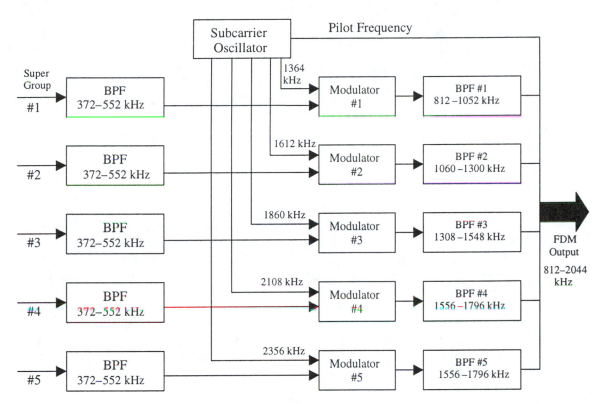

FIGURE 4.5 Three-hundred-voice-channel FDM block diagram (CCITT standard master group).

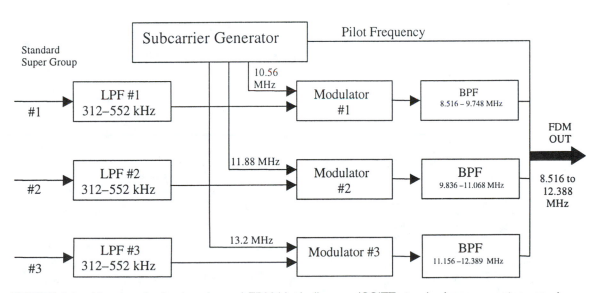

FIGURE 4.6 Nine-hundred-voice-channel FDM block diagram (CCITT standard super master group).

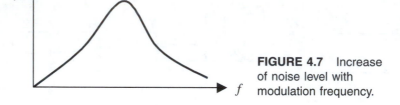

FIGURE 4.7 Increase of noise level with modulation frequency.

noise level will be present, thus reducing the overall SNR, whereas at lower modulation frequencies, the SNR is considerably higher (Figure 4.7). This disadvantage of the FM system is remedied by a **pre-emphasis** circuit incorporated at the input of the FM modulator. Correspondingly, a **de-emphasis** circuit incorporated at the output of the discriminator at the receiver end reverses the pre-emphasis process.

By inserting the pre-emphasis circuit at the input of the FM modulator, the higher modulation frequencies are increased in amplitude, with a corresponding increase of the modulation index (m_f). The pre-emphasis circuit can be designed to increase the modulation index at higher frequencies proportional to the level of the lower modulation frequencies, thus maintaining a constant SNR at the FM receiver. A typical pre-emphasis circuit and its frequency response are shown in (Figure 4.8).

At lower modulation frequencies, ωL is smaller and $1/\omega C$ is higher; at higher modulation frequencies, the reverse is true. With proper circuit design (component selection) of the pre-emphasis circuit, the output level can be maintained constant for both low and high modulation frequencies. This constant output level reflects a highly desirable constant

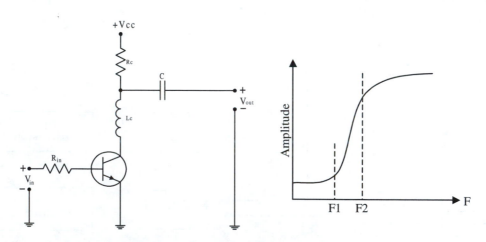

FIGURE 4.8 Pre-emphasis FM circuits.

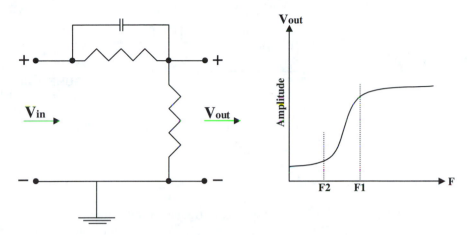

FIGURE 4.9 Pre-emphasis circuit.

signal-to-noise ratio (SNR). For North American FM systems (audio), the ratio of L/R for the pre-emphasis circuit of Figure 4.8 has been set at 75 μs; for Europe and Australia, it has been set at 50 μs. Another pre-emphasis circuit is shown in Figure 4.9. Here

$$f_1 = \frac{1}{2\pi R_1 C} \qquad f_2 = \frac{1}{2\pi R_2 C}$$

where $\quad f_2 - f_1$ = baseband range

The pre-emphasis baseband signal is now ready for FM modulation.

Frequency Modulation (FM)

Frequency modulation (FM) is defined as the process whereby the instantaneous frequency of the carrier is deviated from its center frequency in accordance with the amplitude of the information signal. A common circuit performing the frequency modulation function is the voltage-control oscillator (VCO) (Figure 4.10).

At $e(t) = 0$ V, the output of the VCO is f_c (carrier frequency). Any amplitude variations of $e(t)$ will generate an output signal given by

$$f = f_c \pm Ke(t) \tag{4.1}$$

where
$$\begin{aligned}
f &= \text{instantaneous frequency} \\
f_c &= \text{center VCO frequency} \\
K &= \text{constant of proportionality } (Hz/V) \\
e(t) &= \text{instantaneous input (assumed sine wave)}
\end{aligned}$$

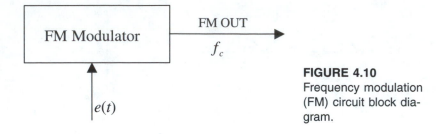

FM OUT

f_c

FIGURE 4.10
Frequency modulation
(FM) circuit block dia-
gram.

Since K is in Hz/V and $e(t)$ is in V

Then: $Ke(t) = (\frac{Hz}{V} \times V)$ corresponds to frequency variation:

$$\Delta f_c = Ke(t)$$

Therefore,

$$\Delta f = f_c + \Delta f \qquad\qquad\text{(4.2)}$$

The input/output waveform of a FM modulated signal is shown in Figure 4.11.

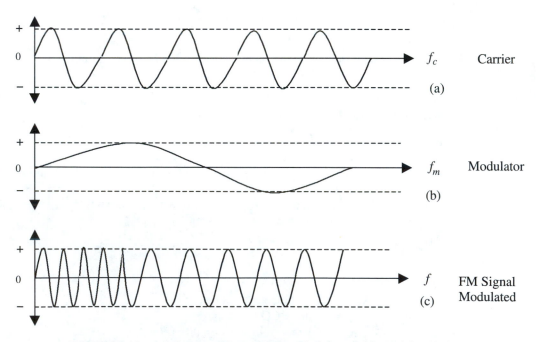

FIGURE 4.11 (a) Carrier, (b) modulating signal, and (c) modulated signal.

FM Modulation Index (m_f)   The **FM modulation index** (m_f) is the ratio of the peak carrier deviation Δf_c to the modulation frequency (f_m):

$$m_f = \frac{\Delta f_c}{f_m} \tag{4.3}$$

A more detailed mathematical interpretation of the FM signal is given as follows: The instantaneous amplitude of the carrier signal is given by

$$e(t) = E \cdot \cos(\omega_t + \phi)t \tag{4.4}$$

where $e(t)$ = instantaneous value
   E = maximum value
   ω = angular velocity
   ϕ = phase

From this expression, if ω is varied by a sinusoidal signal, the resultant combined signal is a frequency-modulated signal expressed by

$$e(t)_{FM} = E_c \cos[\omega_c t + (m_f \sin \omega_m t)] \tag{4.5}$$

where $e(t)_{FM}$ = instantaneous amplitude of the FM signal
   E_c = maximum carrier amplitude
   m_f = modulation index
   ω_f = modulation frequency

The modulation index (m_f) in a FM signal is the equivalent of the percentage of modulation for an AM signal (amplitude modulation) and is used to determine the power spectrum of the FM signal.

Side bands of FM Signals (m_f)

The modulation index (m_f) of a FM signal is also used to determine the side bands and the overall power spectral distribution of the FM signal. Applying the trigonometric identity of Eq. (4.6) into Eq. (4.5),

$$\cos(a + b) = \cos(b) \cdot \cos(a) - \sin(b) \cdot \sin(a) \tag{4.6}$$

We have

$$e(t)_{FM} = E_c[\cos(m_f \sin(\omega_m t)] \cdot \cos\omega_c t - E_c(m_f \sin\omega_m t) \cdot \sin \omega_c t \tag{4.7}$$

Let us first consider a small modulating index, let's say $m_f < 0.25$. Substituting $m_f = 0.25$ into Eq. (4.7) yields

$$\cos(m_f \sin \omega_{m_f})t \approx 1 \tag{4.8}$$

Therefore

$$e(t)_{\text{FM}} = E \cos \omega_c t - E(m_f \sin \omega_{m_f}) \sin \omega_c t$$

or $\tag{4.9}$

$$e(t)_{\text{FM}} = E \cos \omega_c t - \frac{E}{2} \cos(\omega_c \cdot \omega_m)t - \frac{E_{m_f}}{2} \cos(\omega_c \cdot \omega_m)t$$

It is evident from Eq. (4.9) that the transmission of a narrow-band FM signal requires a complete set of side bands plus the transmission of the carrier signal.

For a wide-band FM signal, where $m_f \gg 0.25$, Eq. (4.9) also applies. Because of the substantial difficulties, the amplitudes of the desired frequency components are evaluated according to the Bessel function solutions:

$$
\begin{aligned}
e = E\{ & J_o \sin\omega_c t + J_1[\sin(\omega_c + \omega_m)t - \sin(\omega_c - \omega_m)t] \\
& + J_2[\sin(\omega_c + \omega_m)t - \sin(\omega_c - \omega_m)t] \\
& + J_3[\sin(\omega_c + \omega_m)t - \sin(\omega_c - \omega_m)t] \\
& + J_4[\sin(\omega_c + \omega_m)t - \sin(\omega_c - \omega_m)t] \\
& + J_5[\sin(\omega_c + \omega_m)t - \sin(\omega_c - \omega_m)t] \\
& + J_6[\sin(\omega_c + \omega_m)t - \sin(\omega_c - \omega_m)t] \\
& + J_7[\sin(\omega_c + \omega_m)t - \sin(\omega_c - \omega_m)t] \\
& + J_8[\sin(\omega_c + \omega_m)t - \sin(\omega_c - \omega_m)t] \\
& + J_9[\sin(\omega_c + \omega_m)t - \sin(\omega_c - \omega_m)t] \\
& + J_{10}[\sin(\omega_c + \omega_m)t - \sin(\omega_c - \omega_m)t] \\
& + J_{11}[\sin(\omega_c + \omega_m)t - \sin(\omega_c - \omega_m)t] \\
& + J_{12}[\sin(\omega_c + \omega_m)t - \sin(\omega_c - \omega_m)t]
\end{aligned}
\tag{4.10}
$$

Equation 4.10 shows that, although a pure sine wave frequency modulates a high-frequency carrier, an infinite number of side bands are generated on both sides of the carrier as a result.

Figure 4.12 shows the frequency spectrum of a FM signal and its corresponding amplitudes. The amplitudes of the sidebands are determined by the **Bessel function coefficients** J_0, J_1, J_2, J_3, ..., J_{12}, which in turn are determined by the modulating index m_f (Table 4.1).

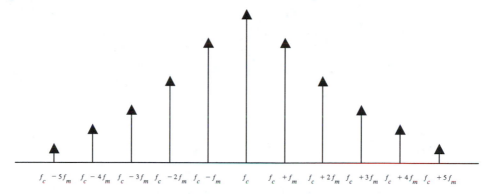

FIGURE 4.12 Frequency Spectrum of a FM signal.

TABLE 4.1 Bessel function coefficients for different modulation indexes (m_f).

m_f	J_1	J_1	J_2	J_3	J_4
0.00	1.00	—	—	—	—
0.25	0.98	0.12	—	—	—
0.50	0.94	0.24	0.03	—	—
1.0	0.77	0.44	0.11	0.02	—
1.5	0.51	0.56	0.23	0.06	0.01
2.0	0.22	0.58	0.35	0.13	0.03

EXAMPLE 4.1

Compute the first two sideband components of a FM signal composed of a 2-MHz carrier frequency and a 15-V_{rms} amplitude, modulated by a sine wave of 1.5 kHz and a peak-to-peak deviation of 0.75 kHz.

Solution
Compute the modulation index (m_f)

$$m_f = \frac{\Delta f_c}{f_m} = \frac{0.75 \text{ kHz}}{1.5 \text{ kHz}} = 0.5$$
$$\therefore m_f = 0.5$$

Determine amplitude levels from the Bessel function table:

$$J_0 = 15 \times 0.94 = 14.1 \text{ V}$$
$$J_1 = 15 \times 0.24 = 3.6 \text{ V}$$
$$J_2 = 15 \times 0.03 = 0.45 \text{ V}$$

Sketch the frequency spectrum.

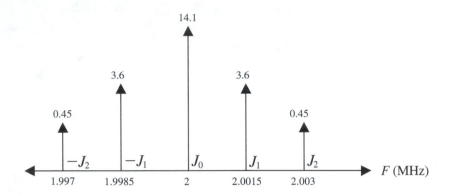

If the amplitude of J_0 is reduced, this energy is not lost but rather is distributed to the sidebands.

FM Bandwidth (BW)

If the index of modulation (m_f) of an FM signal gradually increases beyond 0.25, it will contribute to the generation of a progressively larger number of side bands separated by the modulation frequency (f_m). It is therefore evident that a large modulation index will generate an infinite number of sidebands. Since sidebands determine the FM signal bandwidth, the resultant bandwidth of a wideband FM signal must be infinite. For practical considerations, the **FM bandwidth** is established at the point where the sum of the squares of the J coefficients away from J_0 (inclusive) produce a total power of no less than 98 percent of the total spectral power. This is known as Carson's Rule.

$$BW = 2(f_m + \Delta f_c) \tag{4.11}$$

But

$$m_f = \Delta f_c / f_m \tag{4.12}$$
$$\Delta f_c = m_f \times f_m.$$

Substituting into Eq. (4.11) gives

$$BW = 2(f_m + m_f \cdot f_m)$$
$$= 2f_m + 2m_f \cdot f_m$$
$$= 2f_m(1 + m_f)$$
$$\therefore \text{Bandwidth} = 2f_m(1 + m_f)$$

Finally, a small modulating index generates one set of side bands (narrow-band FM), and a large modulating index generates a large number of side bands (wide-band FM). The bandwidth of the FM signal has a significant effect on the signal-to-noise ratio (the FM improvement factor). Figure 4.13 shows the improvement of SNR in relation to FM signal strength.

At the modulating stage, the FDM baseband signal modulates an IF subcarrier, usually 70 MHz. The output of the FM modulator is fed into the IF amplifier for further amplification. The amplified IF composite signal is fed into the up-converter circuit. This frequency translation is necessary in order to bring the carrier frequency to its predetermined frequency level for transmission. Either a traveling-wave tube (TWT) or a solid-state amplifier further amplifies the carrier frequency (GHz). The transmitter power for line-of-sight microwave links is usually set at 0.1 W, 1 W, or 10 W.

4.1.2 The Receiver

A line-of-sight microwave receiver is composed of an antenna assembly, including the parabolic antenna, the low-noise amplifier (LNA), the waveguide pluming, the down-converter, an IF amplifier, a limiter, the discriminator circuit, and the demultiplexer. From the antenna, the incoming signal is fed to the input of the low-noise amplifier for the first stage of amplification. The output of the LNA through a waveguide assembly is fed into the input of the down-converter/mixer circuit. The output of the local oscillator circuit is also fed into the down-converter/mixer circuit.

The down-converter performs the exact opposite function of the up-converter; that is, it translates the carrier frequency to the 70-MHz or 140-MHz IF frequencies. The 140-MHz IF frequency is used for high-volume voice traffic systems. It also incorporates specialized circuits such as phase equalizers, which are designed to compensate for induced phase distortions of the processed signal. The output of the down-converter is fed into the IF amplifier for further amplification, which is required to elevate the signal amplitude to a level

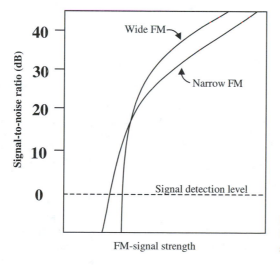

FIGURE 4.13
Improvement SNR with signal strength.

specified by the discriminator circuit characteristics. Prior to the discrimination stage, the IF signal is processed through a limiter circuit. The purpose of the limiter is to reduce the noise level and, consequently, to increase the signal-to-noise ratio, a very important advantage of the FM receiver in comparison with the AM receiver.

The output of the discriminator is the composite FDM baseband signal. This signal is fed into the input of the FDM demultiplexer circuit which separates the combined FDM signal into individual voice channels for routing to their proper destinations. Microwave antennas, low-noise amplifiers, and down-converters are all described in detail in the satellite receiver sections. The only two circuits requiring a brief operational interpretation are the FM discriminator and the FDM demultiplexer circuits.

FM Demodulation

FM demodulation (as with AM) is a process whereby the information signal or baseband is separated from the carrier signal. This process is accomplished through the utilization of a discriminator circuit. The discriminator circuit characteristic is based on the principle that its output voltage is proportional to the variations of the input carrier frequency. Figure 4.14 shows the relationship between the discriminator voltage output variations and input signal variations.

The block diagram of a very simple FM discriminator circuit is shown in Figure 4.15. FM signal detection is achieved through a phase-locked loop (PLL). This circuit is composed of a phase detector, a low-pass filter (LPF), and a voltage-controlled oscillator (VCO) with a specific voltage/frequency sensitivity. When an FM signal is applied at the input of the discriminator circuit, the voltage-controlled oscillator senses and follows any small frequency changes around the carrier frequency. For an input frequency change $f_c + \Delta f_c$, the output voltage of the VCO will increase by an amount proportional to the ratio $\Delta V_o / \Delta f_c$. Similarly, a decrease of the input frequency, $f_c - \Delta f_c$, will generate a lower VCO output voltage.

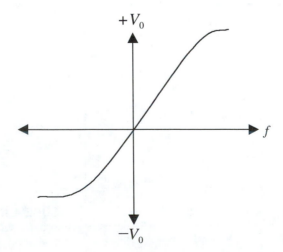

FIGURE 4.14
Frequency versus voltage characteristic current.

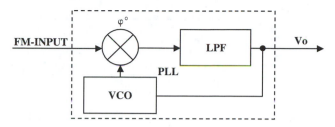

FIGURE 4.15   FM discriminator circuit.

EXAMPLE 4.2

Determine the peak output voltage of a PLL-FM discriminator circuit with sensitivity given by

$$\frac{\Delta f_c}{\Delta V} = 20 \text{ kHz/V} \quad \text{(sensitivity)}$$

The carrier signal is modulated by a sine wave of 2 kHz with an index of modulation $m = 5$.

Solution

The frequency derivation of the carrier is given by

$$\begin{aligned} \Delta f_c &= m_f \cdot f_m \\ &= 5 \times 2 \times 10^3 \text{ Hz} \\ &= 10 \text{ kHz} \\ \therefore \Delta f_c &= 10 \text{ kHz} \end{aligned}$$

The output voltage is

$$V_o = \frac{\Delta f_c}{f_m} = \frac{10 \text{ kHz}}{20 \text{ kHz}} = 0.5 \text{ V}$$

Therefore: $V_o = 0.5$ V. The peak output voltage of the PLL-FM discriminator circuit is 500 mV.

FM improvement threshold

In an analog LOS microwave link, the FM receiver must maintain a specified output SNR in reference to a specified input CNR. This relationship is illustrated in Figure 4.16. The graph of Figure 4.16 identifies four major points: the signal level at point A, the input of the FM receiver is −122 dB with a corresponding output signal to noise ratio equal to 0 dB. Also, at this point the carrier-to-noise ratio at the input of the receiver is 0 dB. At point B,

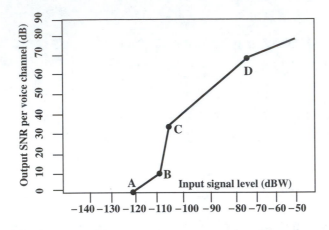

FIGURE 4.16
Relationship between input signal level and output SNR per voice channel of an FM receiver (FM-FDM).

the input signal level is increased to -112 dB, and the carrier-to-noise ratio is 10 dB. This increase of the signal level at the input of the FM receiver by 10 dB reflects an equal increase of the output SNR of 10 dB. It is evident that the portion of the graph between points A and B is linear with a slope equal to one. That is, for a 1-dB increase of the input carrier-to-noise ratio, there is a corresponding increase of the output SNR, also of 1 dB.

Between points B and C, although the graph is still linear, the slope has changed from 1 to 3. That is, for a 1-dB increase of the input CNR ratio, there is a corresponding increase of the output SNR of 3 dB. Therefore, point B is referred to as the **FM improvement threshold** point. It is at this input signal level that the output of SNR increases dramatically, from 10 dB to 30 dB.

Between points C and D, the graph also illustrates a linear increase of the output SNR in reference to the input signal, but here the slope of the line has been decreased back to 1 as between points A and B. Beyond point D, the graph becomes nonlinear. That is, the output SNR does not increase proportionally with the output SNR. Therefore, it is evident that a deterioration of the output SNR occurs and receiver operation beyond this point should be avoided. The FM improvement threshold can be calculated as follows:

$$FM_{th_{dB}} = 10 \log(KT) + 10 \log (BW) + (NF)_{db} + 10_{dB}$$

where

K = Boltzmann's constant $(1.38 \times 10^{-23} \text{ } J\text{-}K)$
T = Noise temperature ambient $(290°)$
BW = Bandwidth (Hz)
NF = Receiver noise figure (dB)

Note: 10 dB is used as a reference input level.

FM Demultiplexing

At the receiver end, the demultiplexing process is exactly opposite to multiplexing (Figure 4.17). The pilot or reference frequency is received and the subcarrier oscillator generates

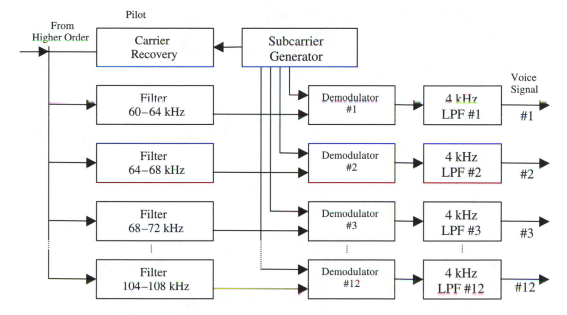

FIGURE 4.17 FDM Demultiplexer circuit (12 voice channels).

the twelve subcarrier frequencies. The demodulation circuit separates the subcarrier from the voice signals, and the output low-pass-filter rejects the unwanted high-frequency components.

EXAMPLE 4.3

Determine the FM improvement threshold level for a FM receiver used in an LOS microwave link operating in FDM mode with a noise figure of 6 dB and a receiver bandwidth of 20 MHz.

Solution

$$\text{FM}_{\text{it}_{\text{dB}}} = 10 \log (KT) + 10 \log (\text{BW}) + (\text{NF})_{\text{dB}} + 10_{\text{db}}$$
$$= 10 \log (1.38 \times 10^{-23}) + 10 \log (290°) + (20 \times 10^{6})_{\text{dB}} + 6 + 10$$
$$= -228.6 + 24.6 + 73 + 6 + 10$$
$$= -113.2 \text{ dB}$$

Therefore, $\text{FM}_{\text{it}} = -115$ dB. The FM improvement threshold is -115 dB.

Knowledge of the FM improvement threshold alternately leads to the calculations of the required transmitter power, provided that **path losses** have already been established. The relationship between required transmitter power, path losses, and FM improvement threshold is given by

$$P_{t_{\text{dB}}} = L_{p_{\text{dB}}} + \text{FM}_{\text{it}_{\text{dB}}}$$

where $\quad P_t$ = transmitter power (dB)

$\quad L_p$ = path losses (dB)

$\quad \text{FM}_{it}$ = FM improvement threshold (dB)

EXAMPLE 4.4

Determine the transmitter power required in an LOS microwave link operating at FDM mode, with a path length of 60 km and a carrier frequency of 2 GHz. The FM receiver's modulation index is 0.5, the modulation frequency is 5 MHz, and the receiver noise figure is 5 dB. The transmitter and receiver antenna diameters are 1 m (parabolic reflectors), and feed losses and branch losses are 5 dB.

Solution

Compute the path losses ($L_{p_{dB}}$)

$$
\begin{aligned}
L_p &= 32.5 + 20 \log(d) + 20 \log(f) \\
&= 32.6 + 20 \log(60) + 20 \log(2 \times 10^3) \\
&= 32.6 + 35.6 + 66 \\
&= 134.1 \\
&\quad \therefore \text{ path losses: } L_p = 134.1 \text{ dB}
\end{aligned}
$$

Compute the FM receiver bandwidth (BW)

$$
\begin{aligned}
\text{BW} &= 2f_m \,(1 + m_f) \\
&= 2(5 \times 10^6)(1 + 0.5) \\
&= 10 \times 10^6 \times 1.5 \\
&= 15 \text{ MHz} \\
\text{BW} &= 15 \text{ MHz} \\
&\quad \therefore \text{ FM receiver bandwidth: BW} = 15 \text{ MHz}
\end{aligned}
$$

Compute the FM improvement threshold ($\text{FM}_{it_{dB}}$)

$$
\begin{aligned}
\text{FM}_{it_{dB}} &= 10 \log (\text{K}) + 10 \log(T) + 10 \log(\text{BW}) + (\text{NF}) + 10 \\
&= -228.6 - 24.6 + 10 \log (20 \times 10^6) + 5 + 10 \\
&= -116 \text{ dB} \\
&\quad \therefore \text{ FM improvement threshold: FM}_{it} = -116 \text{ dB}
\end{aligned}
$$

Compute the antenna gain ($G_t = G_r$)

$$G = \frac{4\pi A_{\text{eff}}}{\lambda^2}$$

$$\lambda = \frac{c}{f} = \frac{3 \times 10^8 \text{ m/s}}{2 \times 10^9 \text{ L/s}} = 0.15 \text{ m}$$

$$\therefore \lambda = 0.15 \text{ m}$$

$$A_{\text{sys}} = KA$$

$$A = \frac{D^2}{4} = \frac{\pi(1)^2}{4} = 0.785 \text{ m}^2$$

$$A_{\text{eff}} = 0.55 \times 0.785 \text{ m}^2 = 0.432 \text{ m}^2$$

$$G = \frac{4\pi(0.432) \text{ m}^2}{(0.15)^2 \text{ m}^2} = 241$$

Therefore, antenna gain $G = G_t = G_r = 23.8$ dB.

Compute $P_{t_{\text{dB}}}$

Incorporating the transmitter and receiver antenna gains plus feed and branch losses into the expression, we have:

$$
\begin{aligned}
P_{t_{\text{dB}}} &= \text{FM}_{\text{it}_{\text{dB}}} - L_{p_{\text{dB}}} - G_t - G_r + L_f + L_{\text{br}} \\
&= -116 - (-134.2) - 23.8 - 23.8 + 2.5 + 2.5 \\
&= -116 + 134.2 - 47.6 + 5 \\
&= -24.4
\end{aligned}
$$

Therefore, transmitter power $P_t = -24.4$ dB or $P_t = 3.6$ mW minimum.

4.2 DIGITAL LINE-OF-SIGHT MICROWAVE LINKS

4.2.1 System Gain

In digital line-of-sight microwave links, the first parameter to be established is the **system gain** (G_{sys}). This parameter incorporates all the gains and losses of the system and also determines the transmitter power required, based on a pre-established **receiver sensitivity** for a given BER, or determines receiver sensitivity based on available transmitter power at a given BER.

The system gain is given by

$$G_{sys} = FM - G_t - G_r + L_p + L_f + L_{br} \qquad (4.13)$$

where
G_{sys} = system gain (dB)
FM = fade margin (dB)
G_t = transmitter antenna gain (dB)
G_r = receiver antenna gain (dB)
L_p = path losses (dB)
L_f = feed losses (dB)
L_{br} = branching losses (dB)

The discrete components of Eq. (4.13) are calculated in the following manner.

Fade Margin

For a 99.99% system availability, the fade margin is given by

$$FM_{dB} = 30 \log(d) + 10 \log(A) + 10 \log(B) + 10 \log(f) - 30 \qquad (4.14)$$

where
A = factor determining the terrain roughness of the path (Table 4.2)
B = factor determining atmosphere impact on the link (Table 4.3)

Path Losses

Path losses for line-of-sight and satellite microwave links are determined as follows. The power density (P_D) at a distance (d) away from the power source in an omnidirectional mode of transmission is given by the expression.

$$P_D = \frac{p_t}{4\pi d^2} \qquad (4.15)$$

where
P_D = power density (W/m^2)
p_t = source power (W)
d = distance (m)

TABLE 4.2 Terrain roughness factor (A).

Terrain	A
Very rough	0.25
Average	1.00
Very smooth	4.00

TABLE 4.3 Atmosphere impact (B).

Atmosphere	B
Very dry	0.125
Average	0.25
Hot and humid	0.5

If the power source is placed at the focal point of a parabolic reflector, the radiated energy will move in the direction pointed to by the antenna and with an angle of directivity relevant to the propagating wavelength and antenna diameter. (for more details see Eq. 6.4.)

Therefore,

$$P_D = \frac{p_t}{4\pi d^2} G_t \qquad (4.16)$$

where G_t = antenna gain (dB)

At the receiver end, the power level of the radiated signal collected is subject to the receiver antenna's effective area A_{eff} (m^2)

Therefore:

$$P_r = P_D G_t A_{\text{eff}} \quad \text{or} \quad P_r = \frac{P_t}{4\pi d^2} G_t A_{\text{eff}} \qquad (4.17)$$

where P_r = received power (W)
A_{eff} = receiver effective area (m^2)

The parabolic reflector antenna gain (receiver) is expressed by

$$G_r = \frac{4\pi A_{\text{eff}}}{\lambda^2} \qquad (4.18)$$

Solving for A_{eff}, we have

$$A_{\text{eff}} = \frac{G_r \lambda^2}{4\pi} \qquad (4.19)$$

Substituting Eqs. (4.18) and (4.19) into Eq. (4.17), we have

$$P_r = \frac{P_t}{4\pi d^2} G_t G_r \left(\frac{\lambda^2}{4\pi} \right)$$

$$= p_t G_t G_r \left(\frac{\lambda}{4\pi} \right)^2$$

Since $\lambda = c/f$, then,

$$p_r = p_t G_t G_r \left(\frac{c/f}{4\pi d} \right)^2$$

where c = velocity of light
           λ = wavelength
           f = frequency

Thus

$$p_r = p_t G_t G_r \left(\frac{c^2/f^2}{(4\pi)^2 d^2/1} \right) = p_t G_t G_r \left(\frac{c^2}{(4\pi)^2 f^2 d^2} \right) \tag{4.20}$$

Assuming the distance for line-of-sight and satellite links is measured in kilometers and the operating frequency in megahertz, Eq. (4.20) is further developed as follows:

$$p_r = p_t G_t G_r \left(\frac{(3 \times 10^8)^2}{(4\pi)^2 (10^6)^2 f^2 (10^3)^2 d^2} \right)$$

$$= p_t G_t G_r \left(\frac{(9 \times 10^{16})}{158 \times (10)^{12} \times (10^6) f^2 d^2} \right)$$

$$= p_t G_t G_r \left(\frac{(9 \times 10^{16})(10^{-18})}{158 f^2 d^2} \right)$$

$$= p_t G_t G_r \left(\frac{0.057 \times 10^{-2}}{f^2 d^2} \right) \tag{4.21}$$

The ratio of the received power to the transmitted power is as follows:

$$\frac{p_r}{p_t} = G_t G_r \left(\frac{0.057 \times 10^{-2}}{f^2 d^2} \right) \tag{4.22}$$

In dB, Eq. (4.22) becomes

$$\frac{p_r}{p_{t_{dB}}} = G_{t_{dB}} + G_{r_{dB}} + \left[10 \log(0.057 \times 10^{-2}) - 10 \log(f^2) - 10 \log(d^2) \right]$$

$$= G_{t_{dB}} + G_{r_{dB}} - \left[32.5 + 20 \log(f) + 20 \log(d) \right]$$

Therefore:

$$\frac{p_r}{p_{t_{dB}}} = G_{t_{dB}} + G_{r_{dB}} - \left[32.5 + 20 \log(f) + 20 \log(d) \right] \tag{4.23}$$

Equation 4.23 defines the ratio of the received power to the transmitted power. This ratio is related to the transmitter and receiver antenna gain, operating frequency (f_{MHz}) and path distance (d_{km}). The section of the equation within the bracket reflects the path losses (L_p). Therefore:

$$L_p = 32.5 + 20 \log(f) + 20 \log(d) \tag{4.24}$$

where f = carrier frequency (MHz)
d = distance (km)

Antenna Gain (G)

Antenna gains for both the transmitter and receiver sections of the link are the same because they use the same operating frequency and the same physical dimensions.

$$G = \frac{4\pi A_{\text{eff}}}{\lambda^2}$$

$$A_{\text{eff}} = K \cdot A$$

$$A = \frac{\pi D^2}{4}$$

where G = antenna gain (dB)
A_{eff} = effective area of the parabolic antenna (m^2)
K = efficiency factor ($\cong 0.55$)
d = antenna diameter (m)

Other Losses

Feed losses at the antenna $\cong 2$ dB
Branching losses (filter circulator) $\cong 2$ dB

Signal fading

In a LOS digital microwave link, **signal fading** is the phenomenon whereby the propagated signal is attenuated beyond the level of the calculated path loss. This signal loss can be in excess of 30 dB and can last from a few seconds to several minutes. Signal fading can be attributed to two main causes. One is atmosphere variation along the transmission path and the other is the physical characteristic of the terrain directly under the signal traveling path.

Multipath Fading

Multipath fading occurs when the transmitted signal follows more than one traveling path, thus arriving at the receiver input at different time intervals and having traveled correspondingly different distances. The deviation of the transmitted signal from its main traveling path is attributed to refractive index variations of the propagation medium, in this case the surrounding atmosphere. Deviation caused by reflection and the scattering of the waves is attributed to the terrain's physical characteristics such as large water-bodies and flat surfaces (Figure 4.18).

Figure 4.18 shows that at the receiver input, the signals arrive at three different time intervals. This reflects time delays for signals not following the **direct path** and results in phase differences and, consequently, phase signal distortion. There are two distinct possibilities: the arriving signals are either in phase or out of phase.

When the signals arrive in phase, the combined signal amplitude increases; when the signals arrive out of phase, the signal amplitude decreases. The increase or decrease of the signal amplitude is frequency dependent (frequency selective fading) (Figure 4.19). This amplitude and phase signal distortion contributes to a proportional increase of the channel intersymbol interference (ISI) and ultimately to undesired system gain and fade margin reductions.

Multipath fading has a greater impact on systems employing multilevel digital modulation techniques than it does on systems employing low-speed modulation techniques. To remedy the impact of multipath fading in digital microwave LOS communications links, especially those operating at high capacity levels, adaptive equalizers are incorporated into the system. Adaptive equalizers sample the amplitudes of the incoming signal at different frequencies within the bandwidth spectrum and, through a filtering process, adjust the sampled levels to a predetermined level.

4.2.2 Line-of-sight digital microwave link system design

The following examples are related to calculations of various important link parameters pertaining to digital baseband transmission.

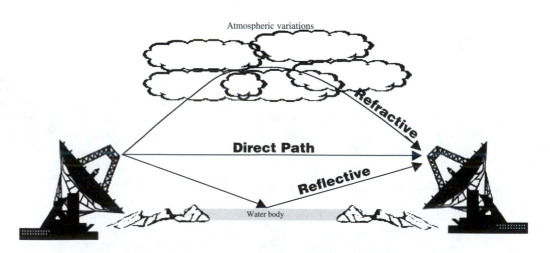

FIGURE 4.18   Multipath fading.

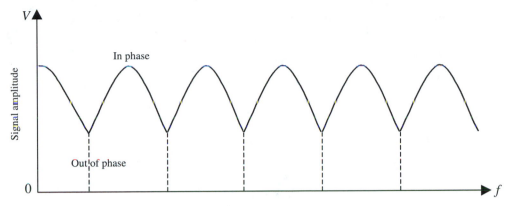

V

Signal amplitude

In phase

Out of phase

0

f

FIGURE 4.19 Frequency selective fading.

EXAMPLE 4.5

Compute the system gain of a line-of-sight microwave link operating with the following system parameters:

Operating frequency: $f = 2\ \text{GHz}$
Receiver and transmitter antenna diameters: $D = 2\ \text{m}$
Feed losses: $L_f = 2\ \text{dB}$
Branching losses: $L_{br} = 1.8\ \text{dB}$
Hop length: $d = 45\ \text{km}$

The link is established over a rough terrain with very dry atmospheric conditions throughout the year.

Solution
The system gain G_{sys} is given by

$$G_{\text{sys}_{dB}} = \text{FM}_{dB} - G_{t_{dB}} - G_{r_{dB}} + L_{p_{dB}} + L_{f_{dB}} + L_{br_{dB}}$$

Compute path losses (L_p) in dB

$$L_p = 32.5 + 20\log(f) + 20\log(d)$$
$$L_p = 32.5 + 20\log(2 \times 10^3) + 20\log(45)$$
$$= 32.5 + 66.02 + 33$$
$$= 131.58$$
$$\therefore L_p = 131.58\ \text{dB}$$

Compute antenna gain ($G_t = G_r$) in dB

$$G = \frac{4\pi A_{\text{eff}}}{\lambda^2}$$

$$A_{\text{eff}} = K \cdot A \quad \text{where } K = 0.55$$

$$A = \frac{\pi D^2}{4} = \frac{\pi (2)^2}{4} = 3.14 \text{ m}^2$$

$$\lambda = \frac{c}{f} = \frac{3 \times 10^8 \text{m/s}}{2 \times 10^9 \text{m/s}} = 0.15\text{m}$$

$$\therefore \lambda = 0.15 \text{ m}$$

$$A_{\text{eff}} = K \cdot A = 0.55 \, (3.14) \text{ m}^2$$

$$G_t = \frac{4\pi (1.727)\text{m}^2}{(0.15)^2 \text{ m}^2} = \frac{21.7}{0.0225} \cong 955$$

Therefore, $G_t = G_r = 29$ dB.

Compute fade margin (FM) in dB
From tables, $A = 0.25$ and $B = 0.125$.

$$\begin{aligned}
\text{FM}_{\text{dB}} &= 30 \log(d) + 10 \log(A) + 10 \log(B) + 10 \log(f) - 30 \\
&= 30 \log(45) + 10 \log(0.25) + 10 \log(0.125) + 10 \log(2 \times 10^3) - 30 \\
&= 49.6 + (-6.02) + (-9.03) + 33 - 30 \\
&= 37.5
\end{aligned}$$

$$\therefore \text{FM}_{\text{dB}} = 37.5 \text{ dB}$$

System gain (G_{sys})

$$\begin{aligned}
G_{\text{sys}} &= \text{FM} - G_t - G_r + L_p + L_f + L_{\text{br}} \\
&= 37.5 - 29 - 29 + 131.58 + 2 + 1.8 \\
&= 114.88 \text{ dB}
\end{aligned}$$

$$\therefore G_{\text{sys}} = 114.88 \text{ dB}$$

EXAMPLE 4.6

A line-of-sight microwave link operates at a frequency of 2.2 GHz and a system gain of 100 dB, and must maintain a reliability factor of 99.99%. Compute the maximum hop length for the following parameters:

1. Antenna diameter for both receiver and transmitter $D = 2.5$ m.
2. Terrain roughness is average ($A = 1$).
3. Atmospheric impact factor is $B = 0.25$.
4. Assume a fade margin of 35 dB.

Solution
System gain:

$$G_{\text{sys}_{dB}} = \text{FM}_{dB} - G_{t_{dB}} - G_{r_{dB}} + L_{p_{dB}} + L_{f_{dB}} + L_{br_{dB}}$$

Solve for L_p (space, or path loss):

$$L_p = G_{\text{sys}} - G_t - G_r + L_f + L_{br} - \text{FM}$$
$$G_t = G_r = G \quad (\text{both transmitter and receiver antenna gain})$$
$$\lambda = \frac{3 \times 10^8 \text{ m/s}}{2.2 \times 10^9 \text{ Hz}} = 0.136 \text{ m}$$
$$\therefore \lambda = 0.136 \text{ m}^2$$
$$A_{\text{eff}} = K \cdot A = 0.55 \times 4.9 \text{ m}^2 = 2.7 \text{ m}^2$$
$$A = \frac{\pi D^2}{4} = \frac{\pi (2.5)^2}{4} = \frac{19.63}{4} = 4.9 \text{ m}^2$$
$$L_p = 100 + 32.6 + 32.6 - 2 - 2 - 35 = 126.2$$
$$\therefore L_p = 126.2 \text{ dB (path losses)}$$
$$G_t = \frac{4\pi A_{\text{eff}}}{\lambda^2} = \frac{4\pi (2.7) \text{ m}^2}{(0.136)^2 \text{ m}^2} = \frac{33.9 \text{ m}^2}{(0.136)^2 \text{m}^2}$$
$$= 1834$$
$$\therefore G_t = 32.6 \text{ dB}$$

Since

$$L_p = 32.5 + 20 \log(f) + 20 \log(d),$$

solve for path length (d):

$$20 \log(d) = L_{p_{dB}} - 20 \log(f) - 32.5$$
$$= 126.2 - 20 \log(2.2 \times 10^3) - 32.5$$
$$= 126.2 - 66.85 - 32.5$$
$$= 26.85$$
$$\frac{20 \log(d)}{20} = \frac{26.85}{20} = 1.3425$$
$$d = \text{antilog}(1.3425)$$
$$\therefore d = 22 \text{ km (hop distance)}$$

Line-of-Sight Microwave Links

EXAMPLE 4.7

Compute the maximum path length and channel capacity (f_b) of a microwave line-of-sight communications link operating under the following system parameters.

Transmitter power: $P_t = 2$ kW
Carrier frequency: $f = 2.2$ GHz
Receiver sensitivity: $P_{r,min} = 10$ nW
Modulation scheme used: QPSK with a 4 MHz bandwidth
System reliability: 99.99%
Antenna diameter: $D = 2.5$ m
Fade margin (assumed): FM $= 37.5$ dB
$E_b/N_o = 10^{-6}$
Receiver noise figure (assumed): NF $= 5$ dB

Solution
System gain: $G_{sys_{dB}}$
System gain is defined as the difference between transmitter power (P_t) and receiver sensitivity ($P_{r,min}$).

$$G_{sys} = P_{t_{dB}} - P_{r,min_{dB}}$$
$$= 33 - (-80)$$
$$= 113 \text{ dB}$$
$$G_{sys} = 113 \text{ dB}$$

Antenna gain: ($G_t = G_r$)

$$\lambda = \frac{3 \times 10^8 \text{ m/s}}{2.2 \times 10^9 \text{ m/s}} = 0.136 \text{ m}$$

$$\therefore \lambda = 0.136 \text{ m}$$

$$A = \frac{\pi D^2}{4} = \frac{\pi (2.5)^2}{4} = \frac{19.6}{4} = 4.9 \text{ m}^2$$

$$A_{eff} = K \cdot A = 0.55 \times 4.9 = 2.7 \text{ m}^2$$

$$G = \frac{4\pi A_{eff}}{\lambda^2} = \frac{4\pi \times 2.7 \text{ m}^2}{(0.136)^2} = 1834$$

$$\therefore G_{dB} = 32.6 \text{ dB}$$

Transmitter and receiver antenna gain $= 32.6$ dB

Path Losses: $L_{p_{dB}}$
System gain:

$$G_{sys} = FM - G_t - G_r + L_p + L_f + L_{br}$$

Solve for L_p:

$$L_p = G_{sys} - FM + G_t + G_r - L_f - L_{br}$$

From tables:

$$L_f = 5 \text{ dB}$$
$$L_{br} = 2 \text{ dB}$$
$$L_p = 113 - 37.5 + 32.6 + 32.5 - 5 - 2 = 133.7$$
$$\therefore L_p = 133.7 \text{ dB}$$

Hop Length: d (km)

$$L_p = 32.5 + 20 \log(f) + 20 \log(d)$$

Solve for d:

$$
\begin{aligned}
20 \log(d) &= L_p + 32.5 - 20 \log(f) \\
&= 133.7 - 32.5 - 20 \log(2 \times 10^3) \\
&= 132.2 - 32.5 - 66.02 \\
&= 35.18 \\
d &= \text{antilog}(1.759) \\
\therefore d &= 57.4 \text{ km (hop distance)}
\end{aligned}
$$

Channel Capacity

The selected digital modulation scheme (QPSK) exhibits a spectral efficiency of 1.9 b/s/Hz while maintaining a $P(e) = 10^{-6}$ at 14 dB CNR.
Therefore, the system bit rate

$$f_b = 1.9 \text{ b/s/Hz} \times 4 \text{ MHz} = 7.6 \text{ Mb/s}$$
$$\therefore f_b = 7.6 \text{ Mb/s}$$

Noise Power: N_{db}

Noise power (N) is given by the following relationship

$$N = KT\text{BW}(\text{NF})$$

where
$$K = \text{Boltzmann's constant } (1.38 \times 10^{-23} \text{ J-K})$$
$$T = \text{temperature-290 K } (273 + 17°C)$$
$$\text{BW} = \text{noise bandwidth (4 MHz)}$$
$$\text{NF} = \text{receiver noise figure: assume (NF)} = 5 \text{ dB}$$

$$(KT)_{dBw} = -174 \text{ dBw}$$

Therefore,

$$N_{dB} = -174 + 10 \log(4 \times 10^6) + 5$$
$$= -174 + 66 + 5$$
$$= -174 + 71$$
$$\therefore N_{dB} = -103 \quad \text{(noise power)}$$

Carrier-to-Noise Ratio (CNR)$_{dB}$

$$CNR_{dB} = P_{r,min_{dB}} - N_{dB}$$
$$= -80 - (-103)$$
$$= 23 \text{ dB}$$
$$\therefore CNR = 23 \text{ dB}$$

Compute Error Power [$P(e)$]

Error power can be estimated as follows:

$$P(e) = e^{-CNR \times \sin^2(\pi/M)}$$

where   CNR = carrier-to-noise ratio (23 dB)
             M = modulation scheme phase levels QPSK (4)
             $\pi/4 = 45°$

$$P(e) = e^{CNR \times \sin^2(\pi/M)}$$
$$= e^{-23 \times \sin^2(\pi/4)}$$
$$= e^{-23 \times 0.5}$$
$$\therefore P(e) = 10^{-5}$$

Voice Channels

The system available bit rate is divided by the T_1 North American digital hierarchy.

$$T_1 = 1.544 \text{ Mb/s}$$
$$f_b = 7.6 \text{ Mb/s}$$
$$\frac{f_b}{T_1} = \frac{7.6 \text{ Mb/s}}{1.544 \text{ Mb/s}} \cong 5$$

There are 24 channels per T_1, and

$$24 \times 5 \cong 120$$

Therefore, number of voice channels $= 120$.

Observations

As we can see here, there is a difference between the CNR established from the graphs for QPSK (14 dB) and the CNR required by the system (23 dB). This 9-dB difference is attributed to system degradations.

The Bit-Energy-to-Noise Density [(Eb/No)dB] degradation, such as cochannel and intersymbol interference (ISI) is given as follows:

$$\left(\frac{E_b}{N_o}\right)_{dB} = \left(\frac{C}{N}\right)\left(\frac{BW}{f_b}\right)$$

$$\left(\frac{E_b}{N_o}\right)_{dB} = \left(\frac{C}{N}\right)_{dB} + 10\log(BW) - 10\log(f_b)$$

$$= 23 + 10\log(4 \times 10^6) - 10\log(7.6 \times 10^6)$$

$$= 23 + 66 - 66.8$$

$$= 20.2$$

where
$$C/N = 23 \text{ dB}$$
$$BW = 4 \text{ MHz}$$
$$f_b = 7.6 \text{ Mb/s}$$
$$\therefore \frac{E_b}{N_o} = 20.2 \text{ dB}$$

4.2.3 Antenna Tower Height

A fundamental element in a line-of-sight microwave link system design is the establishment of repeater path length. For an efficient link and maximum path length, a path profile must be generated in order to provide the required geographical data for antenna height calculations. Ideally, microwave frequencies propagate in a straight line in free space. Since LOS microwave links operate very close to the earth's surface, the propagating waves are subject to atmospheric and other environmental interference such as reflection, refraction, and diffraction (Figure 4.20). Furthermore, even for the modest distance of 50–60 km (average path length), the surface is neither flat nor free from physical obstacles. Physical obstacles and the earth's curvature are important factors in determining antenna tower heights and required transmitted power.

For antenna height calculations in an LOS microwave link, the knowledge of obstacle heights and locations along the path, earth curvature, and Fresnel zone clearance are

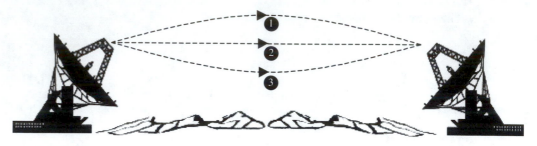

FIGURE 4.20 Bending of the propagated waves due to atmospheric refraction.

absolutely necessary. Under ideal conditions, the propagated wave travels in a straight line. In practice, the wave will bend either toward or away from the earth's surface. The degree of wave bending (refraction) from the straight path is subject to local atmospheric conditions.

The wave bending toward or away from the earth's surface will correspondingly decrease or increase the earth's curvature as a result. The earth's curvature height (h) is given by

$$h = 0.078(d/2)^2 \tag{4.25}$$

where h = curvature height (m)
            d = path length (km)

The relationship of Eq. (4.25) is applied only for a wave propagated in a straight line. When atmospheric refraction and other interference are incorporated into each calculation, the formula is modified as follows:

$$h = \frac{0.078\, d^2}{4K} \tag{4.26}$$

where K = refractivity factor (relating wave bending due to atmospheric conditions and actual earth's radius)

EXAMPLE 4.8

Determine the earth curvature height for a line-of-sight microwave link with a path length of 40 km and a K factor equal to 1.34.

Solution

$$h = \frac{0.078\, d^2}{4\,K} = \frac{0.078(40)^2}{4(1.34)} = 23.28 \text{ m}$$

$$\therefore h = 23.28 \text{ m}$$

The second factor contributing to the antenna height computation is the **first Fresnel zone clearance,** given by

$$r = 8.65\sqrt{d/f} \qquad (4.27)$$

where d = path length (km)
f = operating frequency (GHz)
r = first Fresnel zone clearance (m)

EXAMPLE 4.9

Compute the required antenna tower height of a line-of-sight microwave link operating at 2 GHz with a path length of 50 km and an atmospheric refraction factor of 1.34.

Solution

Compute Earth's Curvature (h)

$$h = \frac{0.078\, d^2}{4K} = \frac{0.078(50)^2}{4(1.34)} = 36.4 \text{ m}$$

$$\therefore h = 36.4 \text{ m}$$

First Fresnel Clearance (r)

$$r = 8.65\sqrt{d/f} = 8.65\sqrt{50/2} = 8.65 \times 5$$
$$= 43.25 \text{ m}$$
$$\therefore r = 43.25 \text{ m}$$

Physical Obstacle Height (20 m, given)

The antenna tower height is equal to the sum total of the earth's curvature, first Fresnel zone clearance, and physical obstacle height.

$$H_{\text{tower}} = h + r + \text{physical condition}$$
$$= 36.4 + 43.25 + 20$$
$$= 99.65$$
$$\therefore H_{\text{tower}} \cong 100 \text{ m}$$

For more complete path length and antenna height calculations, the following are required. First, the geographic location of the link must be identified. A reflectivity contour map should be available to provide the area's reflectivity index. If the link location is in the continental United States, see Figure 4.21.

Since the contour map reflectivity index is at sea level, a topographic survey of the link sight is essential to establish the elevation of physical obstacles along the path.

The reflectivity index N_o obtained from the contour and the site elevation established by the topographical survey are incorporated into a graph shown in Figure 4.22 in order to obtain the K factor.

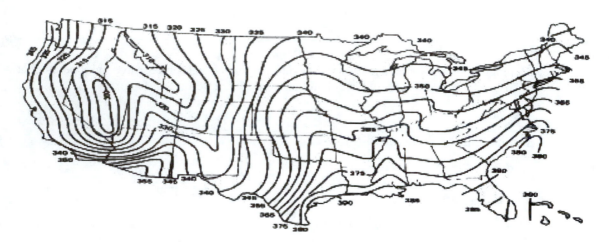

FIGURE 4.21 Index of reflectivity (N_o) for continental United States (sea level).

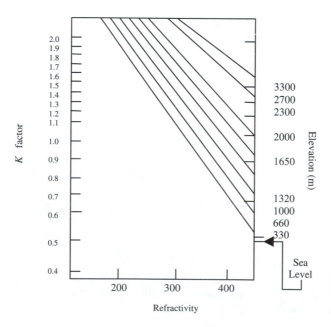

FIGURE 4.22
K factor: based on site reflectivity and elevation.

EXAMPLE 4.10

A line-of-sight microwave link is to be established west of Chicago. The topographic survey of the site indicates a 330-m elevation. Additional survey data show physical obstacles along the selected path, with the tallest at 25 m. Compute the antenna tower height for a 55-km path length and a link operating frequency of 1.8 GHz.

Solution

Establish the Value of K

From the contour map: $\quad N_o = 350$

For site elevation of 330 m: $K = 1.05$

Compute Earth's Curvature (h)

$$h = \frac{0.078 \, d^2}{4K} = \frac{0.078(55)^2}{4(1.05)} = \frac{235.95}{4.2}$$

$$= 56.2 \text{ m}$$

$$\therefore h = 56.2 \text{ m}$$

First Fresnel Clearance (r)

$$d = 55 \text{ km,}$$

$$f = 1.8 \text{ GHz}$$

$$r = 8.65 \sqrt{d/f} = 8.65 \sqrt{\frac{55}{1.8}}$$

$$= 47.8 \text{ m}$$

$$\therefore r = 47.8 \text{ m}$$

Antenna tower height
Physical Obstacle Height = 25 m

$$H_{\text{tower}_m} = h_m + r_m + 25 \text{ m}$$

$$= 56.2 + 47.8 + 25$$

$$= 129$$

$$\therefore H_{\text{tower}} = 129 \text{ m}$$

The antenna tower height is established at 129 m.

DESIGN EXAMPLE 4.11: A COMPLETE LOS DIGITAL MICROWAVE LINK

A line-of-sight microwave link is to be established in Long Island, New York, with the following link parameters.

Operating frequency $f = 2.0$ GHz
Path length $d = 50$ km
Antenna diameter for both receiver and transmitter $D = 1$ m
Receiver sensitivity $P_{r,min} = 10$ pW
Modulation scheme selected: QPSK (spectral efficiency = 1.9 b/s/Hz)
Available bandwidth BW = 20 MHz
Receiver noise figure selected: NF = 5 dB

Sight characteristics:

Terrain roughness: average $(A = 1)$
Atmospheric impact factor: $(B = 0.25)$
Elevation: 50 m above sea level
Tallest physical obstacle: 25 m

Compute

1. Fade margin (FM_{dB})
2. Path losses (L_{pdB})
3. Antenna gains (G_{dB})
4. System gain (G_{sysdB})
5. Channel capacity $(f_{bMb/s})$
6. Noise power (N_{dB})
7. Carrier-to-noise ratio (CNR_{dB})
8. Energy bit-to-noise ratio $[(E_b/N_o)_{dB}]$
9. Error power (P_e)
10. Number of voice channels (N)
11. Transmitter power (P_{tdB})
12. Antenna tower height (H_{tower_m})

Solution

Fade Margin (FM)$_{dB}$

The relationship establishing fade margin (FM) is given by the expression:

$$FM_{dB} = 30 \log(d) + 10 \log(A) + 10 \log(B) + 10 \log(f) - 30$$

where A = factor determining the terrain roughness of the path $(A = 1)$
B = factor determining atmosphere impact on the link $(B = 0.25)$
d = path length (50 km)
f = operating frequency (2 GHz)

Substituting into the equation:

$$FM_{dB} = 30 \log(50) + 10 \log(1) + 10 \log(0.25) + 10 \log(2) - 30$$
$$= 51 + 0 - 6.02 + 33 - 30$$

Therefore, fade margin FM = 48 dB.

Path Losses ($L_{p_{dB}}$)
Path losses are determined by

$$L_p = 32.5 + 20 \log(f) + 20 \log(d)$$

where $\qquad f$ = carrier frequency (MHz)
$\qquad\qquad d$ = path length or distance (km)

$$L_{p_{dB}} = 32.5 + 20 \log(2 \times 10^3) + 20 \log(50)$$
$$= 32.5 + 66 + 34$$
$$= 132.5$$

Therefore: path losses L_p = 132.5 dB.

Antenna Gains (G_{dB})
Antenna gain is given by

$$G_{dB} = 10 \log\left(\frac{4\pi A_{\text{eff}}}{\lambda^2}\right)$$

where $\qquad G$ = antenna gain (dB)
$\qquad\qquad A_{\text{eff}}$ = effective area of the parabolic antenna
$\qquad\qquad d$ = antenna diameter = 1 m

Antenna area $A = \pi D^2/4$

$$\therefore A = \frac{\pi(1)^2}{4} = 0.785 \text{ m}^2$$
$$A_{\text{eff}} = K \cdot A$$
$$K = \text{efficiency factor } (\cong 0.55)$$
$$\therefore A_{\text{eff}} = 0.55 \times 0.785 \text{ m}^2 = 0.432 \text{ m}^2$$

$$\lambda = \frac{c}{f}$$

$$= \frac{3 \times 10^8}{2 \times 10^9} \, \text{m/s}$$

$$\lambda = 0.15 \, \text{m}$$

Substituting into G, we have

$$G_{\text{dB}} = 10 \log\left(\frac{4\pi A_{\text{eff}}}{\lambda^2}\right)$$

$$= 10 \log\left[\frac{4\pi(0.432)}{(0.15)^2 \, \text{m}^2}\right]$$

$$= 23.8$$

Therefore, antenna gain $G = 23.8$ dB.

System Gain ($G_{\text{sys}_{\text{dB}}}$)
The link system gain is calculated as follows:

$$G_{\text{sys}_{\text{dB}}} = \text{FM}_{\text{dB}} - G_{t_{\text{dB}}} - G_{r_{\text{dB}}} + L_{p_{\text{dB}}} + L_{f_{\text{dB}}} + L_{\text{br}_{\text{dB}}}$$

where G_{sys} = systems gain (dB)
             FM = fade margin (dB)
               G_t = transmitter antenna gain (dB)
               G_r = receiver antenna gain (dB)
               L_p = path losses (dB)
               L_f = feed losses (dB)
             L_{br} = branching losses (dB)

Feed and branch losses for both transmitter and receiver consist of the following:

Circulator insertion losses
Connector losses
Directional coupler losses
Coaxial-to-waveguide losses
Rectangular-to-circular-waveguide losses
Other

System Gain
Assuming $L_f = 2.5$ dB and $L_{\text{br}} = 2.5$ dB,

$$G_{\text{sys}_{\text{dB}}} = \text{FM}_{\text{dB}} - G_{t_{\text{dB}}} - G_{r_{\text{dB}}} + L_{p_{\text{dB}}} + L_{f_{\text{dB}}} + L_{br_{\text{dB}}}$$
$$= 48 - 23.8 - 23.8 + 132.5 + 2.5 + 2.5$$
$$= 137.9$$

Therefore, system gain $G_{\text{sys}} = 137.9$ dB.

Channel Capacity $(f_{b_{Mb/s}})$

From tables, the selected digital modulation scheme reflects a spectral efficiency of 1.9 b/s/Hz with a corresponding CNR equal to 16.6 dB while maintaining the required BER of 10^{-8} (Figure 4.23). System bit rate f_b = spectral efficiency × bandwidth, so

$$f_b = 1.9 \times 20 \text{ MHz} = 38 \text{ Mb/s}$$

Therefore, channel capacity $f_b = 38$ Mb/s.

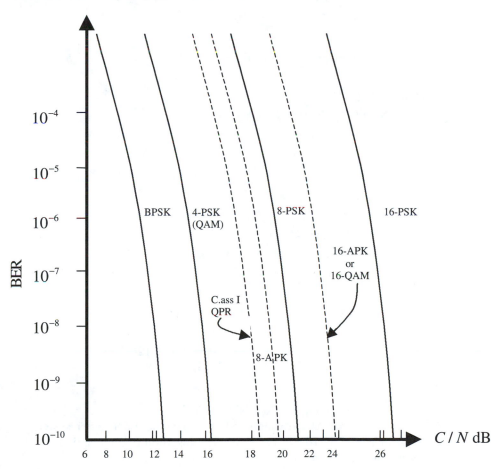

FIGURE 4.23 *P(e)* performance of *M*-ary PSK, QAM, and *M*-ary APK coherent systems. The rms *C/N* is specified in the double-sided Nyquist bandwidth.

Source: Reproduced by permission of HP-Canada.

Noise Power: (N_{dB})

Noise power (N) is given by

$$N = K\,T\,BW(NF)$$

where
K = Boltzmann's constant (1.38×10^{-23} J-K)
T = temperature $290°$ ($273 + 17°C$)
BW = noise bandwidth (20 MHz)
NF = receiver noise figure (assume NF = 5 dB)

$$
\begin{aligned}
N_{dB} &= 10\log(K) + 10\log(T) + 10\log(BW) + NF_{dB} \\
&= 10\log(1.38 \times 10^{-23}) + 10\log(290°) + 10\log(20 \times 10^6) + 5 \\
&= -228.6 + 24.6 + 73 + 5 \\
&= -126
\end{aligned}
$$

Therefore, noise power $N = -126$ dB.

Carrier-to-Noise Ratio (CNR_{dB})

The carrier-to-noise ratio is expressed by

$$
\begin{aligned}
CNR_{dB} &= P_{r,min_{dB}} - N_{dB} \\
&= -110 - (-126) \\
&= -110 + 126 \\
&= 16
\end{aligned}
$$

where
$P_{r,min}$ = receiver sensitivity
N = noise power

$$
\begin{aligned}
CNR_{dB} &= P_{r,min_{dB}} - N_{dB} \\
&= -100 - (-126) \\
&= -100 + 126 \\
&= 26
\end{aligned}
$$

Therefore, carrier-to-noise ratio CNR = 16 dB
Note: It order to offset ISI (intersymbol-interference) and other system degradations, the CNR is measured by 10 dB, to 26 dB. The insertion of an adaptive equalizer into the system will correct the problem.

Bit-Energy–to–Noise Ratio [$(E_b/N_o)dB$]

The Bit-energy–to–noise ratio is given by

$$\left(\frac{E_b}{N_o}\right)_{dB} = \left(\frac{C}{N}\right)_{dB} + 10 \log(\text{BW}) - 10 \log(f_b)$$
$$= 26 + 10 \log(20 \times 10^6) - 10 \log(7.6 \times 10^6)$$
$$= 26 + 73 - 68.8$$
$$= 30.2$$

Therefore, bit-energy–to–noise ratio $E_b/N_o = 30.2$ dB.

Error Power [P(e)]
Error power can be expressed by

$$P_e = e^{\text{CNR} \times \sin^2(\pi/m)}$$

where   CNR = carrier-to-noise ratio (23 dB)
M = modulation scheme phase levels (QPSK)
$\pi/4 = 45°$

$$P_e = e^{\text{CNR} \times \sin^2(\pi/m)}$$
$$= e^{26 \times \sin^2(45°)}$$
$$= e^{26 \times 0.5}$$
$$= e^{-13}$$

Therefore, error power $P(e) \cong 2.2 \times 10^{-6}$.

Number of voice channels
If the microwave link is used for voice traffic, the maximum number of voice channels that can be processed is determined as follows:

$$\text{T}_1 = \frac{f_b}{1.544} = \frac{38}{1.544} \text{ Mb/s} = 24.6$$

Number of voice channels $= 24 \times 24.6 = 590$
Therefore, number of channels $N = 590$.

Transmitter Power ($P_{t_{dB}}$)
Transmitter power is given by

$$P_{t_{dB}} = G_{\text{sys}_{dB}} + P_{r_{min}}$$
$$= 137.9 - 110$$
$$= 27.9 \text{ dB}$$

Line-of-Sight Microwave Links

Therefore, transmitter power $P_t = 27.9$ dB.

Antenna Tower Height (H_{tower_m})

The three main components determining the antenna tower height are:

Earth curvature
First Fresnel clearance
Height of physical objects along the link path

$$H_{tower} = h + r + h_{ph}$$

where H_{tower} = antenna tower height (m)
                   h = earth's curvature (m)
                   r = first Fresnel clearance (m)
             h_{ph} = physical object height (m)

COMPUTE EARTH'S CURVATURE (H)

The earth's curvature is calculated as follows:

$$h = \frac{0.078d^2}{4K}$$

where   d = path length (km)
           K = K factor ($0.5 \leq K \leq 1.5$)

Establish the value of K: From the contour map, the geographic location of Long Island indicates an index of reflectivity of 355. The K-factor is established in relationship to N_o and is found to be equal to 0.85. Substituting into the equation, we have

$$h = \frac{0.078(50)^2}{4 \times 0.85} = 57.35 \text{ m}$$

Therefore, earth curvature $h = 57.35$ m.

FIRST FRESNEL CLEARANCE (R)

The first Fresnel clearance is given by

$$r = 8.65 \sqrt{d/f} = 8.65 \sqrt{\frac{55}{2}} = 8.65 \times 5.244 = 45.3 \text{ m}$$

where $d = 50$ km
$f = 2$ GHz

Therefore, first Fresnel clearance: $r = 43.25$ m

PHYSICAL OBSTACLE HEIGHT

$H_{ph} = 25$ m (given)

$$H_{tower_m} = h_m + r_m + 25 \text{ m}$$
$$= 57.35 + 45.3 + 25$$
$$= 127.65 \text{ m}$$

Therefore, antenna tower height $H_{tower} = 127.65$ m.

QUESTIONS

1. Define the terms *transmission path* and *equipment redundancy*. Why are they required?
2. Define *fade margin*.
3. List the causes for wide-level fade margins.
4. Give the occupied bandwidths for the (a) standard FM/FDM group, (b) standard super group, (c) standard master group, and (d) standard super master group.
5. Sketch the block diagram of the standard group (12-channel FDM).
6. What is the FM improvement factor?
7. Sketch the block diagram of the FM discriminator circuit.
8. Describe multipath fading.
9. Define the terms: (a) *pre-emphasis*, (b) *de-emphasis*. Why is pre-emphasis required?
10. List the main components incorporated into the system gain of a digital microwave link.

PROBLEMS

1. A 10-MHz signal with 10-V_{rms} amplitude is frequency modulated by a 5-kHz sine wave with a peak-to-peak deviation of 1.5 kHz. Determine the first two sideband components of the FM signal.
2. Compute the baseband bandwidth of a FM signal with an index of modulation of 0.87 and a modulating signal of 4 kHz.

3. In an FM/FDM microwave link, the receiver employed exhibits a noise figure of 2.5 dB and an operating bandwidth of 15 MHz. Determine the FM improvement threshold.

4. A line-of-sight microwave link operates with the following parameters:

Mode of operation:	FM/FDM
Carrier frequency:	f_c = 6 GHz
Modulation frequency:	f_m = 20 MHz
Path length:	d = 50 km
Index of modulation:	m_f = 0.55
Receiver noise figure:	NF = 3.5 dB
Antenna (parabolic reflection):	d = 1 m
Feed and branch losses:	5 dB

Compute the required transmitter power (P_t).

5. A digital line-of-sight microwave link operates with the following parameters: (a) operating frequency 2.2 GHz, (b) repeater-to-repeater distance 40 km, (c) both antennas (parabolic reflectors) have 1-m diameters. Feed and branch losses are estimated to be equal to 4 dB (both). Calculate the system gain (G_{syst}). (*Note:* Path terrain is average; atmospheric conditions are average.)

6. Calculate the hop distance of a digital microwave link (LOS) operating at 9 GHz and with a system gain of 125 dB. Both antennas are selected with a gain of 25 dB each. The link is established over an average terrain and average atmospheric conditions. The fade margin is assumed to be equal to 30 dB, and feed and branch losses are equal to 5 dB.

7. A digital (LOS) microwave link operates with a transmitter power of 500 W and carrier frequency of 1.8 GHz. The receiver section employs a QPSK modulation scheme and operates with a bandwidth of 6 MHz and sensitivity of 12 nW. Both receiver and transmitter antennas exhibit a gain of 30 dB. Assuming a fade margin of 32.5 dB and a receiver noise figure of 3.2 dB, calculate the (a) hop distance, (b) link BER, (c) channel capacity, (d) CNR, (e) E_b/N_o, and (f) number of voice channels.

8. Calculate the antenna tower height of an LOS microwave link operating at 1.5 GHz, and a maximum hop length of 45 km. The site survey indicates a 500-m elevation, 20-m height obstacle, and 350 reflectivity number.

5

Communications Satellites

Objectives

- Describe the importance of communications satellites.
- Identify the different classes of satellites.
- Understand the basic principles of planetary mechanics.
- Explain in detail the main sequences of satellite placing into orbit.
- Describe in some detail the different types of satellite antennas.
- List and explain the major modules of a satellite repeater.

Key Terms

INTELSAT
Geosynchronous satellites
Oribiting satellites
Drift orbit
Transfer orbit
Geosynchronous orbit
Apogee
Perigee
Spin axis
Geocentric
Heliocentric

Telemetry
Lens antennas
Low-noise amplifier
Dual-frequency conversion
MMIC
Body stabilization
Spin stabilization
Simple spin
Dual spin
Triple spin
Satellite tracking

Power subsystem
Solar cell
Regulated bus
Unregulated bus
Antenna polarization
Beam-forming network
Angular width
Input/output multiplexer
Group delay equalizer
Parametric amplifier
Image rejection

INTRODUCTION

In the early 1960s a joint venture between NASA, Bell Labs, and JPT resulted in the placement into space of the first communications satellite, at an altitude of approximately 1500 km. This passive repeater was used to reflect signals transmitted from one point on the earth to another point located far beyond the line of signal. In August of the same year the first experimental communications link was established between Holmedel, New Jersey and Goldstone, California. Following this successful experiment, another link was established between New York and France, proving that long-distance communications is possible through artificial satellites.

Since these very early successful experiments, satellite communication has revolutionized the entire communications field, providing efficient and effective global communications by reaching the most inaccessible places on the surface of the earth. The idea for satellite communications was originated from an article published in *Wireless World* in 1945 by Arthur C. Clarke titled "Experimental Relays." In this article the author envisioned the great potential such a system might have in achieving global communications. Three such satellites placed in a geosynchronous or geostationary orbit at a height of 35,780 km and separated by 120° should be able to provide global coverage. Fifteen years later Clarke's vision became reality in the ECHO project. Today communications satellites placed in geosynchronous orbits at a height of 22,000 miles, or approximately 41,000 km, provide voice, video, and data services to millions of customers all over the world. Satellites today are not limited to communications only, as other areas such as remote sensing, search and rescue, and global positioning have completely exploited this fairly new technology.

The U.S. military, sensing the great potential satellite communications might have for strategic and tactical command, initiated an elaborate R&D program to take advantage of this new field. In July 1954 the U.S. Navy was successful in transmitting voice signals using the moon as a reflector. Two years later the same agency established the first communications link between Washington and Hawaii, using the moon as a passive repeater. In 1960 the ECHO project was implemented for civilian use, operating at 960 MHz and 2.29 GHz. The moon and the ECHO project were classified as passive repeaters performing only one function, the reflection of signals. Both systems, although capable of providing multiple access at great distances, suffered from the significant disadvantage of high RF power requirements.

This disadvantage and other problems related to satellite altitude generated the need for active repeaters. Parallel advancements in semiconductor device technology and microelectronics made active communication satellites possible. Today all communications satellites are active repeaters. It would be unwise not to mention the fact that satellites other than those for communication, were placed in space before 1960. These spacecraft were classified as space probes. The first country to place such a probe into Earth's orbit was the Soviet Union with the Sputnik in 1954. This communist success achieved at the peak of the Cold War sent shock waves throughout the industrialized Western world, thus increasing the fear of communist world domination through technological superiority. The free world, led by the United States, mobilized its scientific and engineering resources, and in 1958 the Score satellite was successfully launched carrying a prerecorded message from the U.S. president. Research and development in active communications satellites were accelerated, resulting in the production of the Telstar, Relay and Sincom spacecrafts. The utilization of these types of satellites for communications was limited because they were placed in low orbit and they were not geosynchronous.

The earth tracking stations had the difficult task of knowing the exact orbital path of the spacecraft and its flight time, and communications effectiveness was limited by the horizontal flight time. On July 10, 1962, AT&T launched its first Telstar in an elliptical orbit corresponding to 2.5 hours of orbital period. On May 7, 1963, Telstar II was launched, operating at 6/4 GHz, and capable of carrying one video channel and 600 voice channels.

Parallel to AT&T's launch, RCA developed the Relay spacecraft, similar to Telstar, and at the same time Hughes Aircraft Corporation launched the first synchronous satellite. This active repeater with the capacity of transmitting one video channel or fifty simultaneous voice channels was the first to establish voice communications between Africa and South America and also the first to provide TV coverage between Europe and North America. These early successes generated the incentive for the U.S. administration to introduce legislation for the commercialization of this unique and very important technology.

The Communication Satellite Act was passed by the U.S. Congress in 1962, and a year later, the Communications Satellite Corporation was established. On August 20, 1963, the International Telecommunications Satellite Corporation (**INTELSAT**) was founded by 11 nations in order to oversee the design, development, and maintenance of a global satellite communications network. The first communications satellite to be developed by the new agency was the INTELSAT-I (*Early Bird*), launched in April 1965. This spacecraft was capable of carrying 240 voice channels and had a projected lifetime of 18 months. In reality it lasted over 3 years. INTELSAT-II was launched in 1967, and INTELSAT-III in 1968. Both satellites were placed in geosynchronous orbits, thus providing communications coverage for both the Atlantic and the Pacific Ocean regions. In 1971 INTELSAT-IV was launched. This latest spacecraft was capable of providing global and spot-beam coverage, with a total capacity of 12 video channels or 4000 voice channels. Table 5.1 shows the type and basic characteristics of the INTELSAT series and dates of launching.

TABLE 5.1   INTELSAT Series

Series	Year of Launch	Voice Channel Capacity	Weight (kg)
I	1965	240	38
II	1967	240	86
III	1968	1,200	152
IV	1971	4,000	700
IV[d]	1975	6,000	790
V	1979	12,000	967
VI	1986	32,000	1670

5.1 CLASSIFICATION OF SATELLITES

Satellite systems are classified into four major categories: (a) geosynchronous, (b) polar, (c) mobile, and (d) defense.

Geosynchronous Satellites   **Geosynchronous satellites** are mainly used for international and regional communications. For international communications the INTELSAT family of satellites and the Russian STASINAR systems are utilized, and for domestic service (local television distribution and regional communications), the SATCOM, ANIK, WESTAR, COMSTAR, GALAXY, and MOLNIYA systems are used.

Polar and Orbiting Satellites   The National Oceanic and Atmospheric Administration (NOAA) has developed and operates a system of polar orbiting as well as geosynchronous satellites for meteorological, oceanic and space environmental studies. Both systems are designed to perform specific tasks. For example, the geosynchronous system of satellites is monitoring weather patterns as they develop in the tropics, while the polar satellites orbiting at higher altitudes are used for data collection.

   The first meteorological satellite was placed in orbit on April 1, 1960. From 1960 to 1966, ten Television and Infrared Observation experimental Satellites (TIROS) were in operation. On February 3, 1966, the first fully operational TIROS satellite was placed in orbit. The primary mission of this spacecraft was to provide detailed weather information through photographic means. Nine of these satellites were launched between 1966 and 1970.

   On January 30, 1970, the first of the new series of TIROS satellites was placed in orbit with an accuracy and resolution far exceeding the previous satellites. On June 27, 1979, the TIROS-N series was launched. This spacecraft employed state-of-the-art semiconductor devices and instrumentation far surpassing the earlier spacecrafts. Between 1982 and 1986 more and more advanced TIROS-N series satellites were placed in orbit.

Mobile Satellites   Mobile satellites were specifically designed to provide continuous communications with ships and aircraft, and also to assist in search and rescue missions. Such systems are the AEROSAT, MARISAT, and MAROSAT.

Defense Satellites   Five systems of defense satellites are in operation today: the DSCS, SKYNET, FLT, SATCOM, and NATO systems.

5.2 THE COMMUNICATIONS SATELLITE SYSTEMS

Communications satellites are very complex and highly sophisticated pieces of equipment, utilizing the full spectrum of scientific and engineering know-how in their design and development. For the satellite to fulfill its mission, it must be placed into the appropriate **geosynchronous orbit**. Its orbital path must be closely monitored and if necessary, controlled from the ground. It must also be able to function efficiently for its entire operating life. The mission of the spacecraft and environment in which it operates require the most stringent component and system specifications. For example, the spacecraft's size must be contained within the specified limits dictated by the launching vehicle. Thrusters are required for positioning, altitude control, and reorientation. Solar arrays must be available to provide the necessary power required by the instrumentation and onboard electronics. Dry-cell batteries are needed for maintaining operation during solar eclipses. During the transfer stage from the temporary orbit to the permanent geosynchronous orbit, an **apogee** monitor is required to provide the injection velocity. A number of de-spun highly directional antennas are required for tracking and command as well as to provide global and spot-beam coverage. The communications section employs state-of-the-art electronic components and circuit design. Since the satellite is an active repeater, it receives a very weak signal from the earth's station, amplifies it, translates the signal into a down-frequency, reamplifies it through solid state amplifiers or traveling-wave tube, and finally retransmits the signal back to earth. A more detailed description of satellite subsystems and components will be given in subsequent sections.

5.2.1 Orbits

A communications satellite is designed to operate from a geosynchronous orbit at a height of approximately 22,000 miles or 36,000 km. The placement of the spacecraft to a geosynchronous orbit is achieved through three main sequences: (a) injection of the satellite into space, (b) placement into temporary orbit, and (c) transfer into permanent orbit (Figure 5.1). A more detailed deployment is shown in Figure 5.2.

At a height of approximately 36,000 km the angular velocity of the satellite is 11,068 km/h or 3.07 km/s. This combination of satellite height and angular velocity satisfies the fundamental laws (Kepler's, Newton's) of planetary motion in maintaining orbit without the need for further propulsion requirements, except for minor orbital adjustments. The 24-hour orbital period of the satellite is exactly the same as earth's period around its axis. Thus the satellite seems to be stationary at a reference point on the surface of the earth.

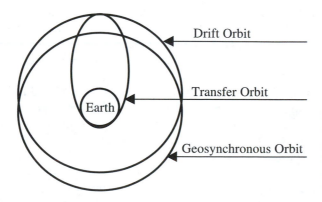

Drift Orbit

Transfer Orbit

Earth

Geosynchronous Orbit

FIGURE 5.1 The complete sequence of placement of an STS/ANIK communications satellite into geosynchronous orbit.

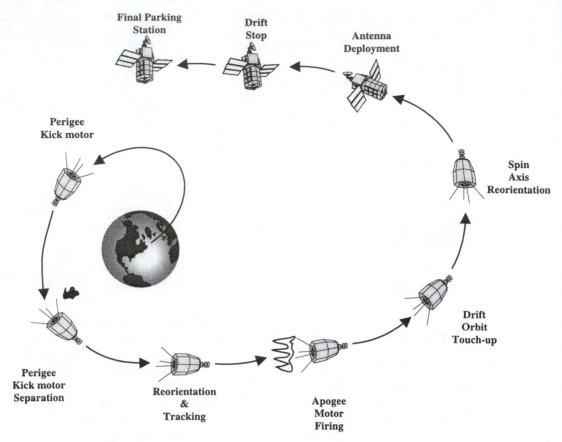

FIGURE 5.2 Stages of Satellites Orbital Placement.

5.2.2 Planetary Mechanics

The successful placement of a spacecraft in a polar or geosynchronous orbit is the direct result of two and a half thousand years of continuous progress in mathematics and planetary science. Greeks in the 7th century B.C. proposed that two kinds of motion are present in the universe: *terrestrial*, which covers the motion of all objects on the surface of the earth, and *celestial*, which covers the motion in space. Greeks also developed the **geocentric** theory, stating that the earth is the center of the universe with the other planets revolving around it (Figure 5.3).

Almost five hundred years later, Ptolemy in Alexandria modified the earlier Greek theory by adding the sun to the circular motion around the earth. Thirteen centuries later the Polish astronomer Nicolaus Copernicus proposed the **heliocentric** theory (*helios: Greek for "sun"*) (Figure 5.4). According to Copernicus, the Sun is the center of the universe and the rest of the planets including earth are revolving around it in a circular motion. The earth is no more than a simple planet, revolving around the sun and also around its axis every 24 hours. Copernicus' heliocentric theory of the universe was very controversial at the time, especially with the powerful church, whose theological doctrine was based on the geocentric concept.

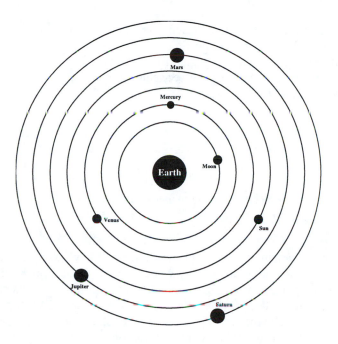

FIGURE 5.3
The Greek geocentric concept of the solar system.

In 1895, Johannes Kepler, a German astronomer, proposed the elliptical motion of the planets around the sun. Kepler felt that in order to better understand the motion of the planets around the sun, the sun must be considered as a reference point. According to Kepler, the orbital motion of the planets is based on three principles, later named after him as Kepler's laws:

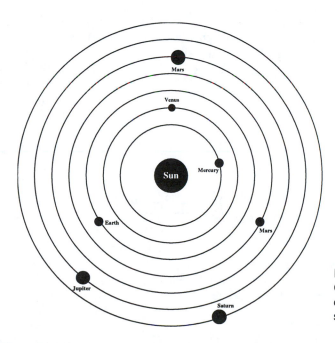

FIGURE 5.4
Conpernicus' heliocentric concept of the solar system.

1. The motion of each planet around the sun is an ellipse with the sun at one focal point.
2. The area covered by the axis connecting the plane and the sun during the elliptical trajectory of the planet is equal at equal time intervals.
3. The ratio of the cube of the distance between a planet and the sun to the square of its period is constant (K):

$$K = \frac{R^3}{T^2}$$

(5.1)

where $\quad R$ = distance between the planet and the sun
$\quad\quad\quad\ T$ = period of the planet

Kepler's three fundamental laws of planetary motion have changed very little since then, so they are as applicable today as they were then. One question Kepler was unable to answer was how the planets were able to maintain their elliptical orbits around the sun. This question was left for Isaac Newton. Born on December 25, 1642, Newton graduated from Cambridge in 1655. His interest in advanced mathematics (calculus), combined with his highly observant nature resulted in the conception of the notion of gravity and gravitational force. Gravitational force effects any object with mass (m). According to Newton, two bodies with mass m_1 and m_2, and separated by a distance (r), exhibit a gravitational force governed by the relationship

$$F = \frac{G \cdot m_1 \cdot m_2}{r^2}$$

(5.2)

where $\quad\quad\quad\quad F$ = intensity of gravitational force (N)
$\quad\quad m_1, m_2$ = mass of each body
$\quad\quad\quad\quad r$ = distance between the two bodies from the center of their universal mass
$\quad\quad\quad\quad G$ = universal gravitational constant (6.67×10^{-18} N)

If a planet is orbiting around the sun in a circular orbit with a constant velocity, its centripetal force is expressed by

$$F = \frac{4\pi^2 \cdot m \cdot R}{T^2}$$

(5.3)

where $\quad m$ = mass of the planet
$\quad\quad\quad R$ = distance between the planet and the sun (centers)
$\quad\quad\quad T$ = orbital period

Referring to Kepler's third law and solving for T^2 we have.

$$T^2 = \frac{R^3}{K} \qquad (5.4)$$

Substituting Eq. (5.4) into Eq. (5.3), we obtain

$$F = \frac{4\pi^2 \cdot m \cdot R \cdot K}{R^3} = \frac{4\pi^2 \cdot m \cdot K}{R^2}$$

Therefore,

$$F = \frac{4\pi^2 \cdot m \cdot K}{R^2} \qquad (5.5)$$

The sun's gravitational attraction on the planet is proportional to its mass (m) and inversely proportional to the square of the distance (R) from the center of the sun. The same relationship can be applied between the earth and its natural satellite the moon, or the earth and any other satellite. Let us consider the moon first. The earth's gravitational attraction on the moon is given by

$$F = \frac{4\pi^2 \cdot m \cdot K_E}{(R_E)^2} \qquad (5.6)$$

where K_E = constant of proportionally for earth
R_E = radius of earth

The gravitational force at the surface of the earth is

$$g_E = \frac{F}{m} \qquad (5.7)$$

Substituting for F gives

$$g_E = \frac{4\pi^2 \cdot K_E \cdot m/(R_E)^2}{m} = \frac{4\pi^2 \cdot K_E}{(R_E)^2}$$

$$g_E = \frac{4\pi^2 \cdot K_E}{(R_E)^2} \qquad (5.8)$$

The acceleration of the moon toward the earth is

$$g_M = \frac{4\pi^2 \cdot K_E}{(R_M)^2} \tag{5.9}$$

where R_M = radius of the moon's orbit from the center of the earth

Dividing Eq. (5.7) by Eq. (5.6), we have

$$\frac{g_M}{g_E} = \frac{4\pi^2 \cdot K_E/(R_M)^2}{4\pi^2 \cdot K_E/(R_E)^2} = \left(\frac{R_E}{R_M}\right)^2$$

Or

$$g_E = g_M(R_E/R_M)^2 \tag{5.10}$$

The acceleration of the moon toward the earth is proportional to earth's gravitational pull, proportional to the square of the earth's radius, and inversely proportional to the moon's orbital radius. Also, the acceleration of the planets toward the sun, or the acceleration of the moon or any other satellite toward the earth is equal and opposite to the centrifugal force. The satellite maintains its orbit with an orbital velocity given by

$$v^2 = g \cdot R_{\text{sat}} \tag{5.11}$$

where g = earth's gravity
R_{sat} = radius from the center of the earth to the center of the satellite

---

EXAMPLE 5.1

A communications satellite is to be placed in an equatorial orbit at a height of 40,000 km. Compute its orbital velocity and period.

Solution
The radius of the satellite from the surface of the earth is

$$R_{\text{sat}} = R_E + \text{height}$$
$$= 6.3 \times 10^6 + 40 \times 10^6 \text{ m}$$
$$= 46.3 \times 10^6 \text{ m}$$
$$\therefore R_{\text{sat}} = 46.3 \times 10^6 \text{ m}$$

The gravitational pull of the satellite is

$$g_{sat} = g_E \left(\frac{R_E}{R_{sat}} \right)^2$$

$$= 9.8 \text{ m/s}^2 \left(\frac{6.3 \times 10^6}{46.3 \times 10^6 \text{ m}} \right)^2$$

$$= 0.18 \text{ m/s}^2$$

The velocity of the satellite is given by

$$v^2 = g_{sat} \cdot R$$

$$v = \sqrt{0.18 \text{ m/s}^2 \times 46.3 \times 10^6}$$

$$= 2.9 \text{ m/s}$$

$$\therefore v = 2.9 \text{ m/s}$$

The orbital period of the satellite is calculated as follows:

$$v \cdot T = 2\pi R$$

$$T = \frac{2\pi R}{v} = \frac{2\pi R \times 40 \times 10^6 \text{ m}}{2.9 \text{ m/s}} = 86.66 \times 10^6 \text{ s}$$

$$T = 24 \text{ hours}$$

Therefore the satellite is in a geosynchronous orbit.

5.2.3 Satellite Launching into Orbit

A communications satellite is launched into space via a launching vehicle. This vehicle can be either a rocket or the Space Shuttle. In both cases, to achieve geosynchronous orbit the satellite must undergo four sequences: (a) ascent, (b) placement into parking orbit, (c) transfer orbit, and (d) placement into permanent circular geosynchronous orbit (Figure 5.5).

The spacecraft is placed on top of the launching vehicle or in the cargo bay of the Space Shuttle, which will eventually carry it into the transfer orbit. During the initial stage, the launching vehicle is programmed to perform absolutely minimum maneuvering in order to survive the enormous aerodynamic stresses generated during flight through earth's atmosphere. The **transfer orbit** is highly elliptical. If a rocket is used as the transportation vehicle, the process of placing the communications satellite into geosynchronous orbit with zero inclination in reference to the equatorial plane involves the following steps: At a launching site close to the equator, the vehicle is programmed to follow an easterly path in order

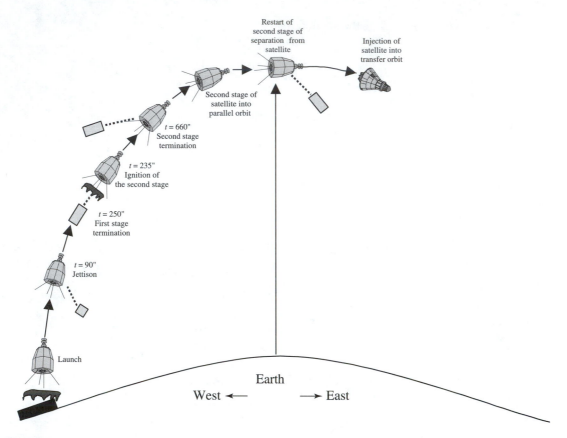

Restart of
second stage of
separation from
satellite

Injection of
satellite into
transfer orbit

Second stage of
satellite into
parallel orbit

$t = 660''$
Second stage
termination

$t = 235''$
Ignition of
the second stage

$t = 250''$
First stage
termination

$t = 90''$
Jettison

Launch

Earth

West ← → East

FIGURE 5.5   The launching of a communications satellite, using a rocket as the transportation vehicle.

to make use of the earth's rotational velocity. After lift-off, the first stage of the rocket is burned for approximately 250 seconds. At this point, it separates from the main vehicle and the second stage is ignited 16 seconds later. Burning for another 400 seconds, the second stage achieves a circular parking orbit at a height of 250 km. Just before the equatorial crossing, the second stage is reignited and the spacecraft is now placed into a highly elliptical transfer orbit, with an apogee of 35,860 km. This apogee is the required height if the final circular geostationary orbit is to be achieved. Placing the satellite in this elliptical orbit also maximizes its bayload carrying capacity. As soon as it reaches its elliptical orbit, the spacecraft is spin-stabilized and its **telemetry** channels are activated for ground control.

Upon ground instructions and when the satellite is at its apogee, the apogee motor is fired and the spacecraft is placed into circular geosynchronous orbit. At this point, the satellite is continuously tracked from the ground control station and guided through its self-propelled system into a specific parking space, ready to start normal operation upon ground command.

The National Aeronautics and Space Administration (NASA) and the European Space Agency (ESA) are the two space authorities capable of placing communications satellites into space. For NASA the main transportation vehicle for placing communications satellites into space is the Space Shuttle, although Delta, Atlas–Centurion, and Titan-IIIc rockets are used as secondary vehicles. The Titan-IIIc rocket has the ability through its upper stage to place the spacecraft into the transfer orbit as well as into the final orbit. The launching site for NASA is Cape Canaveral (Cape Kennedy). This site, located at a latitude of 28.5°N, is not best suited for equatorial orbit, yet it takes full advantage of the earth's rotation for the much-needed extra velocity at this initial stage. The European Space Agency has selected its launching site at Kuru, in the former French Guiana. This site, with a latitude of only 5°N, is best suited for equatorial launch. The agency's main transportation vehicle is the Arean rocket. Other equatorial sites exist in India and off the coast of East Africa.

Of all the transportation vehicles the Space Shuttle has proven to be the most reliable. For example, in 1991 INTELSAT-IV was launched. When the spacecraft was initially placed into its predetermined parking orbit, its booster rocket failed to ignite and the 160-million-dollar satellite was to become the most expensive piece of space junk. This spacecraft was to be placed into circular orbit over the Atlantic to provide full coverage of the 1992 Barcelona summer Olympics. In May 1992, the Space Shuttle *Discovery* in its maiden voyage undertook the difficult task of salvaging the satellite. After several space walks, three astronauts succeeded in pulling the spacecraft into the shuttle's cargo bay. There a new booster rocket was installed, and the satellite was eventually placed into its geosynchronous orbit. The overall salvage operation cost reached the 100-million-dollar mark. It is obvious that this operation would not have been possible without the Space Shuttle.

The utilization of the Space Shuttle as the launching vehicle for communications satellite deployment has other advantages as well. For example, the mass of the satellite can be increased beyond the limits imposed by the launching rockets used. An increase in the size of the spacecraft implies a larger cylinder (spin-stabilization type), a larger solar-cell surface, and therefore an increase in electric power availability.

5.2.4 Stabilization

From the time the satellite is placed into its preassigned parking space (its equatorial geosynchronous orbit), it is anticipated to function efficiently for its entire operating life (approximately ten years) under the most extreme environmental conditions. Efficient operation implies that the spacecraft will maintain an absolutely constant geosynchronous orbit so that its antennas pointing toward the earth will cover the preassigned geographic region. The satellite will operate in a hard vacuum environment, while its various surfaces will be exposed to extreme temperatures dictated by the position of the spacecraft in reference to the earth–satellite–sun axis. Other elements acting upon the spacecraft are gravity, solar radiation, and earth's magnetic field. Earth's nonspherical shape and its nonuniform mass distribution generate a difference in the gravitational force between the center of its mass and the center of the satellite mass. This gravitational difference affects the angle of inclination of the spacecraft, which must constantly be maintained at 0°. The inclination and attitude of a geosynchronous satellite are kept constant by way of spacecraft stabilization and station keeping.

Types of stabilization

There are two basic types of spacecraft stabilization: (a) body stabilization and (b) spin stabilization.

Body Stabilization   Figure 5.6 illustrates body, or three-axis, stabilization. Body stabilization is also subdivided into three categories. In the first category, the spacecraft's attitude is controlled by sensors and gyros while three-axis control for yaw, pitch, and roll is achieved by a number of on-board thrusters. In the second category the spacecraft incorporates in its design a momentum wheel. This wheel provides gyroscopic stiffness for only one axis by maintaining a minimum finite speed. Spin axis alignment is established by a combination of slight variations of the wheel speed and on-board thruster. In the third category, three reaction wheels are responsible for each axis; by varying the speed independently, they control the attitude of the spacecraft in reference to zero momentum.

Spin Stabilization   The most effective form of stabilization is the spinning of the spacecraft while in its final orbit (Figure 5.7). This stabilization method although providing the satellite with attitude gyro-stiffness simultaneously requires de-spun antenna assembly. The de-spun antenna assembly rotates in an opposite direction to that of the spacecraft body, thus maintaining a constant radiation pattern at a specific geographic region on the surface of the earth.

   For spin stabilization, it is necessary that the main body of the spacecraft have a cylindrical shape. The cylindrical body enables the spacecraft to spin around its x-axis with an average of 80 rpm and parallel to earth's polar axis. By rotating around its main axis, the spacecraft achieves gyroscopic stiffness, and attitude corrections are periodically provided by the on-board thrusters in a manner similar to body stabilization. Spin stabilization is also subdivided into three main categories.

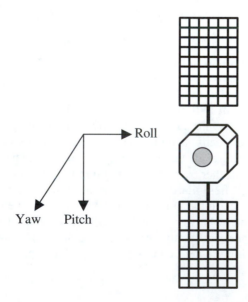

FIGURE 5.6   Body, or three-axis, stabilization.

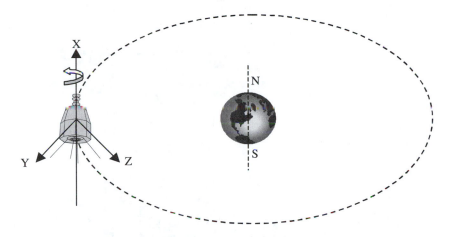

FIGURE 5.7 Spin stabilization.

- **Simple spin**
- **Dual spin**
- **Triple spin**

In simple-spin stabilization (Figure 5.8), the cylindrical body of the spacecraft is spinning around its main axis while an electronically controlled phase array at the top of the cylinder provides beam direction to a fixed geographic area on the surface of the earth. A dual-spin spacecraft is composed of a cylindrical body, very similar to that of a simple-spin spacecraft, plus a de-spun assembly rotating in the opposite direction to the body (Figure 5.9). This arrangement maintains a constant position of the antennas toward a specific

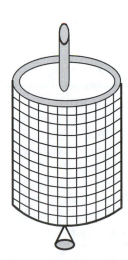

FIGURE 5.8 Simple body spin with phase array.

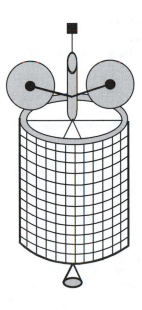

FIGURE 5.9 Dual-spin stabilization with de-spun assembly.

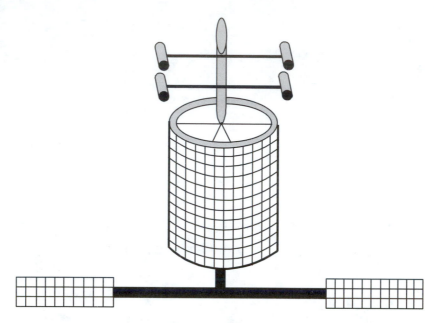

FIGURE 5.10 Triple-spin stabilization with solar array.

geographic region on Earth. The spacecraft with triple-spin stabilization is an extension of the dual-spin spacecraft. In addition to a spinning cylindrical body and a de-spun antenna, a solar array section is also incorporated into the system (Figure 5.10). This section is deployed after the satellite is in its final parking spot. It maintains direct exposure to the sun by revolving around its axis with a circular velocity of one revolution every 24 hours.

5.2.5 Spacecraft Subsystems

Station Keeping, Satellite Tracking, and Command

From the time the satellite is placed in geosynchronous orbit, it must maintain an ideal inclination angle of 0° in order for reliable communications to be established and maintained. This can be achieved relatively easily if only the earth's gravitational force is acting on the spacecraft. In reality the satellite is subject to forces other than earth's gravitational forces, such as those of the moon and sun. Furthermore, earth's nonuniform mass distribution causes the satellite to deviate slightly from the 0° inclination. The deviation of a satellite orbit must be kept within its predetermined nominal longitude for an effective and efficient operation.

The International Telecommunications Union (ITU) requires a maximum deviation of ±1.0°. with a typical spacing between satellites of 2° to 5°. In practice, both the TELE-SAT and INTELSAT systems have maintained a maximum deviation from nominal longitude of ±1.0°. This relatively small degree of east–west deviation allows for an optimum utilization of the available orbiting parking space. The task of correcting any deviations from the 0° inclination of a geosynchronous satellite is achieved through the station-keeping tracking and commands subsystem.

The principal task of this system is to sense spacecraft attitude and provide corrective action when necessary. Two sets of sensory devices are installed on board the spacecraft and are classified in accordance to their source of detection, either as radiation sensory devices or gyroscope sensory devices. The radiation devices are capable of sensing the radiation patterns of the stars, the earth, and the sun; the gyroscope sensors are capable of detecting inertial changes of gyroscopic systems.

For a reliable communications link to be maintained between the satellite and a reference point on the surface of the earth, the satellite antenna must be constantly directed toward that specific point on the surface of the earth. This is accomplished by two infrared sensors capable of detecting the earth's horizon during day and night, thus establishing the local vertical between the satellite and the center of circumference. A longitudinal drift of 0.86°/year is corrected with small rockets which are fired periodically in order to generate the required 52 m/s incremental velocity to offset the longitudinal drift.

Power Subsystem

The primary function of the spacecraft **power subsystem** is the generation, regulation, and distribution of electric power to other subsystems requiring electrical power for their proper operation. Furthermore, the power subsystem must be capable of storing electrical energy during full solar exposure to be used during eclipses, when the primary power subsystem is not functioning properly. The principal device used to generate the required electrical power in a communications satellite is the **solar cell**. This semiconductor device exhibits certain advantageous characteristics relevant to the generation of the required electrical power on board the spacecraft: high reliability, excellent power-to-mass ratio, and relative safety.

Since the solar cell's main function is the conversion of solar energy to electrical energy, the continuous operation of the device requires continuous exposure to sunlight. It is not difficult to satisfy this requirement because the spacecraft is directly exposed to the sun for almost 99% of its operational lifetime. It also offers an excellent power-to-mass ratio and is extremely safe in comparison to its major competitors, such as the nuclear cell. A detailed description of the nuclear cell will be given at the end of this section. In order for the satellite to satisfy all its power requirements, a large number of solar cells are absolutely essential. The class of the satellite determines the arrangement of these cells on board the spacecraft.

For dual-spin stabilization, the solar cells are attached to the cylindrical body. This method is somewhat inefficient because only a portion of the solar cell surface is constantly exposed to sunlight owing to the continuous rotation of the cylindrical body. Power levels of 800 W to 1 kW have been achieved by this method, corresponding to power-to-weight ratios of 10–12 W/g.

For body-stabilized spacecrafts, large panels of solar cell arrays are utilized. These panels, or solar sails, expose all solar cells simultaneously to sunlight. During launch, the solar panels are folded close to the spacecraft's main body. When the satellite has reached its final orbital parking station and has fully stabilized, the solar panels are deployed upon ground command. Special motors adjust the position of the panels every 24 hours in such a way that they are constantly exposed to sunlight. In reference to efficiency, solar panels have achieved an impressive 50 W/kg power-to-weight ratio. Figure 5.11 illustrates the power-to-weight ratio of solar arrays for both spin- and body-stabilization systems.

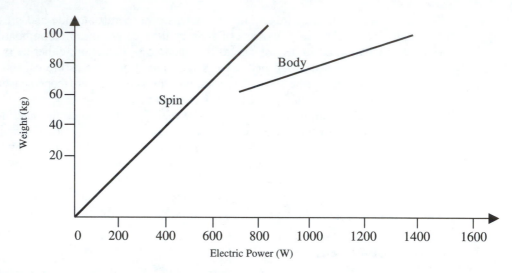

FIGURE 5.11 Solar power-to-weight ratio generation for both spin- and body-stabilization systems.

It is evident from the graph of Figure 5.11 that the utilization of large solar arrays achieves a higher power-to-weight ratio. Unfortunately the complexity and weight of the solar arrays during launch and deployment offset this advantage. Therefore, spin-stabilization satellites are preferable to three-axis systems.

Degradation of Solar Cells Solar cells in space are subject to continuous electron and proton bombardment. This bombardment is more intensified during solar activity. To minimize the damage caused by such solar phenomena, the solar cells are covered with special shields. The gradual degradation of the solar cells and therefore the progressive reduction in power output generation is illustrated in Figure 5.12.

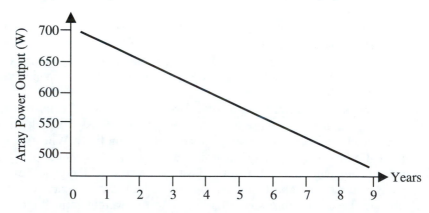

FIGURE 5.12 Solar cell power output degradation in reference to time.

The power output variations during a 1-year period shown in Figure 5.12 are due to intensity variations of solar activity during that period. The gradual solar cell power output degradation is somewhat balanced by a corresponding payload capacity reduction. It is therefore proper to assume that the maximum number of transponder channels is active at the beginning of the spacecraft's life and is gradually reduced toward the end of its operational life. The power output of solar arrays is determined by the system's operating voltage and solar cell irradiation degradation. It is estimated that the solar array voltage during exit from solar eclipse increases the minimum system operating voltage level by threefold. The dramatic increase over a relatively short period of time (10 minutes) can be damaging to system circuitry. It is therefore absolutely necessary to incorporate voltage regulators into the system. These serve a dual function. They store the excessive power generated by the solar arrays during full solar exposure and also maintain a constant voltage output for all systems during total solar eclipse, or full solar exposure. Two basic design concepts are in operation today: the regulated bus (Figure 5.13) and the unregulated bus (Figure 5.14).

In the **regulated-bus** concept, a shut regulator placed in parallel to the solar array maintains a constant operating voltage with a maximum voltage variation of $\pm 2.0\%$. The **unregulated-bus** power system is simpler in its design philosophy than a regulated bus system, the only exception being that it exhibits a better mass-to-power delivery ratio. The main disadvantages of an unregulated-bus system are that it requires a rather complex user power conditioning, exhibits a higher ripple and output impedance, and requires that all the equipment connected to it be fully protected from power supply overload.

Energy Storage   The number of solar eclipses a communications satellite is exposed to in a year of operation is approximately 90, with a maximum delay eclipse time of 72 minutes. To maintain uninterrupted system operation, a secondary electrical power source composed of dry-cell batteries is incorporated into the power source system. NiCd batteries have been used as a secondary energy source, exhibiting an approximate power-to-weight ratio of 2 W/kg and an operating temperature range of between -5 and $+1°C$. Recently other types of batteries have been developed, such as HgH_2 and NiH_2, with much improved power-to-weight ratios.

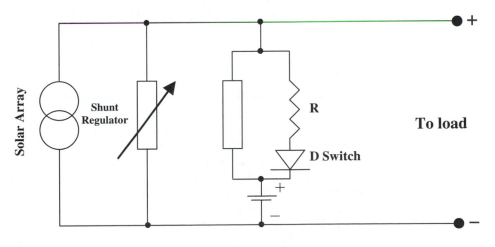

FIGURE 5.13   The regulated bus.

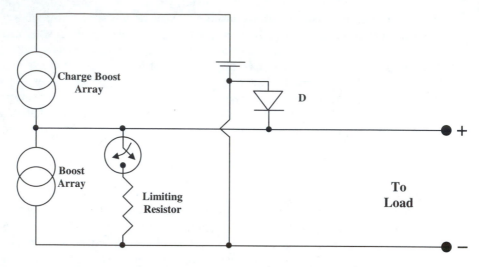

FIGURE 5.14   The unregulated bus.

Alternative Primary Power Source   In an attempt to minimize or completely overcome the solar cell limitations as a primary source in communications satellites, other power sources have been considered, among them the nuclear cell. The utilization of the nuclear power cell as a primary source of power in a geosynchronous communications satellite eliminates two major disadvantages associated with the solar cell array, namely, the need for solar panel with a constant orientation toward the sun, and the need for energy-storing batteries.

Another advantage of the nuclear power cell is its ability to provide constant power to the system for the entire life of the spacecraft. One kilogram of U^{235} can generate the equivalent of 2.5 MW/h of energy. Even though the nuclear power cell is superior to the solar cell, it has not been utilized as a main power source in communications satellites because of the heavy shielding required to protect the payload from radiation exposure during normal orbital operation and the difficulties and risks present during manufacturing and launch.

Satellite Antennas

The satellite antenna subsystem is designed to perform a multitude of functions in accordance with a satellite's primary mission. On board the spacecraft, antennas are either omnidirectional or directional. Omnidirectional antennas are usually designed to facilitate the tracking and command link from earth to satellite and satellite to earth command station during the placing of the spacecraft into its predetermined geosynchronous orbit. Directional antennas are used for both tracking and command after the satellite is positioned into final orbit as well as to provide Earth and spot-beam coverage. For spot-beam coverage, parabolic reflector antennas capable of generating beam widths between 2° and 7° are used, and horn antennas with 17.2° beam widths are utilized for global coverage. Frequency reuse satellite systems employ two transmitting and two receiving directional antennas; one set for each mode of polarization (vertical or horizontal).

During the design process of satellite antennas, a multitude of constraints are taken into account. These constraints arise from such factors as payload mass limitations, attitude stability, and antenna gain requirements. In any event, the primary mission of the communications satellite must always be kept in mind; that is, to provide service to a particular area without interfering with the adjacent areas. Due to the design limitations of satellite antennas, this interference is practically unavoidable. The principal source of such interference is the main lob power overspills to the adjacent area by two well-separated satellites. Each station can significantly reduce this interference problem if the antennas are highly directional.

As previously mentioned, the primary mission of the satellite is to provide reliable communications between two points at the surface of the earth. The stability of the spacecraft plays a crucial role in maintaining reliable communication regardless of the type of stabilization (body or spin). The section on station keeping stated that while in operation the spacecraft is subject to various gravitational field differences. In three-axis stabilization systems, these gravitational field differences affect the angle of inclination. Therefore, the satellite antennas must be capable of compensating for errors induced by these fields, thus maintaining the predetermined radiation patterns. In the early days of satellite development, the antennas had to generate a uniform radiation pattern perpendicular to the satellite rotation axis. Telstar incorporated antennas operating at 6/4 GHz and VHF range with isotropic radiation patterns. The illumination of these parabolic surfaces was achieved through an array of dipoles placed at the reflector focal point. The generated beam was radiated on a plane perpendicular to the axis of the dipole array. A major advantage of this antenna design was that in the case of mechanical failure of the rotating mechanism, the reflector would be ejected. This of course deprived the satellite of its omnidirectivity, but it permitted the system to operate with a reduced load.

Antenna Polarization   The concept of wave polarization is of major importance when designing antennas for communications satellites. In order to increase the system capacity, vertical and horizontal wave polarization is implemented. In vertical wave polarization, the electric vector component of the electromagnetic wave is perpendicular to the wave direction of propagation; in horizontal polarization, the electric vector component is parallel to wave direction of propagation. **Antenna polarization** is usually measured at the far field, thus ensuring that the wave does not change with the increase of the distance from the point of radiation. An empirical relationship defining the minimum distance required to measure the antenna polarization is given by

$$d = \frac{2D^2}{\lambda}$$

(5.12)

where d = distance from point source (m)
     D = antenna diameter (m)
     λ = wavelength (m)

Wave polarization in communications satellite systems was necessary in order to increase the system capacity from 12 to 24 transponders. This doubling of system capacity

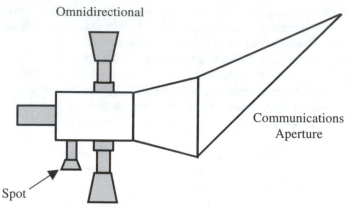

FIGURE 5.15 INTEL-SAT-III antenna assembly.

was achieved as follows: A vertically polarized satellite antenna will transmit and receive only vertically polarized waves. Likewise, a horizontally polarized antenna will transmit and receive only horizontally polarized waves. In both cases, the antenna will reject the opposite wave. INTELSAT-III and ATS-II systems incorporate reflector antennas as shown in Figure 5.15. Employing a reflector counter-rotation mechanism in reference to satellite rotation, a focal axis directed constantly toward a reference point at the earth was achieved. The INTELSAT-III antenna assembly incorporated a rotating flat surface reflector with a 45° angle in reference to the satellite spin axis. The antenna was capable of achieving a 20-dB gain. The ATS-II antenna assembly incorporates two antennas, one operating at 4.15 GHz and the other operating at 6.25 GHz (Figure 5.16).

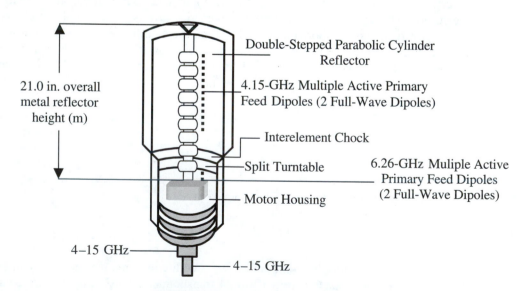

FIGURE 5.16 ATS-II antenna assembly.

There are two other wave polarization methods: circular and elliptical polarization. Circular polarization is also subdivided into two methods: right-hand circular polarization, in which the electric vector component of the electromagnetic wave rotates in a clockwise direction in relation to the direction of wave propagation, and left-hand circular polarization, in which the electric field vector component rotates in an anticlockwise mode in relation to the wave direction of propagation. In elliptical polarization there is a variation of electric field strength as the direct result of the rotation of the electric field.

It is evident from this that all methods of polarization, whether linear, elliptical, or circular, exhibit properties of copolarization and cross polarization. That is, the antennas are capable of transmitting and receiving only similarly polarized waves while rejecting all dissimilarly polarized waves. Each communications satellite employs two sets of cross-polarized antennas; one set for global coverage, with a gain higher than 15 dB, and another set composed of two parabolic reflectors with an average gain of 33 dB and utilizing the carrier frequency twice. This method of frequency reuse was developed by RCA to be used mainly in the SATCOM satellite systems. Today, practically all communications satellites are using the polarization diversity method in order to increase their overall system capacity. Figure 5.17 illustrates a DSCS-II satellite employing two-axis stabilization (spinning). The communications subsystem section and antenna subsystems are mounted in a de-spun base utilizing the horizontal axis stabilization method. In this way, the satellite antennas constantly maintain the required geosynchronous status.

Multibeam Antennas

The progressive technological evolution in spacecraft stabilization enabled antenna subsystem designers to concentrate on innovative designs capable of accommodation changes in performance characteristics once the spacecraft is placed into orbit. These changes are based on postdeployment mission modifications which are most applicable to communications and military satellites. In these cases, the concept of a multibeam antenna (MBA) capable of generating several radiation patterns was utilized.

The main objective of a multibeam antenna is to generate a number of radiation patterns while achieving the maximum possible power gain, typically higher than 25 dB. The block diagram of a multibeam antenna is shown in Figure 5.18. The microwave power is

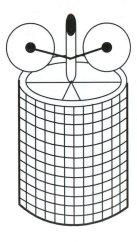

FIGURE 5.17 DCSII satellite.

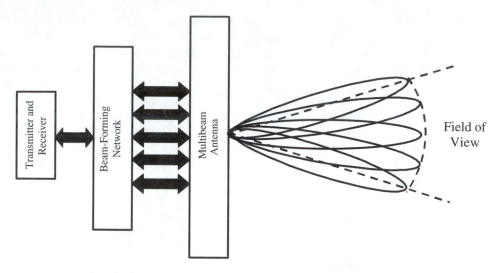

FIGURE 5.18 Multibeam antenna block diagram.

guided to a beam-forming network (BFN) of five ports. Each port is capable of generating a unique radiation pattern. Furthermore, the power distribution to each of the ports can be strictly controlled upon demand. In essence, the complexity and sophistication of a multi-beam antenna system is directly related to the sophistication and complexity of its **beam-forming network**.

A basic beam-forming network is illustrated in Figure 5.19. This network is composed of a number of feed elements and a corresponding number of power dividers and phase shifters. The network allows for a very rapid change in amplitude and phase of the excitation power fed to each horn.

Variable Power Dividers and Phase Shifters   One of the key elements composing a beam-forming network is the power divider (Figure 5.20). A variable power divider is composed of 90° and 180° hybrids connected to phase-shifters. At frequencies below 3 GHz, diode phase shifters are mainly used; at frequencies above 3 GHz range, rectangular waveguides incorporating ferrite phase shifters are utilized. Through the 180° hybrid, the input signal is divided equally at the output port, exhibiting a zero phase difference. Through the 90° hybrid, the input signal is also divided into two at the output port, except that at this point an exact phase difference of 90° is formed between the two output signals. The output signals at port A and port B are expressed as follows:

$$e_1 = \cos\left(\frac{\phi_1 - \phi_2}{2} + \frac{\pi}{4}\right) e^{j\left[\frac{\phi_1 - \phi_2}{2} + \frac{\pi}{4}\right]} \tag{5.13}$$

$$e_1 = \sin\left(\frac{\phi_1 - \phi_2}{2} + \frac{\pi}{4}\right) e^{j\left[\frac{\phi_1 - \phi_2}{2} + \frac{\pi}{4}\right]} \tag{5.14}$$

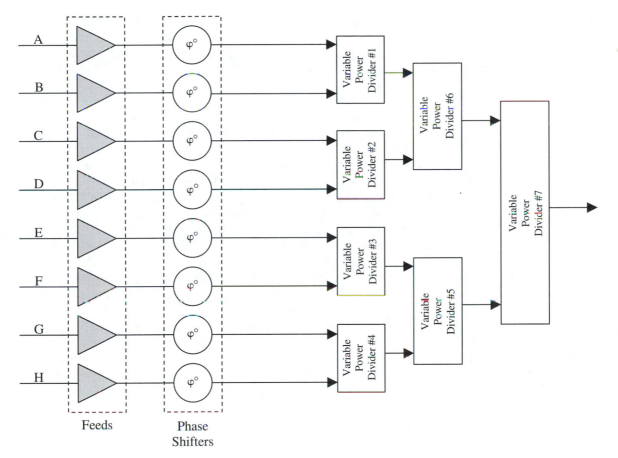

FIGURE 5.19 Beam-forming network (BFN).

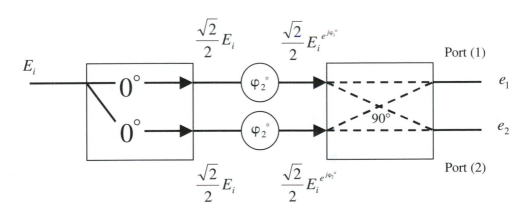

FIGURE 5.20 Schematic of variable power divider.

The insertion phase difference of the phase shifter determines the split of the input signal. Therefore, the variable power divider is capable of dividing the feed signal in terms of amplitude and phase and in accordance with system requirements, in the case of phase, by appropriately selecting the phase difference $(\phi_1 - \phi_2)$.

Lens Antennas   The optics principles are fully applied in the design of a multiple-beam antenna using waveguide lens techniques on board the spacecraft (Figure 5.21). The multiple-feed horn antennas located at an axis perpendicular to the focal axis are fed by a single microwave source. The beam-forming network (BFN) feeds the various length waveguides. If feed horn 3 is located at the focal point of the lens, upon excitation from the microwave source, it will generate a spherical wave propagated parallel to the focal axis of the lens.

Lens antennas exhibit certain advantages over the parabolic reflectors: substantial cross-polarization isolation and light weight in comparison to parabolic reflectors. The only disadvantage is their narrow bandwidth characteristic. If a homogeneous dielectric medium is employed instead of waveguides, an overall bandwidth increase is achieved at the expense of a weight increase. At the output of the lens, the wave is horizontally or vertically polarized and propagated in a direction parallel to the focal axis.

If the feed horn 3 is located at a point other than the focal point and in a direction perpendicular to the focal axis, it will generate a plane wave in a direction opposite and equal to the angle between the incident wave and the focal axis of the lens. The same result will occur for all the individual horn feeds located at the perpendicular plane of the focal axis. The resultant beam at the output of the lens is a multiple beam constituting the overall "field of view" of the lens antenna. Each beam's **angular width** is determined by

$$\phi = \frac{60\lambda}{D}$$

(5.15)

where ϕ = angular width
λ = wavelength
D = lens diameter

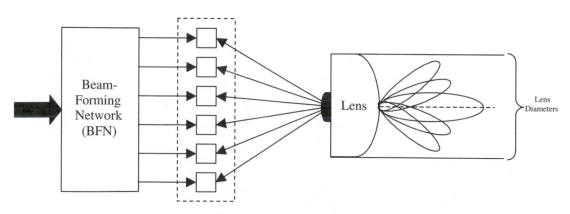

FIGURE 5.21   Lens antennas.

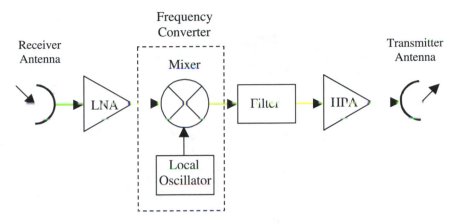

FIGURE 5.22 Block diagram of a simplified satellite repeater.

5.2.6 The Satellite Repeater

Earlier in this chapter it was stated that the primary mission of a geosynchronous communications satellite is to be used as an active repeater, that is, to be able to receive signals transmitted from the earth, amplify these signals, and translate (frequency convert), reamplify, and transmit them back to earth. The block diagram of an oversimplified communications satellite repeater is shown in Figure 5.22. The two main subsystems of a satellite repeater are the antenna and communication subsystems.

Antenna Subsystems

Satellite antennas were discussed at considerable length earlier in "Satellite Antennas." Here a very brief summary will be given. The principal function of an antenna subsystem is to generate microwave beams of a particular shape and transmit them toward the earth at frequencies of 6/4 GHz or 14/12 GHz for the up-link and down-link respectively. They must also be able to accommodate the telemetry tracking and command signals necessary for normal spacecraft operation during its entire operational life. Since most of the communications satellites employ the frequency reuse method of operation in order to increase their information capacity, two parabolic reflectors are utilized, one vertical, the other for horizontal polarization. Normally, a single structure, composed of two parabolic reflectors back to back, is employed (Figure 5.23). The grids of each reflector implement the horizontal or vertical wave polarization and are independently fed by multihorn primary feed arrays, maintaining a polarization isolation of approximately 30 dB.

The Communications Subsystem

The communications subsystem on board the satellite (repeater) consists of the following main units (Figure 5.24):

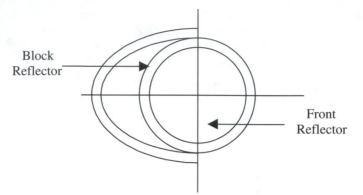

FIGURE 5.23 Vertical and Horizontal polarization antenna.

Low-noise-amplifier (LNA)

Local oscillator (LO)

Mixer

Input multiplexer (MUX)

High-power amplifier (HPA)

Output multiplexer

The information-carrying electromagnetic beam transmitted by the earth station is intercepted by the receiver antenna on board the satellite. This microwave beam has traveled an approximate distance of 40,000 km through space and is heavily attenuated (about 200 dB). The receiver antenna, acting as the interface between the satellite and the earth station provides the first amplification stage of the signal, by an average gain of 20 dB. Then the signal is fed to the input of a **low-noise amplifier** (LNA) through a waveguide bandpass filter (BPF) with a bandwidth of 500 MHz. The presence of an LNA is absolutely necessary because the incoming signal, although amplified by the receiver antenna, is still very weak (on the order of -100 dB). This problem is compounded by the high noise figure of the

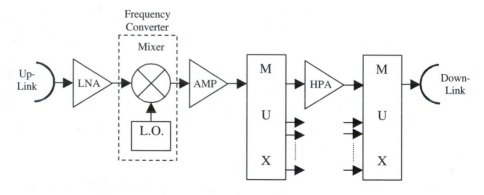

FIGURE 5.24 Satellite repeater block diagram.

mixer stage. Therefore, the LNA must be incorporated between the receiver antenna and the mixer stage. The output of the LNA is fed into the input of the mixer stage. The mixer and the local oscillator circuits perform the frequency translation function of the repeater. The frequency translation from up-link to down-link transmission frequencies can be achieved through one or two stages. If one conversion signal is employed, the process is called "single-frequency conversion." If two stages are employed the process is called "dual-frequency conversion."

Single-frequency conversion is normally used in C-band systems. **Dual-frequency conversion** is usually implemented with satellite repeater systems operating in higher-frequency bands such as the Ku band, with up-link frequencies between 14 GHz and 14.5 GHz, down-converted to 11.7–12.2 GHz using two local oscillators (Figure 5.25). The need for dual-frequency conversion arises from the fact that amplifiers with a high gain and a low noise figure operating in such high frequencies are very difficult to design and construct.

The Input Multiplexer The **input multiplexer**'s main function is to divide the entire available bandwidth of the communications satellite (500 MHz) into a number of transponder channels, determined by satellite type and primary mission. This process is also called "channelization." For example, the early communications satellites such as the INTELSAT-II were designed to accommodate one transponder utilizing the entire available bandwidth of 125 MHz. INTELSAT-III was able to accommodate two transponders with an available bandwidth of 225 MHz each, and INTELSAT-V was able to accommodate 48 transponders in a dual-polarization system with bandwidths of 36 MHz, 41 MHz, 72 MHz, 77 MHz, and 150 MHz.

Input multiplexers are composed of various subcircuits such as filters, amplifiers, **group delay equalizers**, and output circulators (Figure 5.26). The process of channelization of the entire satellite bandwidth to individual transponder channels provides certain advantages contributing to the enhancement of the overall system performance:

1. Reduction of the intermodulation products due to the reduction of the number of subcarriers within the individual channels. The smaller the number of subcarriers in a channel, the smaller the intermodulation distortion resulting from nonlinearities arising from simultaneous subcarrier amplification.

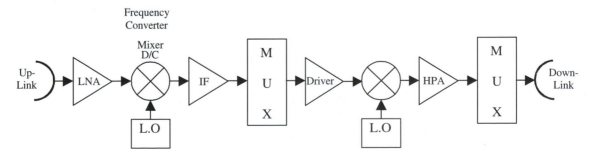

FIGURE 5.25 Ku-band dual-frequency converting satellite repeater.

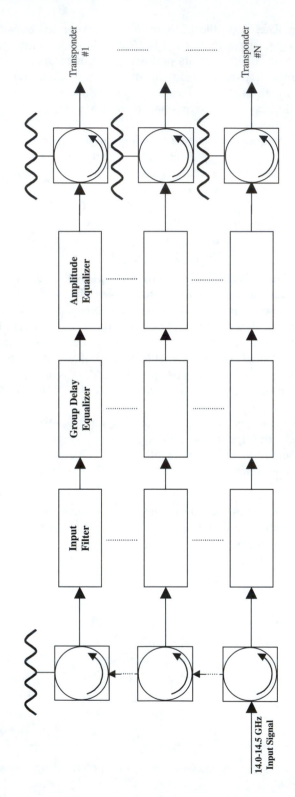

FIGURE 5.26 Input multiplexer.

2. The channelization process also enables the individual channels to be amplified by the high-power amplifiers before the recombination process takes place. Of course, there is a major problem associated with the process of channelization, that of adjacent channel interference. Also, because each individual channel follows a different traveling path, there is a different group delay for each channel, ultimately resulting in an output signal distortion.

Incorporating filter design methods approaching those with ideal frequency response can minimize this problem and its effects. Each channel output from the input multiplexer is applied to the input of a narrow-band driver amplifier. This stage is required in order to increase the amplitude of the signal to the level required by the power amplifier. The output of each driver amplifier is fed into the input of the high-power amplifier, usually a traveling-wave tube (TWT). The traveling-wave tube amplifies each channel to a level required by the down-link transmission.

To achieve its maximum required output power level, this device is driven almost to saturation, resulting in a possible channel performance degradation. The output multiplexer combines the individual channels for final down-link transmission through vertically or horizontally polarized directional antennas. The fact that communications satellites operate at very high frequencies, with very high signal attenuation between spacecraft and earth stations and are subject to noise, interference and extreme space conditions, generated the need for the state of the art in semiconductor devices employed, as well as the implementation of the most sophisticated techniques in circuit and system design. Brief descriptions of such devices and circuits such as LNAs, filters, traveling-wave tubes, and isolators are presented in the following sections.

Parametric Amplifiers   The first stage of a repeater receiver is very crucial to the overall performance of that receiver, because at the input of that receiver, the incoming signal is very weak, while the noise level is comparatively significant. The **parametric amplifier** must be able to detect the signal from the noise level and provide maximum gain over the entire bandwidth. Parametric amplifiers used at the front end of a satellite receiver are capable of satisfying these important requirements by providing an average gain of 13 dB with a gain stability of ± 0.3 dB over the 0–50°C temperature range while maintaining a noise figure of approximately 3 dB.

Low-Noise Isolators   Low-noise isolators are used extensively throughout the transponder as low-, medium-, and high-power isolators, exhibiting losses better than 0.1 dB with power-handling capabilities of up to 250 W.

Graphite–Fiber/Epoxy Filters   One of the most important objectives in designing the narrow-band filters used in the transponder section of a communications satellite receiver is to maintain specified operating characteristics under the extreme temperature variations of space. This important design objective is achieved through the employment of materials exhibiting zero expansion coefficients. Such materials are the graphite–epoxy composites. These composite materials demonstrate almost the same thermal characteristics as the Invar material used in previous designs of microwave filters, with the advantage of a 50% weight reduction. It is imperative to mention here the importance of the critical weight factor of various materials incorporated in the construction of electronic and mechanical modules on board the spacecraft.

FET Amplifiers   Field-effect transistors are used as preamplifiers (drivers) for the traveling-wave tubes. These amplifiers are designed to provide gains of up to 25 dB over the entire bandwidth, while maintaining a stability factor better than 1 dB over the 0–50°C temperature range and a noise figure smaller than 5 dB.

Traveling-Wave Tube (TWT)   Traveling-wave tubes are very important devices used on board the satellite. They are capable of providing the high power gain required for the downlink transmission, usually of 60 dB, with a large dc to RF ratio. Traveling-wave tubes have demonstrated an excellent operating record of up to 50,000 hours, and their weight compares favorably with spacecraft weight restrictions.

MMIC Satellite Transponder Receivers   Advancements in Monolithic Microwave Integrated Circuits (**MMIC**s) have made possible the design and implementation of satellite transponder receivers composed exclusively by MMIC modules. The implementation of MMIC modules for the design and construction of such transponder receivers achieved the ultimate design objectives for reliability, performance, and weight constraints. The block diagram of a MMIC receiver operating in the Ka band is shown in Figure 5.27. This transponder receiver operates at 30 GHz with **image rejection** (IR). It also incorporates in its design a low-noise amplifier and a mixer module, all under the same housing. The utilization of MMICs in this type of design required the solution of some of the most difficult problems encountered in the past, such as the elimination of bandpass filters (BPFs) and the reduction of the excessive phase noise associated with MMIC oscillators.

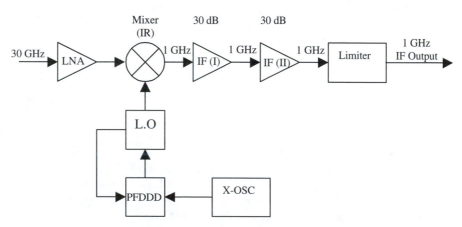

PFCDD = Phase-Frequency-Comparator Digital Divider

FIGURE 5.27   MMIC satellite receiver block diagram (Ka band).

The elimination of the BPFs was absolutely necessary because of their adverse effect on the system weight constraints. The oscillator phase noise is reduced by image rejection and high-speed PLLs (phase-lock loops). Modules incorporated into the MMIC receiver design are the low-noise amplifier, local oscillator, mixer, intermediate amplifiers I and II, limiter, phase-frequency comparator, and crystal oscillators. All but the crystal oscillator are included in the MMIC structure.

Low-Noise Amplifier (LNA)   The basic block diagram of a low-noise amplifier is shown in Figure 5.28. This amplifier is composed of four two-stage microstrips employing GaAsFETs. The electrical characteristics of the LNA are illustrated in Figure 5.29. This figure shows an average 27-dB gain over the 29-GHz to 31-GHz range, with a corresponding noise figure of 8 dB.

Frequency Translator   The block diagram of a frequency translator with an image rejection (IR) module is shown in Figure 5.30. The broadband properties of the frequency doubler are utilized here to achieve the 15-GHz to 30-GHz frequency conversion. The 30-GHz signal is amplified through the two-stage FET amplifier, then fed to the input of the diode mixer. The phase-shifting diode mixer also performs the function of image suppression, thus eliminating the need for an additional narrow-band filter. The graph illustrating the results of the image-rejection capabilities of the mixer is shown in Figure 5.31. Figure 5.31 illustrates the characteristic performance of the frequency-conversion module. An 18-dB image rejection has been achieved with a conversion loss of 1 dB while maintaining a constant phase frequency characteristic of 0.5 GHz to 1.5 GHz over the range.

Local Oscillator   The local oscillator (LO) module is composed of a 15-GHz voltage-control oscillator, a dual-stage and a single-stage 15-GHz amplifier, a power divider, and a two-stage 7.5-GHz amplifier. In the local oscillator circuit of Figure 5.32, a MMIC varactor-tuned voltage-controlled oscillator is used. The VCO output of 15 GHz is divided digitally into 7.5 GHz. The overall performance test results of this MMIC receiver are in compliance with satellite receiver specifications requiring high stability, high image rejection, and a low-noise local oscillator. The design and development of this receiver were performed by Nippon Telephone and Telegraph.

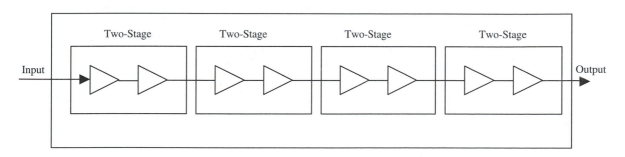

FIGURE 5.28   MMIC low-noise-amplifier block diagram.

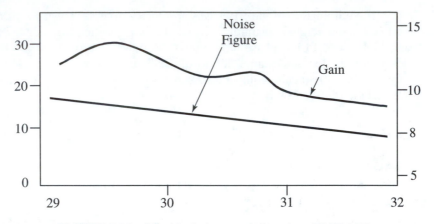

FIGURE 5.29 Electrical characteristics of an MMIC LNA.

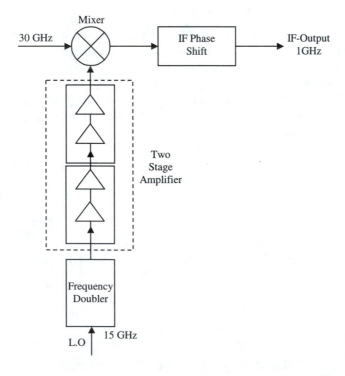

FIGURE 5.30
Frequency translator.

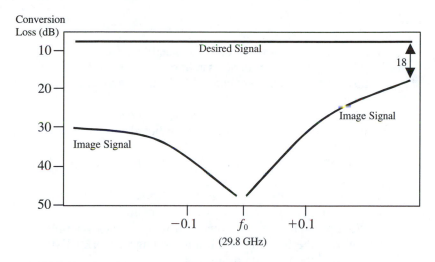

FIGURE 5.31 Image-rejection curves per frequency translator curve.

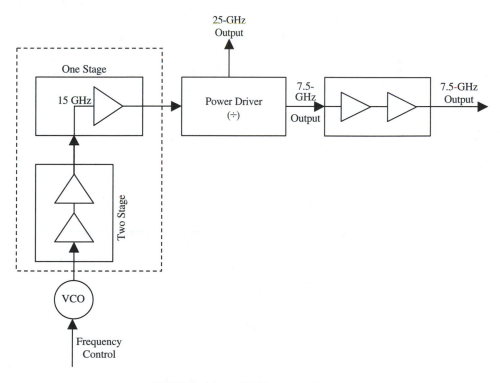

FIGURE 5.32 MMIC load oscillator.

QUESTIONS

1. List the four major categories of satellite systems.
2. What is a geosynchronous satellite?
3. List the three main sequences for placing a geosynchronous satellite into permanent orbit.
4. Define the terms *geocentric* and *heliocentric*.
5. What is the fundamental device on board the spacecraft for electric power generation?
6. Why are voltage regulators required in the power-supply module of a satellite repeater?
7. Define the terms *horizontal polarization* and *vertical polarization*. Why is polarization required in communications satellites?
8. What are multibeam antennas and why are they required?
9. Describe the function of a beam-forming network (BNF).
10. Describe the function of a variable-power divider.
11. Describe the function of a lens antenna and compare it to the parabolic reflector.
12. Sketch the block diagram of a simplified satellite repeater.
13. Describe the function of the input multiplexer in a communications repeater and list the basic components.
14. Sketch the block diagram of an MMIC module. What is an MMIC and why is it employed in a satellite repeater?
15. List the basic characteristics of an LNA amplifier and explain why LNAs are absolutely necessary at the input of a satellite repeater.
16. With the assistance of a block diagram, describe the function of a frequency translator.
17. What is image rejection? Describe its importance in a satellite repeater.

PROBLEMS

1. Determine whether or not a satellite placed in an orbit at a height of 22,000 miles is geosynchronous.
2. Compute the orbital velocity of a satellite placed at a height of 30,000 km with an orbital period of 18 hours.

6

Satellite Earth Stations

Objectives

- Describe the various subsystems of a satellite earth station.
- Define parabolic reflector gain and directivity.
- Calculate the effective isotropic radiated power, figure of merit, and system noise temperature.
- Explain the difference between carrier–to–thermal noise ratio and figure of merit.
- Describe the various subsystems of a satellite earth station transmitter and receiver.

Key Terms

Satellite earth station
Antenna subsystem
Command and control
Diplexer
Pointing motors
Tracking receiver
Electric field
Magnetic field
Horizontal polarization
Vertical polarization
Omnidirectional
Power density
Characteristic impedance
Transmission medium
Permeability
Permittivity
Antenna gain

Focal point
Focal length
Effective area
Efficiency factor
Directivity
EIRP
Figure of merit
System noise temperature
Ambient noise
   temperature
Boltzmann's constant
Space losses
Up-converter
High-power amplifier
Traveling-wave tube
Helix
Transmitter redundancy

Power divider
Guide switch
Carrier combiner
Directional coupler
Down-converter
Low-noise amplifier
Parametric amplifier
Degenerative
Nongenerative
MMIC
HEMT
Nonlinearity
Third-order intercept point
1-dB-gain compression
   point
Image rejection

INTRODUCTION

The main mission of a **satellite earth station** is to establish and maintain continuous communication links with all other earth stations in the system through the satellite repeater. It must also provide and maintain the necessary command and control links with the spacecraft. The main system components of a satellite earth station are the antenna subsystem, transmitter section, receiver section, and command and control section. Figure 6.1 illustrates the block diagram of a satellite earth station.

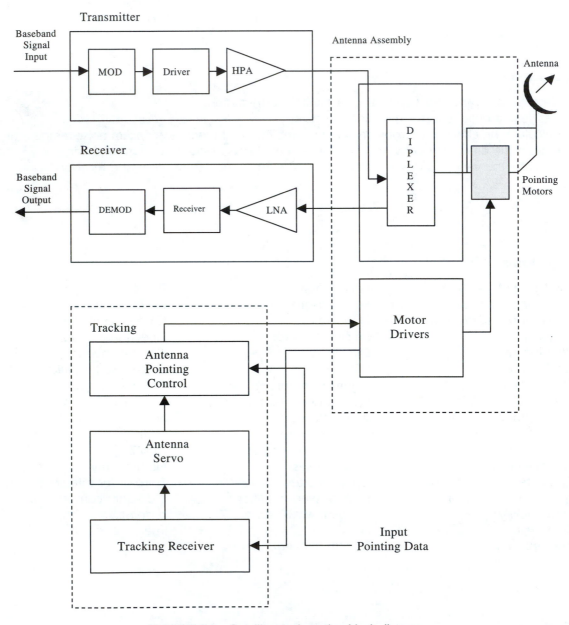

FIGURE 6.1 Satellite earth station block diagram.

6.1 EARTH STATION ANTENNA SUBSYSTEMS

6.1.1 Elements of Electromagnetic Wave Propagation

For long-distance radio communications, electromagnetic waves are used to carry information (voice, video, and data) from one point to another through the atmosphere or space. The electromagnetic waves are transverse waves, composed of an **electric field** and a **magnetic field** perpendicular to each other and both perpendicular to the direction of propagation (Figure 6.2).

Electromagnetic waves are generated when electric current of high frequency flows through a conductor. This current generates an electric field with a corresponding potential difference developed across the conductor. The electric field generates a magnetic field, with an alternating transfer of energy from electric field to magnetic, and from magnetic to electric. If all the energy of the magnetic field is unable to return to the conductor, then this energy is said to propagate in free space in the form of an electromagnetic wave with the velocity of light (3×10^8 m/s). If electric power in the form of an electromagnetic wave is purposely released into space, then an effort must be made to maximize it. The electrical aperture capable of maximizing the transmission of electric power generated by a transmitter into free space is the directional antenna.

Electromagnetic waves are classified in accordance to their polarization as vertically polarized or horizontally polarized. The polarization of a wave is determined by its electric field. If the electric field is vertical to the direction of propagation, the wave is vertically polarized. If the electric field is horizontal to the direction of propagation, the wave is horizontally polarized. This concept has found useful applications in line-of-sight and satellite communications, where both vertically and horizontally polarized waves are utilized in order to maximize system capacity.

The propagation of an electromagnetic wave through space can be accomplished by either the omnidirectional or directional mode of transmission. In the **omnidirectional** mode, a point source of electromagnetic radiation is assumed (Figure 6.3). The resulting waveform is a sphere with radius d expanding with the velocity of light. The **power density** at the surface of the sphere can be determined by

$$P_D = \frac{P_t}{4\pi d^2} \tag{6.1}$$

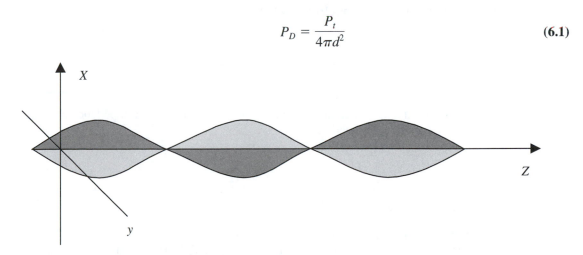

FIGURE 6.2 Electromagnetic waves.

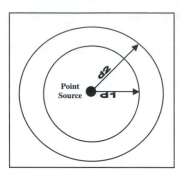

FIGURE 6.3
Omnidirectional mode
of transmission.

where $\quad P_D$ = power density (W/m^2)
$\qquad P_t$ = power at the point source (W)
$\qquad d$ = radius of the sphere (m)

From Eq. (6.1) it is evident that the power density away from the source is proportional to the power of the source and inversely proportional to the square of the distance (inverse square law).

The power density (P_D) of an electromagnetic wave away from the source of radiation is considered to be equivalent to the power of an electric circuit. Similarly, field intensity is considered to be equivalent to the voltage generated across that circuit. Therefore, field intensity (ε) is expressed by

$$\varepsilon = \sqrt{P_D Z} \qquad \qquad \textbf{(6.2)}$$

where $\quad P_D$ = power density (W/m^2)
$\qquad Z$ = **characteristic impedance** of the **transmission medium**

Substituting Eq. (6.2) into Eq. (6.1) we have

$$\varepsilon = \sqrt{\frac{P_t}{4\pi d^2} Z} \qquad \qquad \textbf{(6.3)}$$

If the propagation takes place in free-space, Z is the characteristic impedance of free space, denoted as (Z_0) and calculated as follows:

$$Z_0 = \sqrt{\frac{\mu_0}{\varepsilon_0}} \qquad \qquad \textbf{(6.4)}$$

where $\quad \mu_0$ = **permeability** of free space (1.26×10^{-6} H/m)
$\qquad \varepsilon_0$ = **permittivity** of free space (8.85×10^{-12} F/m)

Substituting into Eq. (6.4) for μ_0 and ε_0 we have

$$Z_0 = \sqrt{\frac{1.26 \times 10^{-6}\ \text{H/m}}{8.85 \times 10^{-12}\ \text{F/m}}} = 377\ \Omega \quad \text{or} \quad 120\pi\ \Omega$$

$Z_0 = 377\ \Omega$ is defined as the characteristic impedance of free space.

6.1.2 Antenna Parameters

In satellite communications systems, highly directional antennas are used to provide the much needed **antenna gain** both for the ground segment and on board the spacecraft. This very high signal amplification is required in order to compensate for the substantial losses the signal has suffered while propagating through space. The power density of an electromagnetic wave away from the source in a directional mode of transmission is expressed by

$$P_D = \frac{P_t}{4\pi d^2} G_t \tag{6.5}$$

where $\quad P_t$ = transmitter power at the source (W)
$\quad\quad\quad\ d$ = distance from the source (m)
$\quad\quad\quad G_t$ = transmitter antenna gain (dB)

Since microwave frequencies exhibit the same properties as optical rays, the transmitter and receiver antennas must be in the line of sight. This is a requirement for the receiver antenna to intercept and collect the maximum possible transmitted RF power. Similar to the transmitter, the receiver antenna is designed for maximum directivity and gain. These two fundamental requirements of directional antennas are subject to their physical properties and system operating frequencies. For example, the physical size of high-gain antennas must be several times longer than their operating wavelengths. Operating in microwave frequencies, high-gain satellite system antennas can easily be designed and manufactured because the diameter of the smallest parabolic reflector is several times longer than that of its operating wavelength. In satellite communications systems, the most commonly used antennas are the parabolic reflectors.

6.1.3 Parabolic Reflectors

Operating in microwave frequencies, parabolic reflector characteristics are governed by the same principles as those of visible light (Figure 6.4). In Figure 6.4, the distance
$FA + Aa = FB + Bb = FC + Cc = K$

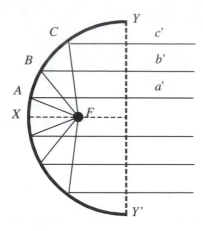

FIGURE 6.4 Parabola.

where
K = constant
F = **focal point** of the parabola
FX = **focal length**
YY' = diameter of the parabola

The ratio of the focal length to the diameter is called the aperture of the parabola. Therefore, the aperture (A) is given by

$$A = \frac{FX}{YY'}$$ (6.6)

If the parabola is rotated in its focal length axis by 360°, a parabolic reflector will be developed.

Assume a light source is placed at the focal point of the parabolic reflector. All the optical rays originating from the focal point will be reflected by the inner surface of the parabola, and they will travel the same distance to the mouth of the reflector, provided that the inner surface of the reflector is free of physical anomalies. It is therefore evident that an omnidirectional source of light placed at the focal point of a parabolic reflector generates a highly directional beam of light. Alternatively, if a parabolic reflector intercepts the beam of light, the intercepted beam of energy will be concentrated at the focal point of the reflector. It is also evident that the energy collected from a large surface will be concentrated in a very small area, that of the focal point. This is indicative of parabolic reflector amplification characteristics. Since electromagnetic waves in the microwave range behave the same way as those of visible light, then parabolic reflectors exhibiting excellent gain and direction characteristics are appropriately used in satellite communication systems operating at microwave frequencies.

Parabolic Reflector Power Gain (G_p)

The power gain of a parabolic reflector is expressed by the relationship

$$G_p = \frac{4\pi A_{\text{eff}}}{\lambda^2}$$ (6.7)

where
G_p = power gain (ratio)
A_{eff} = **effective area** of the reflector (m^2)
λ = operating wavelength (m)

The effective area (A_{eff}) is given by

$$A_{eff} = KA \tag{6.8}$$

where
K = **efficiency factor,** with a value between 0.55 and 0.65
A = area of the reflectors

The reflector maximum area (A) is expressed by

$$A = \frac{\pi D^2}{4} \tag{6.9}$$

where
D = diameter of the reflector (m)

Substituting Eqs. (6.8) and (6.9) into Eq. (6.7), we have

$$G_p = \frac{K4\pi(\pi D^2)}{\lambda^2 4} = K\left(\frac{\pi D^2}{\lambda}\right)$$

Therefore,

$$G_p = K\left(\frac{\pi D^2}{\lambda}\right) \tag{6.10}$$

Equation (6.10) indicates that the gain of a parabolic reflector is proportional to its diameter and inversely proportional to its operating wavelength.

Directivity

The **directivity** of the reflector antenna is given by

$$\phi = \frac{70\lambda}{D} \tag{6.11}$$

where
ϕ = directivity in degrees
λ = wavelength (m)

Effective Isotropic Radiated Power (EIRP)   Most often in satellite communications the concept of **EIRP**(effective isotropic radiated power or equivalent isotropic radiated power) is used to specify the required channel power capable of saturating the input of the receiver. The product of the RF transmitter power and its antenna gain, EIRP is considered to be the total radiated power from a directional antenna, in reference to ideal isotropic sources of radiation with a gain equal to 0 dB or 1 W. Therefore,

$$\text{EIRP} = P_t G_t \qquad\qquad (6.12)$$

This expression can be better illustrated by the following examples.

EXAMPLE 6.1

Determine the EIRP of a microwave transmitter of 500 W coupled to a directional antenna with a gain of 30 dB.

Solution

$$G_t = 30 \text{ dB}$$
$$G_t = \text{antilog}(3.0) = 1{,}000$$
$$\text{EIRP} = P_t \times G_t$$
$$= 50 \times 1000$$
$$= 500{,}000$$
$$= 5 \times 10^5 \text{ W}$$
$$= 57 \text{ dB}$$

EXAMPLE 6.2

Determine the transmitter antenna gain in dB required for a microwave link with an EIRP of 4×10^4 W and transmitter power equal to 500 W.

Solution

$$\text{EIRP} = P_t \times G_t$$
$$G_t = \frac{\text{EIRP}}{P_t} = \frac{4 \times 10^3 \text{ W}}{500 \text{ W}} = 8$$
$$G_t = 10 \log(8) = 9$$
$$\therefore G_t = 9 \text{ dB}$$

EXAMPLE 6.3

The EIRP of a microwave link is 2×10^4 W. If the system employs a highly directional antenna with a gain of 40 dB, determine the required transmitter power.

Solution

$$\text{EIRP} = P_t \times G_t$$

$$P_t = \frac{\text{EIRP}}{G_t}$$

$$G_t = 40 \text{ dB} = 10,000 = 1 \times 10^4$$

$$P_t = \frac{2 \times 10^4}{1 \times 10^4} = 2 = 2.0 \text{ W}$$

$$\therefore P_t = 2 \text{ W}$$

Figure of Merit (G_r/T)

The ability of an earth station to receive satellite signals is measured by the ratio of its receiver antenna gain (G_r) to the **system noise temperature** (T). This ratio is referred as **figure of merit** (G_r/T). Standards for INTELSAT systems have set the figure of merit to be equal or higher than 40.7 dB

$$(G_r/T) = 40.7 \text{ dB} \tag{6.13}$$

where
G_r = receiver antenna gain (dB)
T = system noise temperature at the input of the LNA (dB)

In order for a figure of merit to be equal to or better than 40.7, an excellent combination of receiver antenna gain and low system noise temperature must be achieved.

EXAMPLE 6.4

Determine the system noise temperature (T_{sys}) of a satellite receiver station in order to maintain a constant figure of merit equal to 40.7 dB with a receiver antenna gain of 55 dB.

Solution
In decibels,

$$\frac{G_r}{T_{\text{sys}}} = 40.7 \text{ dB}$$

$$G_{r_{\text{dB}}} - T_{\text{sys}_{\text{dB}}} = 40.7$$

$$-T_{\text{syst}_{\text{dB}}} = 40.7 - G_{r_{\text{dB}}}$$

$$T_{\text{syst}_{\text{dB}}} = G_{r_{\text{dB}}} - 40.7 \text{ dB}$$

Since $G_r = 55$ dB,

$$T_{\text{syst}} = 55 - 40.7 = 14.3 \text{ dB}$$

In degrees Kelvin,

$$T_{sys} = \text{antilog}(1.43) \cong 27$$
$$\therefore T_{sys} = 27 \text{ K}$$

EXAMPLE 6.5

The system noise temperature at the input of the LNA of a satellite receiver station operating in the C band is 450 K. If the diameter of the receiver antenna (parabolic reflector) is 30 m, compute the earth station figure of merit.

Solution
Compute G_r in dB

$$A = \frac{\pi D^2}{4} = \frac{\pi (30)^2}{4} \text{ m}^2 = 707 \text{ m}^2$$
$$A_{eff} = 0.55\, A$$
$$= 0.55 \times 707 \text{ m}^2 = 388 \text{ m}^2$$

For C-band down-link:

$$f = 4 \text{ GHz}$$
$$\lambda = \frac{c}{f} = \frac{3 \times 10^8 \text{ m/s}}{4 \times 10^9}$$
$$\therefore \lambda = 0.075 \text{ m}$$
$$G_r = \frac{4\pi A_{eff}}{\lambda^2}$$
$$= \frac{4\pi (388) \text{m}^2}{(0.075)^2 \text{ m}^2}$$
$$= 866 \times 10^3$$

In dB,

$$G_r = 10 \log(866 \times 10^3) = 59$$
$$\therefore G_r = 59 \text{ dB}$$

The figure of merit (G_t/T_{sys}) is given by

$$\frac{G_t}{T_{sys}} = 59 - 26.5 \text{ dB} = 32.5 \text{ dB}$$

$$T_{sys} = 10 \log(450 \, °K) = 26.5 \text{ dB}$$

$$\therefore \frac{G_t}{T_{sys}} = 29.5 \text{ dB} \quad \text{(figure of merit)}$$

System Noise Temperature (T_{sys})

System noise temperature is the sum of the antenna and receiver noise temperatures:

$$T_{sys} = T_{ant} + T_{rec} \qquad \text{(6.14)}$$

where T_{sys} = system noise temperature
T_{ant} = antenna noise temperature
T_{rec} = receiver noise temperature

Antenna noise temperature (T_{ant}) is defined by

$$T_{ant} = \frac{T_0(L_a - 1) + T_s}{L_a} \qquad \text{(6.15)}$$

where T_0 = **ambient noise temperature** (290 K)
L_a = antenna losses at base (dB)
T_s = sky noise temperature (K)

Receiver noise temperature (T_{rec}) is defined as follows:

$$T_{rec} = T_0(L_{in} - 1) + T_{LNA}L_{in} + \frac{T_{pm} \times L_{IN}}{G_{LNA}} \qquad \text{(6.16)}$$

where T_{rec} = total noise temperature at the input of the LNA (K)
T_0 = ambient temperature (290 K)
L_{in} = total losses from reference to the input of the LNA (dB)
T_{LNA} = noise temperature of the LNA (K)
T_{pm} = preamplifier and mixer noise temperature (K)
G_{LNA} = LNA gain (dB)

EXAMPLE 6.6

Compute the system noise temperature and figure of merit (G_r/T_{sys}) for the earth receiver station of a satellite system given the following data:

Operating frequency: 14/12 GHz
Receiver antenna diameter: 10 m

Losses at the base of the antenna: 0.15 dB
Sky noise temperature: 70 K
Total waveguide losses: 0.25 dB
Direction coupler insertion losses: 0.1 dB
$T_{LNA} = 120$ K
$G_{LNA} = 40$ dB
Preamplifier noise temperature $= 550$ K

Solution
Antenna Temperature

$$L_a = \text{antilog}(0.15/10) = 1.035$$
$$T_{ant} = \frac{T_0(L_a - 1) + T_{sys}}{L_a} = \frac{290 \cdot (1.035 - 1)}{L_a} = 77.5$$

Therefore,

$$T_{ant} = 77.5 \text{ K}.$$

System Noise Temperature

$$T_{rec} = T_0(L_{in} - 1) + T_{LNA} \times L_{in} + \frac{T_{pm} \times L_{in}}{G_{LNA}} \qquad \textbf{(6.16)}$$

The total sum of losses due to waveguide pluming at the input LNA is as follows:

Waveguide losses: 0.25 dB
Directional coupler losses: 0.10 dB

$$L_{in} = \text{antilog}(0.35/10) = 1.08$$
$$T_{rec} = 290(1.08 - 1) + 120 \times 1.08 + \frac{550 \times 1.08}{1 \times 10^4}$$
$$= 23.2 + 129.6 + 0.05$$

Therefore, $T_{rec} = 153$ K.
System noise temperature (T_{sys})

$$T_{sys} = T_{ant} + T_{rec} = 77.5 + 153 = 230.5°\text{K}$$

Therefore, $T_{sys} = 230.5$ K.
In dB,

$$T_{sys} = 10\log(230) = 24$$

Therefore, system noise temperature $T_{sys} = 24$ dB.

Receiver figure of merit (G_{rec}/T_{sys})

$$A = \frac{\pi D^2}{4} = \frac{\pi \times 10^2}{4} = 78.5 \text{ m}^2$$
$$A_{eff} = 0.55 \times 78.5 \text{ m}^2 = 43.2 \text{ m}^2$$
$$\lambda = \frac{3 \times 10^8 \text{ m/s}}{12 \times 10^9 \text{ m/s}} = 0.025 \text{ m}$$

Compute G_{rec}:

$$G_{rec} = \frac{4\pi A_{eff}}{\lambda^2} = \frac{4\pi(43.2) \text{ m}^2}{0.025^2 \text{ m}^2} = \frac{543 \text{ m}^2}{0.025^2 \text{ m}^2} = 868,587$$
$$\therefore G_{rec} = 59 \text{ dB}$$

Compute figure of merit (G_{rec}/T_{sys}):

$$\frac{G_{rec}}{T_{sys}} = 59 - 24 = 35$$

Therefore, $G_r/T_{sys} = 35$ dB.

EXAMPLE 6.7

Determine the maximum system noise temperature allowed in the system of Example 6.6 if a figure of merit equal to 40.7 dB is to be maintained.

Solution

$$\frac{G_{rec}}{T_{sys}} = 40.7 \text{ dB}$$
$$T_{sys} = 59 - 40.7 = 18.3 \text{ dB}$$
$$T_{sys} = \text{antilog}(1.83) = 67$$

Satellite Earth Stations

Therefore, the system noise temperature must not exceed 67 K:

$$T_{\text{sys}} \leq 67 \text{ K}$$

Carrier–to–Thermal Noise Ratio (C/T)

The carrier-to-thermal noise ratio is derived from the carrier–to–noise power ratio as follows:

$$\frac{C}{N} = \frac{C}{KT} \tag{6.17}$$

where K = **Boltzmann's constant** $= 3.8 \times 10^{-23}$
BW = channel bandwidth (Hz)
T = thermal noise (K)

In dB,

$$\frac{C}{N_{\text{dB}}} = \frac{C}{T_{\text{dB}}} - 10 \log(1.38 \times 10^{-23}) - 10 \log(\text{BW})$$

Solving for C/T,

$$\frac{C}{T_{\text{dB}}} = \frac{C}{N_{\text{dB}}} + 10 \log(1.38 \times 10) + 10 \log(\text{BW})$$

The carrier–to–noise temperature ratio (C/N) is expressed as dB/K and is related to the figure of merit (G/T) as follows.

Relationship Between *G/T* and *C/T*   There is a relationship between the figure of merit and the carrier–to–thermal noise ratio, given by

$$\left(\frac{C}{T}\right)_{\text{dB}} = (P_t \times G_t)_{\text{dB}} - L_{\text{dB}} + \left(\frac{G}{T}\right)_{\text{dB}} \tag{6.18}$$

where $P_t \times G_t$ = EIRP (dB)
L = **space losses** (dB)
G/T = figure of merit (dB)

EXAMPLE 6.8

Determine the carrier–to–noise temperature ratio of a satellite system with an isotropic radiated power of 20 dB, space losses of 200 dB, and which is required to maintain a figure of merit (G/T) of 40.7 dB.

Solution
In dB,

$$\left(\frac{C}{T}\right)_{dB} = (P_t \times G_t)_{dB} - L_{dB} + \left(\frac{G}{T}\right)_{dB}$$

$$= EIRP - L_{dB} + \left(\frac{G}{T}\right)_{dB} = 20 - 200 + 40.7 = -139.3$$

Therefore,

$$\frac{C}{T} = -139.3 \text{ dB}$$

EXAMPLE 6.9

In a satellite communications down-link, the effective isotropic radiated power of the transponder antenna is 22 dB and the space losses are 198 dB. If the carrier–to–noise temperature ratio (C/T) is 136 dB, calculate the antenna gain for a system noise temperature of 87.8 K.

Solution
In dB

$$\left(\frac{C}{T}\right)_{dB} = EIRP_{dB} - L_{dB} + \left(\frac{G_{rec}}{T}\right)_{dB}$$

$$-136 = 22 - 198 + \left(\frac{G_{rec}}{T}\right)_{dB}$$

$$\left(\frac{G_{rec}}{T}\right)_{dB} = -136 + 198 - 22 = 40 \text{ dB}$$

Since the system noise-temperature is given as 87.8 K then

$$T_{dB} = 10 \log(87.8)$$
$$= 19.43$$
$$\therefore T = 19.43 \text{ dB}.$$

Satellite Earth Stations

Substituting into the flow-of-merit equation, we have

$$G_{\text{rec}} = 40 - 19.43 \cong 20.6$$

Therefore, the antenna gain is equal to 20.6 dB.

6.2 The Earth Station Transmitter

The principal function of the earth station transmitter of a communications satellite system is to generate the composite baseband signal by multiplexing all the incoming transponder signals (usually digital), then modulating the result with an IF subcarrier and up-converting it to a carrier frequency by the frequency translation circuit. The carrier frequency is then amplified through a **high-power amplifier** and finally transmitted to satellite transponder via the antenna subsystem. The basic block diagram of an earth station transmitter is shown in Figure 6.5.

The multiplexer network, through the frequency-division multiplexing (FDM), or time-division multiplexing (TDM) process, multiplexes the individual transponder signals. The number of transponder signals combined through the multiplexing process is determined by the system available bandwidth. For the C band and Ku band, the available bandwidth is 500 MHz. After multiplexing, the composite signal modulates a subcarrier frequency and then up-converts it to either 6 GHz or 14 GHz, for the C band or Ku band respectively. The block diagram of an **up-converter** circuit is shown in Figure 6.6.

The IF signal is mixed with the local oscillator carrier frequency. At the output of the mixer circuit there are a number of frequencies such as the IF, the local oscillator sum and differences, and a number of harmonics. The bandpass filter (BPF) is designed to process only the desired frequency band (up link) and to reject all the other frequencies. The output of the BPF is then fed into the input of the high-power amplifier (HPA). This signal amplification through the HPA is absolutely necessary because in conjunction with the antenna, it generates the required EIRP dictated by the system specifications.

6.2.1 Up-Converter Circuit

The principal function of the up-converter circuit is to translate the modulated IF output frequency to up-link frequency levels of 6 GHz or 14 GHz for the C Band or Ku Band re-

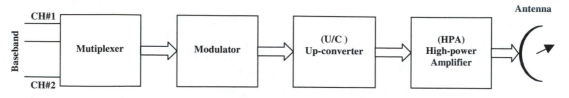

FIGURE 6.5 Simplified block diagram of an earth station transmitter.

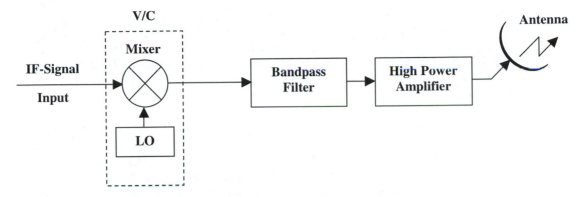

FIGURE 6.6 Up-converter block diagram.

spectively. The ever-increasing demand for higher system capacities and better error per-formance proportionally increases up-converter design constraints. Figure 6.7 shows an up-converter circuit incorporated into a digital radio system, utilizing the 64-QAM modulation scheme.

This up-converter (U/C) is composed of two balanced field-effect transistors (FETs), an input transformer, two IF chocks and one 180° hybrid. The input transformer provides the IF phase inversion required before both signals are applied to the gates of the balanced

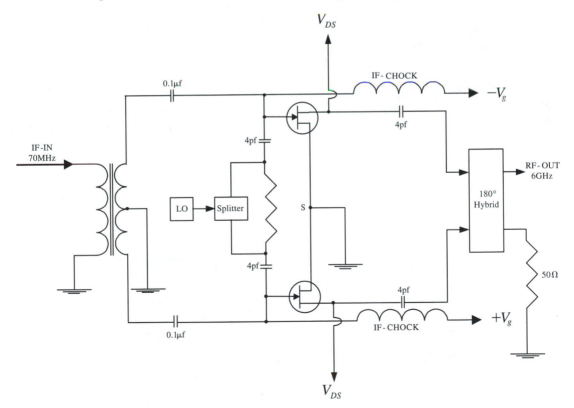

FIGURE 6.7 A 70-MHz–to–6-GHz balanced FET U/C.

FETs. The local oscillator output is applied to each gate of the FET through the power splitter. The drains of both transistors are connected to the 180° hybrid. The combination of the power splitter and hybrid achieves the required local-oscillator signal suppression. Each set of 4-pF capacitors and IF chocks connected between the gates of the FETs and power supply eliminates the possibility of an IF short circuit through the FETs and power supply. The FET devices are selected with performance characteristics in compliance with up-converter circuit requirements. Figure 6.8 illustrates the relationship between input and output power levels.

It is evident from Figure 6.8 that a linear conversion gain of approximately 1.7 dB can be achieved. The up-converter circuit has also demonstrated an operating power output level of 13 dB. The local oscillator (L.O.) output level required to drive the up-converter circuit is 18 dB. The minimum acceptable L.O. leakage at the input of the filter is -15 dB (Figure 6.9), and the local oscillator power-level suppression is at 33 dBm.

The bit-error-rate (BER) performance curve is illustrated in Figure 6.10. Figure 6.10 shows that an increase of the circuit power output level from 1 dBm to 3.7 dBm corresponds to an increase of the BER from 10^{-31} to 10^{-21}. In all, this converter demonstrates certain advantages over the standard diode type, such as a conversion loss of 2 dB instead of 6 dB for diode circuits. It also shows an increase of the output level and contributes to a proportional reduction of the overall transmitter power requirements.

6.2.2 The High-Power Amplifier (HPA)

The two fundamental devices used for high-power amplification in satellite earth stations are the **traveling-wave tube** (TWT) and the gallium–arsenide FET (GaAs FET). Traveling-wave tubes are used when high-power amplification is required, and GaAs FETs are used when low-power amplification is required.

The amplification action of a traveling-wave tube is based on the interaction between a RF field and an electron beam. The physical layout of a TWT device is shown in Figure 6.11. In order for interaction between the RF field and the electron beam to take place, the

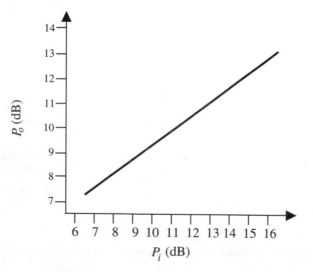

FIGURE 6.8 P_o versus P_i.

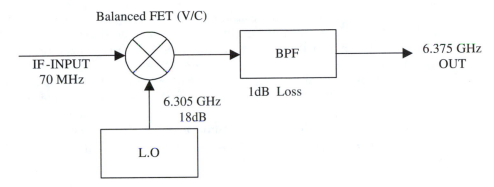

FIGURE 6.9 Balanced FET U/C block diagram.

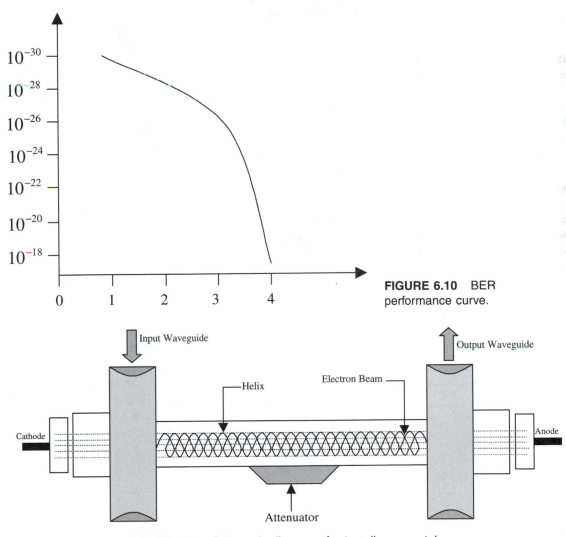

FIGURE 6.10 BER performance curve.

FIGURE 6.11 Schematic diagram of a traveling-wave tube.

velocities of the RF field and electron beam must be equal. Since the velocity of the RF field is ten times higher that that of the electron beam, the velocity of the RF field must be reduced to the level of the electron beam if energy is to be transferred from the electron beam to the RF field.

The **helix** inside the tube is used for that very reason, that is, to slow down the forward propagation of the RF fields to the propagation velocity of the electron beam. The interaction between the RF field and electron beam results in the transfer of energy from the beam to the RF signal. Hence, amplification of the RF signal can be achieved. A wide range of TWTs exists, designed to generate various levels of output power within the microwave frequency spectrum.

6.2.3 Earth Station Transmitter Redundancy

In satellite communications systems, continuous and uninterrupted transponder service is the most important objective. This highly desirable system efficiency is achieved by incorporating into the design of the earth station a redundancy mechanism capable of providing continuous service even if a high-power amplifier has completely failed to operate. This redundancy mechanism is shown in Figure 6.12.

For dual polarization systems, two up-converters are required. The output of each converter is fed into the input of the **power divider.** The rest of the operation is the same as that of Figure 6.13. The only exception, the output of high-power amplifier 1, is fed into the vertically polarized feed of the transmitter antenna, while the output of high-power amplifier 2 is fed into the horizontally polarized feed of the transmitter antenna. Figure 6.14 illustrates another method of power output redundancy. This method incorporates three high-power amplifiers. The redundancy mechanism of Figure 6.14 can also be utilized when two carrier frequencies of the same polarization mode are to be transmitted concurrently through a third carrier combiner before the antenna feed.

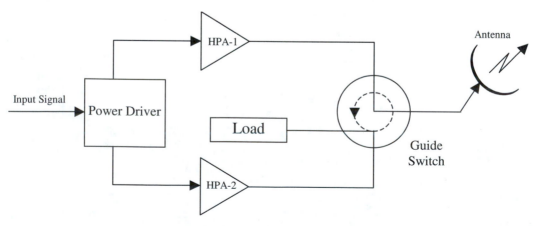

FIGURE 6.12   HPA redundancy scheme.

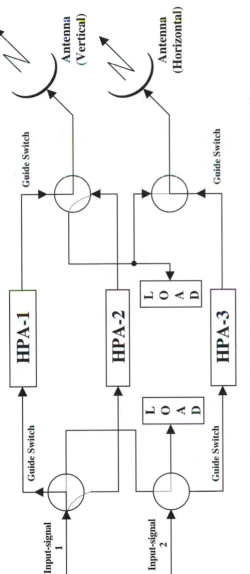

FIGURE 6.13 Power output redundancy incorporating three HPAs.

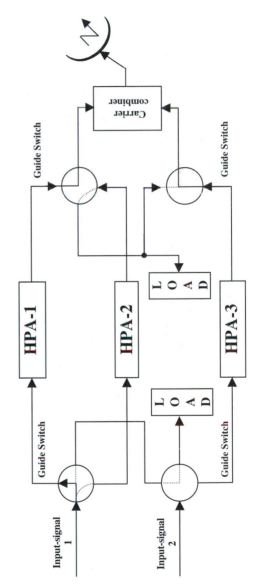

FIGURE 6.14 Power output redundancy with carrier combiner.

FIGURE 6.15   Carrier combiner through a directional coupler.

Combining Carriers

There are two methods of carrier combining. One utilizes a **directional coupler,** the other a dual hybrid filter. When two carriers are combined through a directional coupler (Figure 6.15), each carrier suffers a power loss almost equal to the directional coupler's insertion loss.

If x carriers are to be combined, the required carrier power loss is given by

$$L_\lambda = 10 \log(x) \tag{6.19}$$

where L_λ = total power loss
           x = number of carriers to be combined.

The number of directional couplers required is $x - 1$. It is evident from this that the method of combining carriers is relatively expensive in terms of carrier power loss. Figure 6.16 illustrates the block diagram of a **carrier combiner** with more than two carriers. To reduce the carrier power losses associated with the directional combiner method, the dual-hybrid-filter method is utilized (Figure 6.17). The two bandpass filters are identical. The mid-frequency and bandwidth are that of carrier-1. Hybrid-1 divides carrier-1 equally into two, and through the bandpass filters the two signals are again recombined to generate carrier output-1.

Hybrid 2 also divides carrier 2 equally into two signals, but these signals are reflexed by both bandpass filters, because the center frequency is different from that of carrier 2. Theses two signals are recombined by the second hybrid and appear at the same output port with carrier 1. The power loss for both carriers is much smaller using this method of carrier combining rather than the directional coupler method. For example, the typical inser-

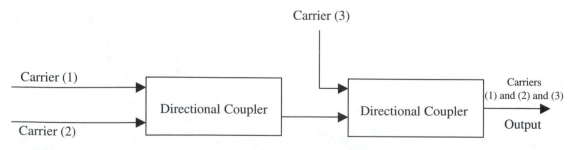

FIGURE 6.16   Carrier block diagram for more than two carriers through directional compilers.

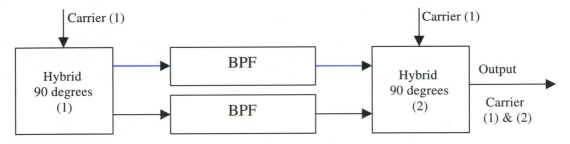

FIGURE 6.17 Dual-hybrid-filter method of carrier combiners.

tion loss of a 90° hybrid is approximately 0.1 dB and the filter loss is approximately 0.6 dB. Carrier 1 processed through both 90° hybrid and filter suffers a total loss of 0.8 dB [(0.1 × 2) + 0.6], and carrier 2 suffers a total loss of 0.2 dB (0.1 × 2).

Combining Power Amplifiers

There are three compelling reasons for combining power amplifiers in the earth station transmitter section. The first reason is related to the fact that higher power output in an earth transmitter can be obtained by combining a number of lower-power amplifiers. The second reason is economics. Since lower-power amplifiers are more economical to design and develop, the total cost of the combined number of low-power amplifiers is less than the cost of an equivalent single HPA. The third reason is related to reliability. When a power combiner mechanism is employed that is composed of a number of low-power amplifiers, failure of one or two of these amplifiers, although it may reduce the total transmitter power, does not completely incapacitate the system. Such a power combiner mechanism is shown in Figure 6.18. The power combiner mechanism of Figure 6.18 employs two 90° hybrid and two high-power amplifiers with equal gains and power output. The input signal is equally divided by the hybrid 1 then fed into the inputs of the two identical high-power amplifiers. Since the output power of both amplifiers is P_0 with gain G_p, the input power (P_i) must be restricted to the level determined by P_0/G_p for each amplifier. Hybrid 2 combines the outputs of each HPA, thus generating a total power output of $P_{ot} = 2P_o$. It is therefore apparent that the circuit of Figure 6.18 performs the function of power combining.

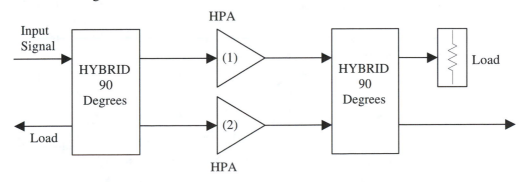

FIGURE 6.18 Power combiner mechanism.

6.3 THE EARTH STATION RECEIVER

The receiver section of a satellite earth station performs the exact opposite function to that of the transmitter. Figure 6.19 illustrates an oversimplified version of an earth satellite receiver. The microwave beam intercepted by the receiver antenna is collected at the focal point of the secondary antenna and, through a waveguide assembly, is fed to the input of a **low-noise amplifier** (LNA).

In most receiver systems the LNA is located at the focal point of the parabolic reflector antenna. This way, unnecessary power loss between antenna and LNA is minimized. (Receiver and transmitter antenna characteristics are dealt with in separate sections.) In summary, both antennas must exhibit a very high gain and a narrow bandwidth. Although the antenna gain is in the order of 50 dB, the output signal from the antenna is extremely weak and very compatible with the input noise power. Therefore, the presence of an LNA is absolutely necessary.

6.3.1 Low-Noise Amplifier (LNA)

The previous section mentioned the importance of the LNA as the first stage of active amplification of the incoming satellite signal. The two basic LNA types are the **parametric amplifier** and the GaAs FET amplifier. Parametric amplifiers were exclusively used in the early stages of satellite system development. These amplifiers were capable of providing reasonable gains over the entire bandwidth with substantially low noise temperatures. Recent developments in advanced semiconductor device technology gave birth to such devices as the GaAs FETs and high-energy-mobility semiconductor devices. With such devices it is possible to achieve high gains with very low noise temperatures, high stability, and reliability, coupled with a substantial increase in cost effectiveness.

6.3.2 The Parametric Amplifier

The principal performance characteristics of a parametric amplifier are high gain and low noise figure (NF). The amplification action of a parametric amplifier takes place when exchange of energy is used to amplify the input signal. In practical parametric amplifier designs, the capacitive reactance (X_C) is utilized through a varactor device to provide the required energy exchange mechanism. A varactor is a semiconductor diode whose junction capacitance can be varied in accordance with a reverse dc bias voltage. The capacitance across the pn junction is varied in accordance with

$$C_v = \frac{C}{(V_b - V)^M} \qquad (6.20)$$

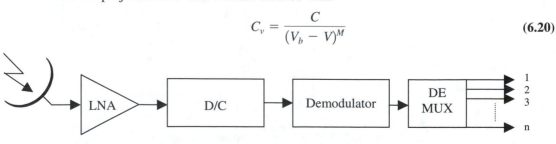

FIGURE 6.19   Oversimplified block diagram of a satellite earth station receiver.

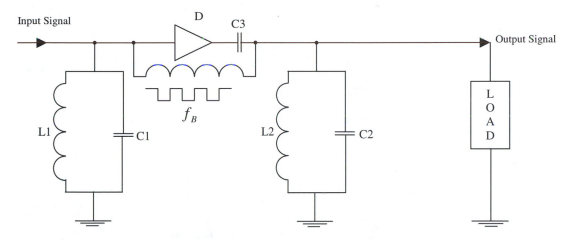

FIGURE 6.20 Degenerative parametric amplifier.

where C_v = varactor capacitance at specific reverse-bias voltage
C = varactor capacitance at *zero* bias voltage
V_b = junction barrier potential (normally 0.7 V)
M = constant, characteristic of the semiconductor material used
$(0.3 < M < 0.5)$

When varactor diodes are used in parametric amplifiers, their typical reverse-bias voltages are in the order of 4 V to 12 V, corresponding to a maximum operating frequency of 20 GHz. Ideally, parametric amplifiers must generate a zero noise figure (NF), because an ideal reactive element does not have resistive elements to induce power dissipation. In practice, a very small equivalent resistive element is present as a result of semiconductor material imperfections. These material imperfections reflect a nonzero, yet very small, noise figure by comparison. There are two basic modes of parametric amplifier operations: **degenerative** mode, and **nongenerative** mode.

The Degenerative Parametric Amplifier

The block diagram of a degenerative parametric amplifier is shown in Figure 6.20. The varactor diode is connected in series between the input and the output impedance of the circuit. For amplification to take place, stored energy from the varactor diode must be transferred to the load precisely when the input signal is at the positive or negative peak. This implies that the capacitance of the varactor at these points must go to its minimum. This is achieved by precisely selecting the exact phase relationship between the input signal and the varactor bias frequency (f_b). Even the smallest phase difference between the two signals will reduce the circuit's amplification capability.

Nongenerative Parametric Amplifiers

The critical phase relationship required for the generative parametric amplifier between the input signal and the varactor dc biasing voltage is somewhat relaxed with the modified circuit of Figure 6.21. In this circuit, a third resonant element, with a resonance frequency equal to the sum of or difference between the input and the biasing signals, is added. The

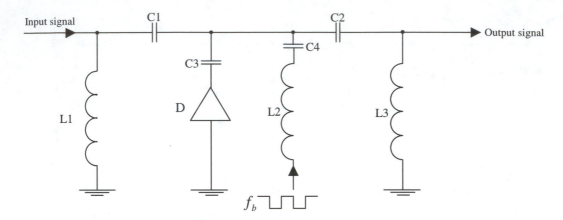

FIGURE 6.21 Nongenerative parametric amplifier.

rest of the operation is the same as with the previous circuit, except that any phase drift between the input and biasing signals does not degrade the overall amplification capability of the parametric amplifier. It is also evident from the description of the circuit that it performs the up- or down-conversion of the input signal:

$$f_o = f_i + f_b \quad \text{or} \quad f_o = f_i - f_b \tag{6.21}$$

6.3.3 MMIC LNAs

The ever-increasing demand for high device performance together with circuit complexity, space, and weight reductions, generated the need for the development of monolithic mi-

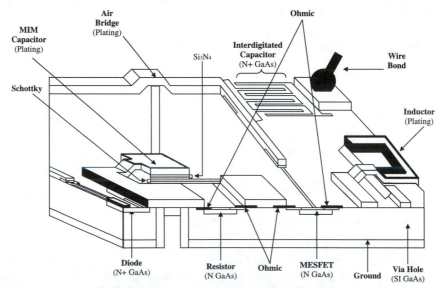

FIGURE 6.22 A three-dimensional MMIC device.

crowave integrated circuits (**MMICs**). These are integrated circuits incorporating a large number of active and passive solid state components operating in microwave frequencies. Such components as FETs, diodes, resistors, capacitors, and transmission lines are all packaged in a single GaAs chip with dimensions in the millimeter range.

A three-dimensional MMIC device is shown in Figure 6.22. Today, MMICs are widely used in satellite military and instrumentation applications, performing such functions as amplification, signal translation, modulation, and multiplexing. The principal active device employed in MMIC structures is the GaAs MESFET(Metal–Semiconductor Field-Effect Transistor), because the overall behavior of this device operating in microwave frequencies is well understood. An MMIC amplifier operating at 10 GHz is shown in Figure 6.23. The frequency response of the amplifier of Figure 6.23 is shown in Figure 6.24.

High-Energy-Mobility Devices

High-energy-mobility semiconductor devices employed in MMIC structures are superior to MESFET devices, especially in reference to the noise figure. A discrete **HEMT** device is capable of providing a gain of 6 dB at a frequency range higher than 90 GHz. At lower frequencies (e.g.,10 GHz) gains of 10 dB are obtained with a corresponding noise figure of 1 dB. These figures strongly support the utilization of the HEMT semiconductor devices in MMIC structures, and particularly in low-noise amplifiers. The schematic diagram of an MMIC HEMT LNA is illustrated in Figure 6.25.

This MMIC LNA employs a 0.25-μm HEMT device exhibiting 6 dB of gain over the 20-GHz to 30-GHz frequency range and a noise figure of 5 dB. The circuit of Figure 6.25 is designed to operate at 20 GHz to 40 GHz using transmission lines and shunt-shorted stubs as matching elements for both input and output ports. A special computer program was used to manipulate a large number of different parameters incorporated into the circuit design. For loss minimization and other practical considerations, the matching elements were selected between 30 Ω and 90 Ω. The measured performance of this LNA circuit in comparison with the theoretical performance is shown in Figure 6.26. The 6-dB gain was obtained with circuit biasing voltages $V_d = 3$ V, $V_g = -0.3$ V, and $I_{d_s} = 15$ mA. The

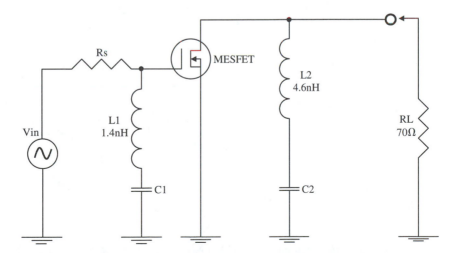

FIGURE 6.23   Schematic diagram of an MMIC amplifier operating at 10 GHz.

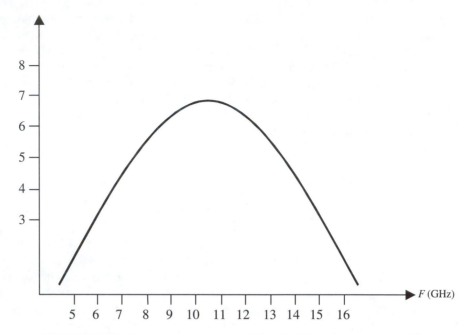

FIGURE 6.24   Frequency response of the 10-GHz microwave amplifier.

equivalent circuit of this amplifier is illustrated in Figure 6.27. The measured noise figure is shown in Figure 6.28.

The substantially lower noise figure of HEMT devices in relation to MESFET devices is attributed to higher cutoff frequencies (g_m/C_{gs}) and a higher correlation coefficient. The very fact that HEMT devices have a lower noise conductance is a determining factor in the overall noise figure reduction (the noise figure is less susceptible to source impedance variations). The higher gain of the HEMT device is attributed to higher transconductor and lower output conductance. The fact that the AlGaAs used in HEMT devices can facilitate higher doping than the GaAs used in MESFET devices compounded by an increase in the saturation velocity of the electron–gas layer, results in a higher transconductance for HEMT devices Figure 6.29. The lower output conductance of an HEMT device is the result of a thinner epitaxial layer.

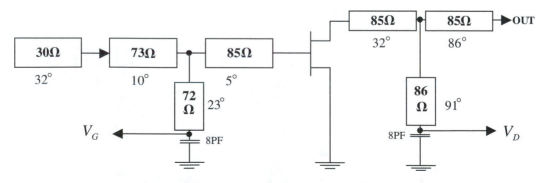

FIGURE 6.25   25 MMIC LNA operating in the Ka band.

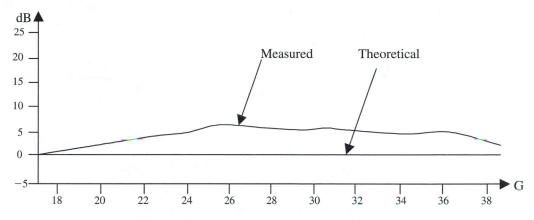

FIGURE 6.26 Frequency response of the MMIC LNA.

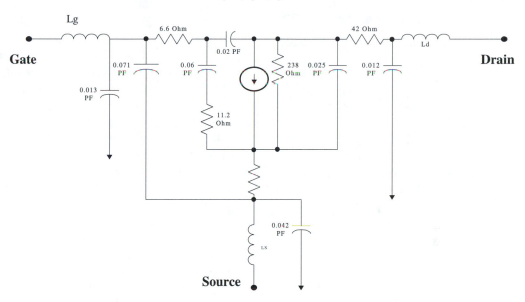

FIGURE 6.27 MMIC LNA equivalent circuit.

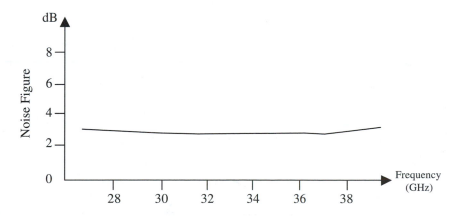

FIGURE 6.28 Measured noise figure versus frequency for the MMIC LNA.

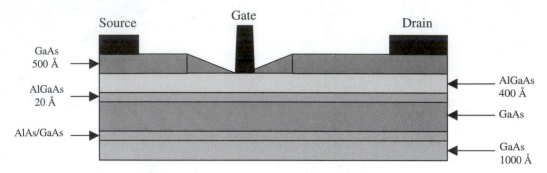

FIGURE 6.29　HEMT device cross section.

LNA Redundancy

Redundancy considerations are as important for the receiver section of the earth station as they are for the transmitter, if continuous link operation is to be maintained. A low-noise-amplifier redundancy block diagram is shown in Figure 6.30. In Figure 6.30 the LNA redundancy block diagram consists of two LNAs. Under normal operating conditions only one LNA is actively connected to the receiver section. The second LNA is in stand-by mode. When LNA 1 fails, the second LNA is automatically activated between the receiver antenna feed and the input of the down-converter through two microwave switches. A more elaborate scheme is implemented using three LNAs for cross-polarization transmission (Figure 6.31). If either LNA 1 or LNA 3 fails, LNA 2 will be switched on automatically through the appropriate microwave switches (1 and 3 or 2 and 4). The redundant LNA 2 is capable of performing in the cross-polarization mode of operation.

Nonlinearity

So far the importance of the LNA as the first stage in the process of down-link signal detection has been discussed. Also discussed was the importance of strictly maintaining LNA specifications such as gain (over the entire 500-MHz band width), and noise figure.

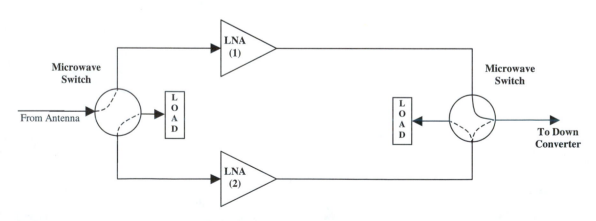

FIGURE 6.30　LNA redundancy block diagram.

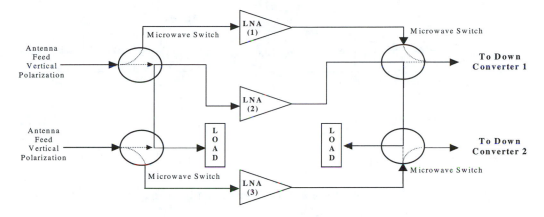

FIGURE 6.31 LNA redundancy for cross-polarization mode of operation.

Furthermore, one additional critical parameter related to LNA performance must be taken into consideration. This parameter is related to amplifier **nonlinearities**. Low-noise-amplifier nonlinearity determines the power levels of various spurious signals present within the operating bandwidth. The degree of LNA nonlinearity is expressed in terms of the **third-order intercept point** and 1 dB gain compression. Both these parameters determine the level of nonlinearity of a low-noise amplifier. By examining the LNA power-transfer characteristic curve (Figure 6.32), the third-order intercept point can be determined.

Third-Order Intercept Point   Figure 6.32 illustrates the input/output power-transfer characteristic curve of an LNA. From this curve the third-order intercept point is established as follows: The third-order intercept point is the power output level generated by the receiver due to the intermodulating process when two signals of different frequencies f_1 and f_2 are applied at the input of the receiver. The receiver nonlinearity will produce an output signal of frequency $2f_1 - f_2$. This signal will be compared with either f_1 or f_2 of the input signals. It is evident from the characteristic curve of Figure 6.32 that, the higher the third-order intercept point, the better the receiver nonlinearity.

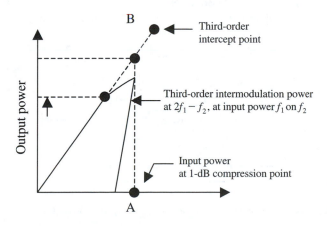

FIGURE 6.32   LNA power-transfer characteristic curve.

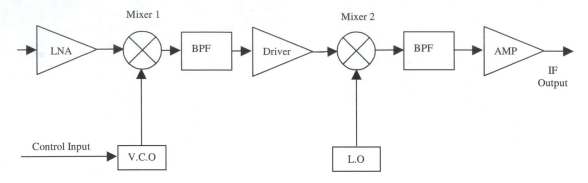

FIGURE 6.33 Down-converter block diagram.

1 dB-Gain Compression Point The **1-dB-gain compression point** is established when a continuous increase of the input signal generates an output signal that is 1 dB less than the output power generated by the amplifier operating at the linear portion of its characteristic curve. From the curve of Figure 6.32, it is also evident that the smaller the input power required generating the 1-dB-gain compression, the smaller the receiver non-linearity.

Down Converter (D/C)

The output of the LNA is fed into the input of the **down-converter.** The block diagram of a down-converter module is shown in Figure 6.33. The output of the LNA is fed into the mixer circuit and, combined with the output of the voltage-controlled oscillator, generates the first IF frequency. For example the LNA output bandwidth is between 3.7 GHz and 4.2 GHz. The VCO circuit is designed to cover the frequency band between 2.82 GHz and 3.32 GHz. The first IF output frequency is then 3.7 GHz $-$ 2.82 GHz, equal to 0.88 GHz, or 4.2 GHz $-$ 3.32 GHz, equal to 0.88 GHz. The output of the BPF with a bandwidth of 0.88 GHz is fed to the input of the second mixer through a driver circuit. Since a 70-MHz IF output signal is required, the local oscillator is designed to provide a fixed output frequency of 950 MHz (880 MHz $+$ 70 MHz).

The block diagram of another converter circuit covering the 6/4 GHz band is shown in Figure 6.34. This broadband down-converter has demonstrated an overall BER im-

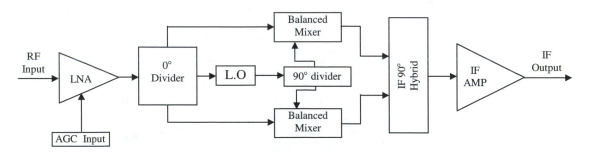

FIGURE 6.34 Block diagram of a 6/4-GHz down-converter.

provement of 0.5 dB. It is composed of a low-noise amplifier, a power driver, two double-balanced mixers (Schottky diodes), an RF 90° coupler, an IF 90° hybrid, and an IF amplifier. This down-converter operates over the frequency range of 3.5 GHz to 6.5 GHz, exhibiting an overall conversion gain of 37 dBm for an input signal larger than −42 dBm. The IF output is maintained at a constant level of −5 dBm through the AGC circuit. This down converter also exhibits an **image rejection** of better than 20 dB over the operating range and a noise figure less than 2.5 dB.

The Low-Noise Amplifier (LNA)   The first stage of a down-converted module is the low-noise amplifier. This amplifier is composed of two stages. The first stage is designed to maintain very low noise impedance; the second stage is designed to provide the required high gain. Test results have demonstrated a gain of 17 dB with a noise figure of 1.7 dB over the entire frequency range (Figure 6.35).

The Local Oscillator   A bipolar transistor is used as the active device in the design of the local-oscillator circuit employed in the down converter module. The design objective of this local oscillator is to provide a high power output at the oscillating frequency with a corresponding low phase noise. A dielectric resonator is used as the tuned circuit in this particular design.

The 90° Divider   The 90° divider exhibits a phase difference of less than 1.6 dB between the two outputs and 1 dB in magnitude variations over the 3-GHz bandwidth.

The Image Rejection Mixer   The image rejection mixer is composed of a power divider, two balanced mixers, the 90° divider, and an IF hybrid. This circuit demonstrates a greater than 20-dB image rejection with a conversion loss of 6.8 dB over the entire bandwidth (Figures 6.36 and 6.37).

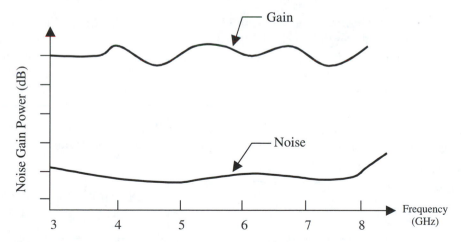

FIGURE 6.35   Performance curves of a LNA circuit incorporated at the input of a down-converter module.

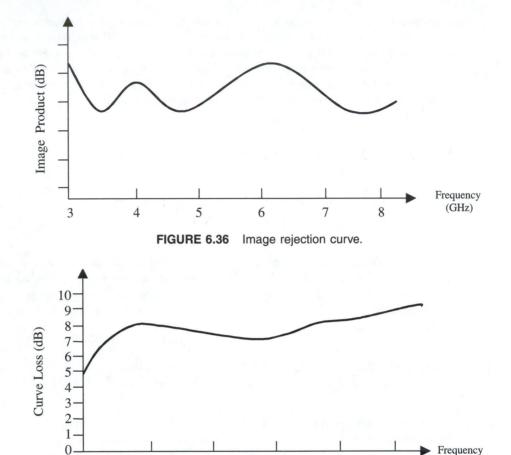

FIGURE 6.36   Image rejection curve.

FIGURE 6.37   Conversion loss curve.

The 90° Hybrid   The 90° hybrid has a bandwidth of 20 MHz (60 MHz–80 MHz), an output port phase difference of 1°, and a magnitude of 0.5 dB over the 20-MHz bandwidth. In summary, the down-converter described exhibits an excellent noise figure, low phase noise, and high image rejection. The D/C IF output is further amplified in order to maintain an appropriate signal level as required by demodulation circuit input sensitivity.

The Demodulator   The block diagram of a demodulator circuit is shown in Figure 6.38. The demodulator circuit is composed of a limiter and a phase-lock loop (PLL) discrimina-

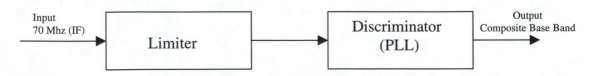

FIGURE 6.38   Demodulator block diagram.

tor. The limiter circuit is necessary to maintain the required signal-to-noise ratio at the output of the circuit. This improvement in the signal-to-noise ratio is usually referred to as the "FM-improvement factor."

The Demultiplexer   The composite baseband signal retrieved by the demodulator is fed into the input of the demultiplexer. At the demultiplexer output, the individual transponder signals are regenerated and further processed to the appropriate destinations.

QUESTIONS

1. List the main components of a satellite earth station. With the assistance of a block diagram, briefly explain its function of operation.
2. Define an electromagnetic wave.
3. Briefly describe the fundamental properties of parabolic reflectors.
4. What are the two components determining effective isotropic radiated power (EIRP)?
5. Define *figure of merit* (*G/T*).
6. Define (a) system noise temperature and (b) carrier–to–thermal noise temperature.
7. List the main components of an earth station transmitter. With the assistance of a block diagram, briefly explain its function of operation.
8. What is the function of an up-converter? Sketch the block diagram and briefly explain its operation.
9. Give the reasons why high-power amplifiers (HPAs) are required in an earth transmitter station.
10. Describe the two fundamental devices used as high-power amplifiers in a satellite earth station.
11. With the assistance of a schematic diagram, briefly explain the function of the operation of a traveling-wave tube (TWT).
12. Explain why HPA redundancy is required in a satellite earth station. List the different redundancy methods.
13. List the main components of an earth station receiver. With the assistance of a block diagram, briefly explain its function.
14. Describe briefly the various types and operating characteristics of low-noise amplifiers (LNAs).
15. Why is LNA redundancy required in an earth receiver station?
16. With the assistance of a block diagram, describe the function of the LNA redundancy scheme employed for a cross-polarization mode of operation.
17. Define the terms, "*nonlinearity*" and *1-dB-gain compression point*.
18. With the assistance of a block diagram, describe the function of a satellite earth station down-converter.

PROBLEMS

1. The transmitter of a satellite earth station generates 1kW of RF power at its output. Compute the effective isotropic radiated power if a parabolic reflector with a 50-dB gain is used as the transmitter antenna. Assume a feed loss from transmitter to antenna of 2 dB.

2. A satellite earth station transmitter generates an effective isotropic radiated power equal to 85 MW. If the transmitter power output is 750 W, calculate the antenna gain in dB.

7

Satellite Access

Objectives

- Define satellite access.
- Identify the various types of satellite access.
- Describe the concepts of FDMA and TDMA.
- Explain the difference between FDMA and TDMA.

Key Terms

Access	CDMA	Signaling
INTELSAT	MCPC FDMA	Switching
Single access	FDM-FM-FDMA	DASS
Multiple access	SCPC	Channel encoding
FDMA	PCM-PSK-FDMA	Frame structure
TDMA	The SPADE	Frame synchronization

INTRODUCTION

The unique feature of a geosynchronous communications satellite lies in its ability to provide reliable communications links for the transmission of voice, video, and data for a large number of earth stations separated by long distances. This ability of the system is achieved through the process of satellite **access.** In the early phases of satellite communications, only two earth stations were capable of establishing communications links (Figure 7.1). The **INTELSAT-I** system incorporated two earth antennas, four on-board antennas, two transponders, and four frequencies. The first transponder utilized the 6.30102-GHz up-link and the 4.081-GHz down-link frequencies; the second utilized the 6.3899-GHz up-link and 4.16075-GHz down-link frequencies. Since a single station was able to access and utilize the entire transponder bandwidth, INTELSAT-I was referred to as a **single-access** system. The ever-increasing demand for more earth stations generated the need for the design of wide-band repeaters involving **multiple-access** techniques. The concept of multiple access was first introduced in the INTELSAT-II satellite systems (Figure 7.2).

The single-access mode was unable to satisfy the fundamental requirements established by the INTELSAT systems. These requirements were related to the maximization of system capacity, efficient utilization of the available bandwidth, interconnectivity and cost efficiency. Fundamentally, there are three modes of multiple access.

1. Frequency-division multiple access **(FDMA)**
2. Time-division multiple access **(TDMA)**
3. Code-division multiple access **(CDMA)**

Frequency-division multiple access and time-division multiple access are the most common access modes used in satellite communications. Accessibility to a satellite transponder can

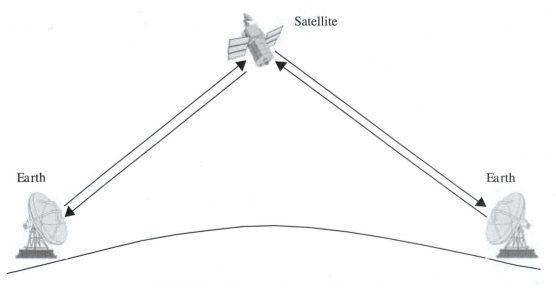

FIGURE 7.1   Single-access satellite system.

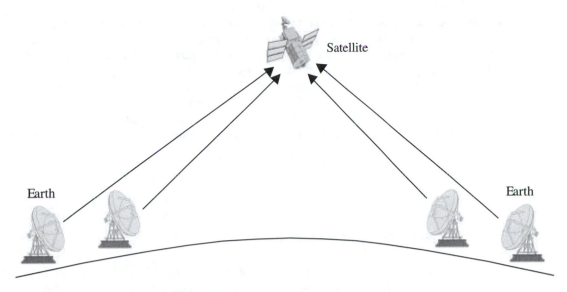

FIGURE 7.2 Multiple-access satellite system.

also be classified in terms of the degree of channel availability to the user, or to all users at all times upon demand (demand-assigned multiplex access.) This mode of access is most applicable to low- and medium-capacity links. Heavy traffic routes are better served by a single-access mode of operation requiring very large earth station antennas. When the FDMA mode of operation is utilized, the transponder available bandwidth is subdivided into a number of channels separated by guard bands (Figure 7.3).

The earth station transmitter is capable of transmitting to a number of these channels, and the earth station receiver is designed to receive the entire RF spectrum and, through filtering, to tune in to the desired channel. When the TDMA mode of access is utilized, a single carrier is used by all earth stations in a time-sharing mode of operation. When an earth station transmits a burst of information at a preassigned time to a satellite, the entire bandwidth can be utilized by that earth station. The satellite receives the transmitted burst,

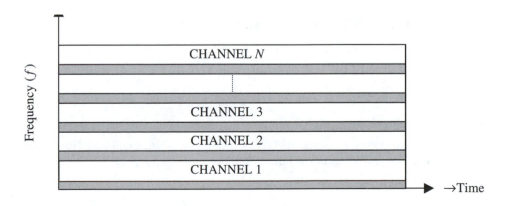

FIGURE 7.3 Frequency-division multiple-access bandwidth channelization.

and amplifies, down-converts, and retransmits it back to earth. The next burst, perhaps originating from another earth station, follows exactly the same pattern. The fact that bursts are transmitted in sequence separated by narrow guard bands necessitates optimum synchronization techniques so that burst overlap can be avoided.

In code-division multiplex access, each earth station transmits coded information to the satellite regardless of any overlap with other stations that may be transmitting simultaneously. At the receiver end, the separation of the transmitted information by each station is achieved through the detection of the individual earth station's transmitted identification code.

7.1 FREQUENCY-DIVISION MULTIPLE ACCESS

Starting with INTELSAT-II, the multiple-access mode of operation was necessitated by an ever-increasing number of earth stations requiring access to a common transponder. Frequency-division multiple access can be divided into two basic categories: fixed (multiple channels per carrier) and demand (single channel per carrier).

7.1.1 Multiple Channels Per Carrier (MCPC) FDMA

In this FDMA mode of operation, a number of voice channels are frequency multiplexed to form a single baseband signal. This signal is then frequency modulated (suppressed carrier), up-converted, amplified, and then transmitted to a satellite transponder. At the receiver end, the reverse process is performed. After amplification the down-link RF frequency is translated to IF frequency, then fed to an FM discriminator input for the composite baseband signal to be recovered. It is imperative to mention that, at the input of the FM discriminator circuit of the earth receiver, there also exists a noise signal. Therefore, the signal-to-noise ratio at this point is critical in establishing overall system performance.

FDM-FM-FDMA Carrier-to-Noise Ratio

The block diagram of an FM demodulator is given in Figure 7.4. The FM signal is given by

$$e_{\text{FM}} = E_c \cos[\omega_c t + (m_f \sin \omega_m t)] \tag{7.1}$$

The noise power consists of AWGN (Additive White Gaussian Noise) and other interferences. The carrier-to-noise ratio (CNR) is expressed by

$$CNR = \frac{C}{N} = \frac{S}{N}\left(\frac{b}{B}\right)\left(\frac{f_m}{f_{rms}}\right)^2 \tag{7.2}$$

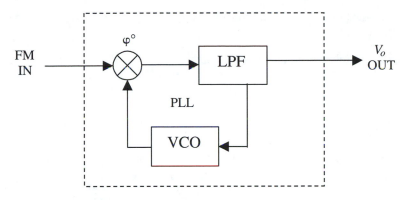

FIGURE 7.4   FM discriminator block diagram.

where C = carrier power (W)
            N = noise power (W)
            BW = system bandwidth
            f_m = modulation frequency
            f_{rms} = rms frequency duration $\left(\dfrac{\Delta f}{\sqrt{2}}\right)$

From Eq. (7.2), the CNR can be calculated if the voice-channel signal-to-noise ratio and the number of voice channels are established. The value of CNR must at least be maintained at a 10-dB level. This level is referred to as the "FM demodulation threshold level." The bandwidth (BW) is determined by

$$BW = 2(f_m + \Delta f_c) \qquad \text{(Carson's Rule)} \tag{7.3}$$

For FDM-FM-FDMA, the value of Δf_c is given by

$$\Delta f_c = glf_{rms} \tag{7.4}$$

where g is the peak-to-rms voltage ratio of a specific number of active voice channels per carrier. (It is estimated that the active number of channels comprises 25% of the total number. For the specific application of FDM-FM-FDMA, the g-factor is 10 dB.)

(l) is a loading factor with a value related to voice channels given by the following: For a number of channels less than 240,

$$l = \text{antilog}\left(\frac{-1 + 4\log X}{20}\right) \tag{7.5}$$

where x = number of channels

For a number of channels more than 240,

$$l = \text{antilog}\left(\frac{-15 + 10\log X}{20}\right) \tag{7.6}$$

For FDM-FM-FDMA satellite systems, the baseband frequency including the guard band for specific numbers of voice channels is shown in Table 7.1.

TABLE 7.1   Baseband frequency and voice channels for FDM-FM-FDMA satellite system.

Number of channels	24	30	60	72	96	132	192	252	312	432	612	792	972	1092
Baseband frequency (Hz)	108	156	252	300	408	552	802	1052	1300	1796	2540	3284	4028	4812

7.1.2 Single-Channel-Per-Carrier (SCPC) Demand Assignment

The fixed-assignment mode of satellite access is excellent for high-traffic routes, because a single carrier utilizes all the available transponder bandwidth for transmission. INTELSAT-V, for example, utilizes the FDM-FM-FDMA fixed mode for the transmission of 1400 voice channels in its international routes, and TELESAT–Canada carries up to 900 voice channels (one-way) for its heavy routes using the same FDM-FM-FDMA mode of operation. The fixed-assignment mode of operation presents certain difficulties when interconnectivity is required for low-traffic multiuser links such as those required by national systems. If the FDM-FM-FDMA system was to be implemented in a low- or medium-traffic multiuser setting, it is estimated that only 37.5% of the available capacity could be utilized. This low efficiency is translated into a substantial revenue loss for the system operator. The other major problem relates to interconnectivity.

The FDM-FM-FDMA mode of operation does not offer the flexibility of frequency reconfiguration required by a multiuser setting with an ever-increasing demand and complexity. To overcome these difficulties, the demand-assignment mode of operation was developed. With demand assignment, the available transmission bandwidth is divided into a number of single channels and each channel is assigned to individual users (carriers). The carrier information can either be analog or digital in form. INTELSAT has developed two basic systems operating on demand assignment. PCM-PSK-FDMA **(SPADE),** a 12- to 24-channel system, and **PCM-PSK-TDM** (MAT-1), a 12- to 120-channel system.

The SPADE System

The acronym SPADE is derived from *S*ingle-Channel-Per-Carrier *P*CM multiple *A*ccess *D*emand *A*ssignment *E*quipment. This system was developed with the following basic philosophy in mind. The 36-MHZ INTELSAT transponder bandwidth is to be divided into 800 discrete channels assigned to a common pool accessible on demand to all earth stations. The most important component for the SPADE system is the common signaling channel.

It is evident that the SPADE system is a modified frequency-division multiple-access system, divided into three discrete operating subsystems: (a) signaling, (b) switching, and (c) transmission. Figure 7.5 shows the block diagram of the SPADE system.

Operation   From Figure 7.6, channels 1 and 2 are not included in the pool of usable frequencies. Instead these are used to provide the isolation band required between the central signaling channel (CSC) and the rest of the voice channels. Therefore, the total available number of channels in the SPADE system is 794.

The DASS Unit   The most important unit in the SPADE system is the *D*emand-*A*ssignment *S*ignal and *S*witching, or **DASS,** unit. The main function of this unit is to continuously route information to all system subscribers, relating request calls and channel availability. Upon receipt of a request call, the DASS unit will instantaneously select a pair of available frequencies from the pool of frequencies and convey the call request and the selected response channel to the appropriate destination station. It will also inform the rest of the stations in the network and remove the newly assigned frequency pair from the pool of frequencies.

The Control Unit   As soon as the selected frequencies are assigned to a particular call, the control unit generates the selected frequencies through the frequency synthesizer circuit upon the command of a digital code. Each control unit in the network is capable of generating all 800 frequencies. The voice signal to be transmitted is converted to a binary stream of information through the process of pulse-code modulation (PCM). The channel synthesizer circuit performs timing and framing of the PCM signal. The composite digital signal is digitally modulated (QPSK) with a preassigned intermediate frequency, up-converted, amplified, and then fed to the antenna subsystem for transmission to the satellite. The opposite process takes place at the receiver end. The received digitally coded and modulated voice signal is down-converted, demodulated, decoded, and finally processed to the other party through the terrestrial network. Upon completion of the call, the DASS unit disconnects the frequencies and returns them to the common pool, informing all the other users in the network of the change.

SPADE Development

The technical characteristics of the SPADE system are as follows:

Channel encoding: 7-bit PCM
Companding: A-law
Modulation: QPSK
Bit-rate: 64 kb/s
Channel bandwidth: 38 kHz
Spacing between channels: 45 kHz
BER: 10^{-4}

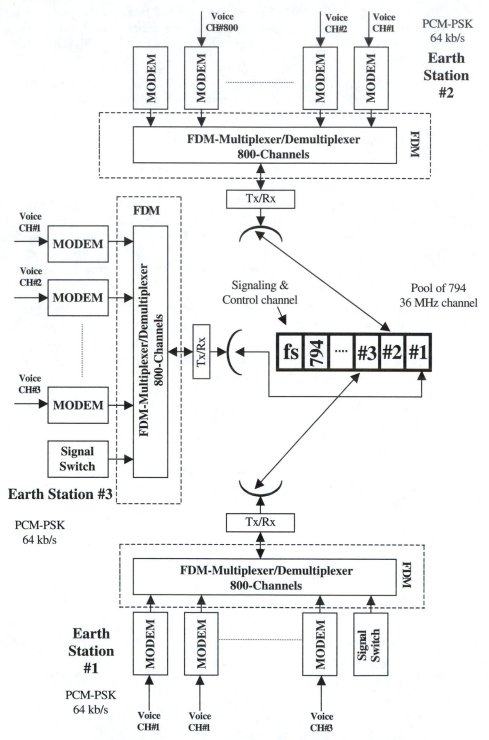

FIGURE 7.5 The SPADE system block diagram (FDMA).

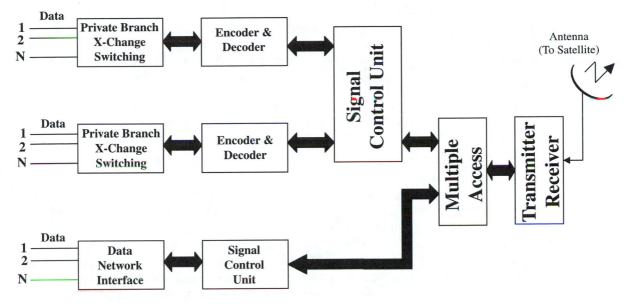

FIGURE 7.6   TDMA system for voice and data transmission.

The substantial advancements made in digital technology development were instrumental in the design and implementation of the SPADE systems. Since the voice channels must be digitally encoded, the CCITT recommended a 7-bit or 8-bit voice signal quantization. The implementation of A-law companding (European), instead of μ-law companding (North American) was based on the fact that the A-law companding exhibits better linear characteristics than the μ-law companding and therefore is easier to implement (Figure 7.7).

The selection of QPSK as the baseband digital modulation technique was based essentially on two factors. First, the required effective isotropic radiated power (EIRP) for a satellite transponder with 36 MHz of available bandwidth was established at 22 dBm. At

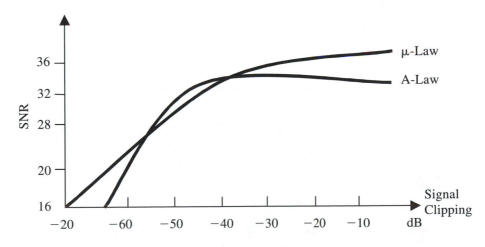

FIGURE 7.7   Comparison of performance between μ–low and A-law companding.

this transponder bandwidth, 400 digitally coded and BPSK-modulated voice channels could be accommodated. Second, it was also evident from related studies that slightly fewer than 40% of all channels in the transponder bandwidth were simultaneously active. This gave rise to the possibility of an increase in channel bandwidth from the original 40 kHz to 45 kHz. Therefore, by utilizing the QPSK modulation scheme (higher spectral efficiency), the 800-voice-channel target was achieved while maintaining the required EIRP of 22 dBm.

The performance characteristics of an early SPADE modem employing a QPSK modulation scheme are shown in Figure 7.8. For a continuous mode of modem operation, a 7.4-dBm satellite EIRP with an earth station figure of merit (*G/T*) of 40.7 dB/K was required. One of the crucial factors limiting system capacity was the intermodulation product noise.

The Signaling Channel It was mentioned earlier that the key unit required of establishing inter-connectivity between earth stations and satellite is the demand assignment **signaling** and **switching** unit. The signaling channel is an integral part of the DASS unit. Its main circuitry includes the synchronization circuitry and the PCM-TDM modem. The synchronization function is for burst control; the PSK modem function is to process the related information between the earth stations and between the earth station and the satellite. Each earth station employs a small computer capable of storing and processing signals received from all the other stations and the satellite. The access time of all the earth stations in the network and the reference station is approximately 1 ms. Since the reliability and signal quality of these units is very critical to the overall operation of the SPADE system, the QPSK modulation scheme with a burst rate of 128 kb/s is employed, and the transmission power is kept to a level of 7 dB above the voice-channel level. This requires a BER equal to 10^{-7}.

Frequency Stability One of the main difficulties encountered in the early stages of the SPADE system development was the maintenance of frequency stability between all the earth stations in the network. The primary sources of frequency drift beyond the ± 2-KHz

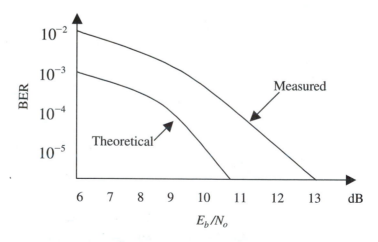

FIGURE 7.8 Performance characteristics of an early SPADE modem employing the QPSK modulation scheme.

maximum allowable were the up- and down-converters. To overcome this major difficulty, one of the earth stations was assigned the task of transmiting a pilot frequency via the satellite to all other earth stations in the network. Upon receipt of the pilot frequency, each station centered the entire 800-channel bandwidth through an automatic-frequency-control loop within its filter bandwidth. In this way, intersymbol interference (ISI) was kept to an absolute minimum.

7.2 TIME-DIVISION MULTIPLE ACCESS (TDMA)

One of the principal objectives of a communication satellite is to maintain maximum system capacity with an acceptable signal-to-noise ratio (C/N) at the receiver end over its entire operational life. With the utilization of the FDM-FM-FDMA mode of operation, maximum system capacity was practically impossible to achieve. The main factor preventing the system from operating at its maximum efficiency was the traveling-wave tube. To minimize distortion these high-power amplifiers are operating below maximum output power.

The implementation of the SPADE system increased the overall system capacity through a more efficient utilization of the available transmitter power and a substantial reduction of the intermodulation noise. The time-division multiple-access mode of operation (Figure 7.7) further diminishes the problems associated with power amplifier nonlinearities and system capacity. The TDMA mode of operation allows all earth stations linked to the network to transmit bursts of digitally coded information in such a synchronized manner that no two burst time frames overlap. The TDMA system also allows for the utilization of the entire transponder bandwidth. Thus, maximum system capacity can be achieved (Figure 7.9).

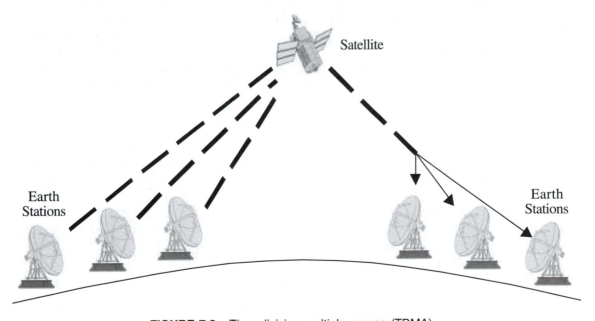

FIGURE 7.9 Time-division multiple access (TDMA).

Upon receipt of a burst of digital information, the satellite transponder amplifies, down-converts, reamplifies, and retransmits the burst to all stations covered by the satellite footprint. All earth stations within the footprint receive the entire burst sequence, and each station recognizes and extracts its own burst. The availability of digital technology in the mid-1960s supported the development of TDMA systems. In 1963, COMSAT initiated the development of a three-terminal system with a 6 Mb/s bit rate, translating to a 72-voice system capacity.

In late 1966, this system was tested through INTERSAT-I with positive results. Most encouraging were the results in the critical area of burst synchronization, which showed that synchronization down to few nanoseconds was possible. Following this success, Nippon Telephone and Telegraph developed a 13.664 Mb/s system, corresponding to a 244-voice-channel capacity, utilizing the 7-bit PCM voice signal quantization.

The highly encouraging results of the first TDMA system generated the incentive for the development of the second generation of TDMA systems by COMSAT Labs. These systems exhibited an impressive 50 Mb/s bit rate, corresponding to a 700-voice signal capacity, utilizing the 8-bit PCM voice signal quantization. In 1970, a compatible TDMA system of 50 Mb/s incorporating the pulse-code modulation (PCM), *T*ime-*A*ssignment-*S*peech-*I*nterpolation (TASI) method was developed and tested by INTELSAT-III with very promising results.

7.2.1 TDMA Frame Synchronization

Figure 7.10 illustrates a typical TDMA **frame structure.** A time-division multiple-access frame structure (Figure 7.10) is normally composed of two reference bursts and traffic bursts separated by a guard band. A particular earth station in the satellite network is assigned the task of transmitting one reference pulse to the satellite transponder and to all stations in the network. Another earth station is assigned the task of automatically switching over and transmitting reference burst 2, in the event of reference burst 1 transmission failure. In this way the overall system reliability is further enhanced. Reference bursts do not contain traffic information but merely provide the transmission reference required by all earth stations in the network and the satellite transponder. The length of the frame may be different for

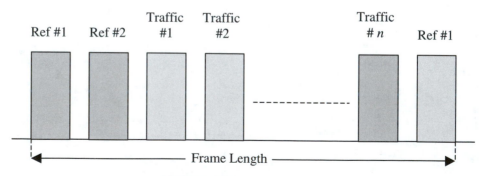

FIGURE 7.10   TDMA frame.

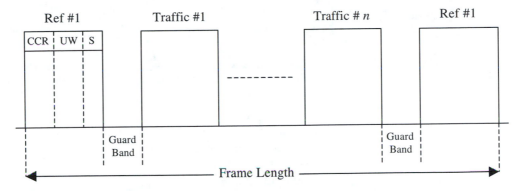

FIGURE 7.11 TDMA frame with burst structure.

different systems. For example, SATCOM-1 utilized a frame length of 20 ns, whereas INTELSAT-V is operating with a frame length of 2 ms. A small quad band is allowed between reference bursts and traffic bursts as well as between traffic and traffic bursts. Guard bands are absolutely necessary to avoid burst overlap. Their length must be equal to the time required for the transmission and reception of the TDMA reference pulse.

A more detailed explanation of the frame structure is shown in Figure 7.11. The reference burst consists of a carrier clock recovery (CCR) word or unique word (UW) and a signaling channel (S). The carrier and clock recovery word at the start of the TDMA burst sequence enables the receiver demodulator to recover the phase of the carrier, thus maintaining a BER within the predetermined limits. The length of the CCR is directly related to the bit rate of the TDMA system used.

For example, a long CCR word is required for a high-bit-rate system, whereas a short CCR word is required for a low- to medium-rate TDMA system. In a typical 120 Mb/s TDMA system, a carrier clock recovery word of 400 bits is required. The inclusion of the unique word in the reference burst assists the receiver in locating the traffic burst and frame synchronization. The signaling channel is also composed of three subchannels: (a) the order-wire subchannel, (b) the housekeeping or management channel, and (c) the transmission-timing subchannel.

The order-wire burst contains the coded information available to all earth stations in the network, indicating whether a voice or data signal is to be transmitted. The housekeeping or management channel contains coded information relating to time-slot boundaries for each earth station, burst plan alterations, and status reports from the reference station.

The traffic burst follows the reference burst. This time subframe is also divided into smaller time slots. The information contained in these time slots is relevant to the type of service to be established (voice, video, data), which ultimately will determine the length of the traffic slot.

7.2.2 TDMA Frame Efficiency and Channel Capacity

Frame efficiency is defined as the ratio of the time occupied by the information bits to the total time occupied by the frame. A practical formula expressing frame efficiency quantitatively is given by

$$\eta = 1 - \frac{P(2 + n)}{T_f R} \tag{7.7}$$

where
η = frame efficiency
R = link total bit rate
n = traffic burst number within the frame
P = the sum of preamble bits in the frame
T = period of the frame

In practical systems, frame efficiency of the order of 90–95% is required. It is evident from Eq. (7.8) that the time slot allocated for quad and preamble bits must be kept very small, in comparison to traffic time, if acceptable frame efficiency is to be accomplished. These are limitations for both quad and preamble time lengths, based on synchronization accuracies and signaling and detection requirements. Frame length limitations, on the other hand, are based on transmission delays and other economic considerations.

In systems such as the INTELSAT employing a TDMA mode of operation, the maximum number of voice channels that can be accommodated within the system is given by

$$K = \frac{R}{\eta r} \tag{7.8}$$

where
K = number of voice channels
R = total satellite bit rate
n = frame efficiency
r = voice-channel bit rate

EXAMPLE 7.1

A communications satellite system is used exclusively for voice signal transmission. It employs an 8-PSK modulation scheme with a spectral efficiency of 3.356 b/s/Hz, corresponding to an error power $P(e)$ of 10^{-8} while maintaining a C/N equal to 21.5 dB. If earth stations transmit with TDMA one burst per frame for a frame period of 2 ms, determine the maximum number of voice channels the transponder bandwidth can accommodate if the total number of guard and preamble bits is 680.

Solution

Transponder bit rate (R)

$$R = \text{Bandwidth} \times \text{spectral efficiency}$$
$$= 36 \text{ MHz} \times 3.356 \text{ b/s/Hz} = 120.816 \text{ Mb/s}$$

Therefore, $R = 120.816$ Mb/s.

Frame efficiency (η)

$$\eta = 1 - \frac{P(2 + n)}{T_f R} = 1 - \frac{680(2 + 12)}{2 \times 10^{-3} \times 120.816 \times 10^6} = 1 - \frac{9520}{241.7 \times 10^3} = 0.96$$

Therefore, $\eta = 96\%$ frame efficiency.

Number of voice channels (K)

$$K = \frac{R}{\eta r} = \frac{120.832 \times 10^6}{0.96 \times 64 \times 10^3} = 1966$$

Therefore, number of voice channels $K = 1966$.

Therefore the maximum numbers of voice channels this satellite system can effectively accommodate is 1966.

QUESTIONS

1. Define satellite access. Describe the difference between single and multiple access.
2. Name and define the three modes of multiple access.
3. Define the CNR relationship of a FDM-FM-FDMA demodulator circuit and list all the parameters incorporated into the relationship.
4. Define the SPADE system and list the most important components of the system.
5. With the assistance of a block diagram, describe the function of operation of the SPADE system.
6. Define briefly the operation of a time-division multiple-access (TDMA) scheme.
7. Define the significance of the DASS unit and briefly describe its operation.
8. What is frame efficiency of a TDMA system? List all the components determining frame efficiency and indicate their relationships.

PROBLEM

Compute the maximum number of voice channels a communications satellite transponder bandwidth can carry, based on the following satellite operating parameters.

TDMA system (one burst per frame, for a frame period of 2.5 ms).
Modulation: QPSK.
Modulation efficiency: 2.2b/s/hz
BER $= 10^{-8}$
$C/N = 18$ dB.
Guard and preamble bits $= 700$

8

Satellite Links

Objectives

- Identify the various parameters incorporated in satellite system design.
- Define **system availability** and **bit error rate** in digital satellite link design.
- **Compute:**

  Effective isotropic radiated power.

  Transmitter antenna gain.

  Up-link losses.

  Power at the input of the LNA.

  Carrier-to-noise ratio.

  Figure of merit.

  Receiver noise figure.

  Transponder voice signal capacity.

Key Terms

Power amplifier back-off loss
Feed loss
EIRP
Atmospheric loss
Free-space loss
Transponder capacity

Equivalent noise temperature
Figure of merit
Carrier–to–equivalent noise temperature ratio
Carrier–to–thermal noise density ratio

Bit energy–to–thermal noise density ratio
Transponder gain
LNA noise figure
The earth segment
The space segment

8.1 SATELLITE LINK ANALYSIS AND DESIGN

8.1.1 System Parameters

As with the design of any terrestrial communications link, the design of digital satellite links requires the establishment of specific parameters necessary for effective and efficient link operation. Such parameters are listed here for both up-link and down-link operation.

Up-Link
The Earth Segment

Transmitter power (P_t, dB)
Transmitter antenna gain (G_t, dB)
Power amplifier back-off loss (L_b, dB)
Effective isotropic radiated power (EIRP, dB)
Atmospheric loss (L_a, dB)
Free-space loss (L_s, dB)
Modulation scheme used
Transponder capacity (bit rate)

The Space Segment (on board the satellite)

Received power (P_r, dB)
Receiver antenna gain (G_r, dB)
Feed loss (L_f, dB)
Equivalent noise temperature (T_e, K)
Figure of merit (G/T_e, dB)
Carrier–to–equivalent noise temperature ratio (C/T_e, dB)
Carrier–to–thermal noise density ratio (C/N_o, dB-Hz)
Carrier-to-noise ratio (CNR, dB)
Energy bit–to–thermal noise density ratio (E_b/N_o)
Transponder gain (G_{sat}, dB)
LNA noise figure (NF, dB)

Down-Link
The Space Segment

Transmitter power (P_t, dB)
Antenna gain (G_t, dB)
Power amplifier back-off loss (L_b, dB)
Feed losses (L_f, dB)
Effective isotropic radiated power (EIRP, dB)
Free-space losses (L_s, dB)

The Earth Segment

Received power (P_r, dB)

Antenna gain (G_r, dB)

Feed losses (L_f, dB)

Equivalent-noise temperature (T_e, K)

Figure of merit (G/T_e, dB)

Carrier–to–thermal noise density (C/N_o, dB)

Carrier–to–noise ratio (CNR, dB)

Energy-bit–to–thermal noise ratio (E_b/N_o)

LNA noise figure (NF, dB)

Some of these parameters for both up-links and down-links will be assumed based on system requirements; the rest will be calculated. A brief summary discussion of some of these parameters follows.

Effective Isotropic Radiated Power (EIRP)

The effective isotropic radiated power (**EIRP**) was previously defined as the product of the transmitter power (P_t) and the transmitting antenna gain (G_t) given by

$$\text{EIRP} = P_t \times G_t$$

In dB,

$$\text{EIRP}_{dB} = 10 \log(P_t) + 10 \log(G_t)$$
$$\text{EIRP}_{dB} = P_{t_{dB}} + G_{t_{dB}} \tag{8.1}$$

It is important to mention that when the total transmitter power P_t is taken into consideration for the calculation of EIRP in Eq. (8.1), this power is always less than the maximum power output generated by the traveling-wave-tube (TWT) amplifier incorporated into the final stage of the transmitter assembly.

This power amplifier **back-off loss** (L_b) is necessary to keep high-power amplifier (HPA) nonlinearity to a minimum by operating the device in the linear section of its operating characteristic curves. This mode of operation of the HPA eliminates the possibility of intermodulation distortion, which occurs when such amplifiers operate in the nonlinear section of their characteristic curve. This sacrifice of power output is a trade-off for distortion minimization. The back-off loss must be incorporated into the calculations for determining the final EIRP. Another element that must be incorporated into the final calculations of the EIRP is the **feed loss** (L_f). This loss is encountered between the antenna and the final stage of the power amplifier. If both losses are taken into account, the final formula determining the EIRP is given by

$$\text{EIRP}_{dB} = P_{t_{dB}} + G_{t_{dB}} - (L_{b_{dB}} + L_{f_{dB}}) \tag{8.2}$$

8.2 DIGITAL SATELLITE LINK DESIGN

In any digital satellite communications link design, the two fundamental parameters determining an efficient and effective system operation are system availability and bit error rate (BER). These two system properties are interrelated by definition. System availability is defined as the percentage of the time the system is operating while maintaining a pre-established bit error rate. The system availability for communications satellite links is set at 99.95% of the time with a bit error rate not exceeding the 10^{-7} level under no-fading conditions.

The main source of signal fading is rain. Characteristic curves relating different levels of rainstorms, operating frequencies, and attenuation are shown in Figure 8.1. Figure 8.1 is based on an antenna elevation of approximately 15°. For different antenna elevations, characteristic curves are shown in Figure 8.2.

The dimensions of a typical rainstorm are estimated to be 15 km wide by 5 km deep. The total signal attenuation due to rainfall is calculated as follows: For both up-link and down-link power-budget calculations, signal attenuation due to a rainstorm can be extrapolated from graphics. For example, for a system operating at the C band (6/4 GHz), a rainstorm estimated to be 15 km by 5 km, for a linear total of 20 km, reflects an attenuation equal to 20 km $\times$ 0.5 dB/km, or a total signal attenuation of 10 dB (up-link). For a down-link operating at 4 GHz, an attenuation of approximately 0.1 dB/Km is encountered, for a total power attenuation of 20 $\times$ 0.1 dB/km = 2 dB.

In a practical system, the up-link attenuation due to rainfall is offset by a corresponding increase of the transmitter power output. By increasing the transmitter power beyond its normal operating power, the high-power amplifier of the transmitter is driven closer to a nonlinear operation, resulting in a possible increase of intermodulation distortion and a corresponding increase of BER. In this case, an attempt is made to maintain a balance between signal fading and BER.

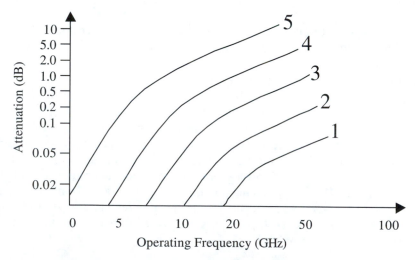

FIGURE 8.1   Attenuation due to rain for different rainfall intensities. (Key: 1 = 0.25 mm/h; 2 = 1 mm/h; 3 = 4 mm/h; 4 = 16 mm/h; 5 = 100 mm/h.)

Chapter 8

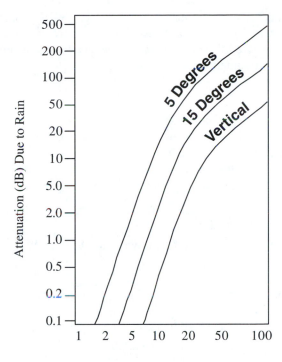

FIGURE 8.2
Attenuation due to rain for different antenna elevations.

DESIGN EXAMPLE 8.1

A transponder link must be established in a communications satellite system operating in the C band, with the following transponder parameters:

Operating frequency: 6/4 GHz (C band)
Transponder bandwidth: 36 MHz
BER $= 10^{-9}$ (clear weather) but not necessarily exceeding 10^{-7} under heavy rain
System availability: 99.95%
Modulation scheme: QPSK
Earth station antennas: parabolic reflectors with diameter equal to 30 m
Transponder gain on board the satellite: $G_{sat} = 100$ dB
Down-link: 3.730 GHz
Transmitter power: 2 kW
Up-link 5.945

Compute

Up-Link

MAXIMUM TRANSPONDER CHANNEL CAPACITY (f_b, Mb/s)
The transponder channel capacity (f_b) is based on the digital modulation scheme used and the transponder bandwidth:

$$f_b = \text{Digital modulation scheme efficiency} \times \text{Channel bandwidth} \qquad (8.3)$$

Satellite Links

267

The QPSK modulation scheme reflects an efficiency of approximately 1.7 b/s/Hz while maintaining BER $= 10^{-9}$ with a corresponding $E_b/N_o = 13$ dB. Therefore, $f_b = 1.7$ b/s/Hz $\times$ 36 Mb/s $= 61.2$ Mb/s. Allowing for system losses, a more conservative 60 Mb/s channel capacity is assumed.

$$\therefore f_b = 60 \text{ Mb/s} \qquad \text{(channel capacity maximum per transponder)}$$

EFFECTIVE ISOTROPIC RADIATED POWER (EIRP)

EIRP is the product of the earth transmitter power in dB added to the transmitter antenna gain in dB, minus the back-off losses $L_{b_{dB}}$ and feed losses $L_{f_{dB}}$. Therefore,

$$\text{EIRP}_{dB} = P_{t_{db}} + G_{t_{dB}} - L_{b_{dB}} - L_{f_{dB}} \tag{8.4}$$
$$P_t = \text{transmitter power output} = 2 \text{ kW}$$

In dB,

$$P_{t_{dB}} = 10 \log(2 \times 10^3) = 33$$
$$L_{b_{dB}} = 2 \text{ dB} \quad \text{(branch losses)}$$
$$L_{f_{dB}} = 2 \text{ dB} \quad \text{(feed losses)}$$

EARTH STATION TRANSMITTER ANTENNA GAIN (G_t)

The parabolic reflector antenna gain is calculated as follows:

$$G_t = \frac{4\pi A_{\text{eff}}}{\lambda^2} \tag{8.5}$$

where
$\quad A_{\text{eff}} = $ effective antenna area (m^2)
$\quad \lambda = $ operating wavelength (m)
$\quad G_t = $ antenna gain

$$A_{\text{eff}} = KA \tag{8.6}$$

where
$\quad K = $ efficiency parameter (0.55)
$\quad A = $ antenna aperture

$$A = \frac{\pi D^2}{4} \tag{8.7}$$

where
$\quad D = $ antenna diameter (m)

Therefore,

$$A = \frac{\pi(30)^2}{4} = 706 \text{ m}^2$$

Substituting A into A_{eff},

$$A_{\text{eff}} = 0.55 \times A = 0.55 \times 706 \text{ m}^2 = 388 \text{ m}^2$$

Substituting A_{eff} into G_t,

$$\lambda = \frac{c}{f} = \frac{3 \times 10^8 \text{ m/s}}{5.945 \times 10^9} = 0.05 \text{ m}$$

$$G_t = \frac{4\pi(388) \text{ m}^2}{\lambda^2} = \frac{4\pi(388) \text{ m}^2}{(0.05 \text{ m})^2} = \frac{4875}{2.5 \times 10^{-3}} = 1.95 \times 10^6$$

where c = speed of light in vacuum = 3×10^8 m/s
$\quad\quad\quad G_t = 1.95 \times 10^6$

In dB,

$$G_t = 10 \log(1.95 \times 10^6) \cong 59$$
$$\therefore G_t \cong 63 \text{ dB}$$

Substituting values in the equation we have (in dB)

$$\text{EIRP}_{\text{dB}} = P_{t_{\text{dB}}} + G_{t_{\text{dB}}} - L_{b_{\text{dB}}} - L_{f_{\text{dB}}} = 33 + 63 - 2 - 2 = 88 \text{ dB}$$
$$\therefore \text{EIRP} = 92 \text{ dB}$$

UP-LINK PATH LOSSES (L_{ul})

The EIRP generated by the earth station transmitter antenna assembly suffers a substantial loss through the propagation path of 42,000 km between the earth station antenna and the satellite receiver antenna. Up-link path losses are expressed by

$$L_{ul_{\text{dB}}} = 32.5 + 20 \log(d)_{\text{dB}} + 20 \log(f)_{\text{dB}} \quad\quad\quad\quad (8.8)$$

where d = Path length in km (42,000 km)
$\quad\quad\quad f$ = Up-link frequency in MHz (5.945 GHz)

Substituting into the equation for d and f

$$L_{ul_{dB}} = 32.5 + 20 \log(42{,}000) + 20 \log(5.945 \times 10^3)$$
$$= 32.5 + 20(4.623) + 20(3.77)$$
$$= 32.5 + 92.46 + 75.4$$
$$= 200.45$$
$$\therefore L_{ul_{dB}} \cong 201$$

POWER RECEIVED AT THE SATELLITE LNA INPUT (P_r)

The EIRP generated by the earth station has suffered an attenuation of 201 dB traveling the path distance of 42,000 km. This signal is intercepted by the satellite receiver antenna, amplified by the antenna, and then fed into the receiver low-noise amplifier (LNA) input. The signal strength at the input of the LNA is given by

$$P_{r_{dB}} = \text{EIRP}_{dB} - L_{ul_{dB}} + G_{r_{dB}} - L_{f_{dB}} \qquad (8.9)$$

where
P_r = received power at the input of the LNA on board the satellite (dB)
EIRP = effective isotropic radiated power from earth station (equal to 92 dB)
L_{ul} = path losses (equal to 201 dB)
L_f = feed losses (2 dB)
G_r = satellite receiver gain

$$G_r = \frac{4\pi A_{\text{eff}}}{\lambda^2}$$
$$A_{\text{eff}} = KA$$
$$A = \frac{\pi D^2}{4}$$

where
D = antenna diameter (1 m)
K = efficiency factor (0.55)
λ = operating wavelength (m)
C = speed of light (3×10^8 m/s)

$$A = \frac{\pi(D)^2}{4} = \frac{\pi \times 1^2}{4} = 0.78 \text{ m}^2$$
$$A_{\text{eff}} = KA = 0.55 \times 0.78 \text{ m}^2 = 0.429 \text{ m}^2$$
$$\lambda = \frac{c}{f} = \frac{3 \times 10^8 \text{ ms}}{5.945 \times 10^9} = 0.5 \times 10^{-1}$$

Substituting into Eq. (8.5) gives

$$G_t = \frac{4\pi(0.429)}{\lambda^2} = \frac{4\pi(0.429)}{(0.5 \times 10^{-1})^2} = \frac{5.39 \times 10^2}{0.25} = 2156$$

$$G_{r_{dB}} = 10 \log(2156)$$

$$\therefore G_{r_{dB}} \cong 33$$

Substituting into Eq. (8.4),

$$P_{r_{dB}} = \text{EIRP}_{dB} - L_{ul_{dB}} + G_{r_{dB}} - L_{f_{dB}}$$
$$= 92 - 201 + 33 - 2$$
$$= -203 + 121$$
$$= -78$$

Therefore, $P_{r_{dB}} = -78$.

In watts,

$$P_r = \text{antilog}\left(-\frac{78}{10}\right)$$

$$\therefore P_r = 15.8 \times 10^{-9} \text{ W} \quad \text{or} \quad P_r = 15.8 \text{ nW}$$

CARRIER-TO-NOISE RATIO (CNR)$_{ul}$
The relationship determining the up-link carrier-to-noise ratio (CNR) is as follows:

$$CNR_{ul_{dB}} = \left(\frac{C}{N_o}\right)_{dB} - 10 \log(\text{BW}) \qquad (8.10)$$

where C/N_o = carrier-to-noise power density
          BW = transponder bandwidth (36 MHz) (System bandwidth
              for C band equal to 500 MHz)

The carrier-to-noise power density (C/N_o) is given (in dB) by

$$\left(\frac{C}{N_o}\right)_{dB} = \text{EIRP}_{dB} + \frac{G_r}{T_{sys_{dB}}} - L_{ul_{dB}} - L_{sys_{dB}} - K_{dB} - L_{envir_{dB}} \qquad (8.11)$$

where   EIRP $= 92$ dB
$L_{ul} = 201$ dB
$L_{sys} =$ system losses $\cong 2$ dB
$L_{envir} =$ environmental losses $\cong 2$ dB
$T_{sys} \cong 1000$ K
$K_{dB} = -228.6$ dB
$G/T =$ **Figure of merit** to be calculated

FIGURE OF MERIT (G_r/T_{sys})

$$\frac{G_r}{T_{sys_{dB}}} = G_{r_{dB}} - 10 \log(T_{sys}) = 33 - 30 = +3 \text{ dB} \qquad (8.12)$$

$$\therefore \left(\frac{G_r}{T_{sys}}\right)_{dB} = +3$$

Substituting for values in the equation,

$$\left(\frac{C}{N_o}\right)_{dB} = \text{EIRP}_{dB} + \frac{G}{T_{sys_{dB}}} - L_{ul_{dB}} - L_{sys_{dB}} - K_{dB} - L_{envir_{dB}}$$
$$= 92 + 3 - 201 - 2 - (-228.6) - 2$$
$$= 118.6$$
$$\frac{C}{N_o} \cong 118 \text{ dB}$$

The carrier-to-noise ratio for the up-link is

$$CNR_{ul} = \frac{C}{N_o} - 10 \log(\text{BW}) = 118 - 10 \log(500 \times 10^6) = 118 - 87 = 31$$
$$\therefore CNR_{ul} = 31 \text{ dB}$$

BIT ENERGY–TO–THERMAL NOISE DENSITY RATIO (E_b/N_o)
The **bit energy–to–thermal noise density** ratio was given by

$$\frac{E_b}{N_o} = \frac{C}{KT_e} \times \frac{1}{f_b} \qquad (8.13)$$

where $C =$ carrier power or $P_t = -78$ dB
$K =$ Boltzmann's constant $(1.38 \times 10^{-23}$ J-K or -228.6 dB/W$)$
$f_b =$ bit rate per transponder (60 Mb/s). For 12 transponders (C Band), the system bit rate is 12×60 Mb/s $= 720$ Mb/s.

E_b/N_o at the satellite is calculated as follows:

$$\frac{C}{KT_e} = \frac{C}{N_o} = 118 \text{ dB}$$

$$\frac{E_b}{N_o} = \frac{C}{N_o} - 10 \log(f_b)$$

$$= 118 - 10 \log(720 \times 10^6)$$

$$= 118 - 88.6$$

$$= 29.4 \text{ dB}$$

$$\therefore \frac{E_b}{N_o} = 29.4 \text{ dB}$$

In any digital satellite communications system, the CNR must be higher than the E_b/N_o. In this case, CNR = 31 dB and E_b/N_o = 29.4 dB, so the CNR is higher than the E_b/N_o by +3 dB.

Down-Link

EFFECTIVE ISOTROPIC RADIATED POWER (EIRP)
For the down-link, the effective isotropic radiated power is calculated the same way as for the up-link:

$$\text{EIRP}_{dB} = P_{t_{dB}} + G_{t_{dB}} + G_{\text{sat}}$$

where
P_t = transmitter power on board the satellite (dB)
G_t = antenna gain on board the satellite (dB)
G_{sat} = transponder gain (100 dB)

downlink f = 3.73 GHz

$$\lambda = \frac{c}{f} = \frac{3 \times 10^8 \text{ m/s}}{3.73 \times 10^9}$$

$$= 0.8 \times 10^{-1}$$

$$A_{\text{eff}} = KA$$

$$K = 0.55$$

$$A = \frac{\pi D^2}{4}$$

$$= \frac{\pi \times 1^2}{4}$$

$$= 0.78 \text{ m}^2$$

$$\therefore A_{\text{eff}} = 0.55 \times 0.78 \text{ m}^2$$

$$= 0.43 \text{ m}^2$$

Therefore,

$$\text{EIRP}_{\text{dB}} = P_{t_{\text{dB}}} + G_{t_{\text{dB}}} + G_{\text{sat}} = -78 + 29 + 100 = 51$$
$$\therefore \text{EIRP}_{\text{dB}} = 51 \text{ dB}$$

DOWN-LINK PATH LOSSES (L_{dl})

Down-link path losses are calculated in the same way as up-link losses. The only difference is the frequency component for the down link ($f_{dl} = 3.73$ GHz).

$$G_t = \frac{4\pi A_{\text{eff}}}{\lambda^2} = \frac{4 \times \pi \times 0.43}{(0.8 \times 10^{-1})^2} = 844$$

Therefore,

$$L_{dl_{\text{dB}}} = 32.5 + 20 \log(d) + 20 \log(f)$$

where $d = $ distance in km (36,000)
            $f = $ frequency in MHz (3.73×10^3)

$$L_{dl_{\text{dB}}} = 32.5 + 20 \log(42{,}000) + 20 \log(3.73 \times 10^3)$$
$$= 32.5 + 91 + 71.5$$
$$= 195$$
$$\therefore L_{dl_{\text{dB}}} = 195$$

RECEIVED CARRIER POWER (P_r)

The carrier power received at the input of the LNA of the earth receiver antenna is the sum total of all gains and losses through the down-link path.

$$\therefore \text{ in dB, } G_{t_{\text{dB}}} = 10 \log 844$$
$$G_{t_{\text{dB}}} \cong 29 \text{ dB}$$

Therefore,

$$P_{r_{\text{dB}}} = \text{EIRP}_{\text{dB}} + G_{r_{\text{dB}}} - L_{dl_{\text{dB}}} - L_{f_{\text{dB}}} \tag{8.14}$$

where $P_r = $ received carrier power (dB)
            $G_r = $ earth station antenna gain (58 dB)
            $L_{dl} = $ down-link path losses (197 dB)
            $L_f = $ feed losses (2 dB)

For the receiver carrier power calculations, only the earth receiver antenna gain is unknown.

$$K = 0.55$$
$$A_{\text{eff}} = KA = 0.55 \times 706 = 388 \text{ m}^2$$
$$\lambda = \frac{c}{f} = \frac{3 \times 10^8 \text{ m/s}}{3.73 \times 10^9} = 0.8 \times 10^{-1}$$
$$G_r = \frac{4\pi A_{\text{eff}}}{\lambda^2} = \frac{4\pi(388) \text{ m}^2}{(0.8 \times 10^{-1})^2} = \frac{4.8 \times 10^5}{0.64}$$
$$G_{r_{\text{dB}}} \cong 58$$

In dB,

$$P_{r_{\text{dB}}} = 51 + 58 - 197 - 2 = -90$$

Therefore, $P_{r_{\text{dB}}} = -90$ dB is the total received carrier power at the input of the earth receiver LNA.

FIGURE-OF-MERIT (G/T)

In a previous section it was mentioned that the figure of merit for an earth station antenna operating at the C band must satisfy the condition $G/T \geq 40.7$ dB. For the system under consideration $(G_r = 58$ dB), the receiver **equivalent noise temperature** must not exceed

$$10 \log T = 40.7 - G_t = 40.7 - 58 = 17.3$$
$$T = \text{antilog}(1.73)$$
$$= 53 \text{ K}$$
$$\therefore T = 53 \text{ K}$$

RECEIVER NOISE FIGURE (NF)

The receiver noise figure in dB is given by

$$\text{NF}_{\text{dB}} = 10 \log\left(\frac{T_e}{T_o} + 1\right) \tag{8.15}$$
$$= 10 \log\left(\frac{53}{290} + 1\right)$$
$$= 0.72$$

CARRIER-TO-THERMAL NOISE DENSITY $(C/N_o, \text{dB})$

For a received carrier power

$$P_r = -90 \text{ dB}$$
$$N_o = KT = 1.38 \times 10^{-23} \times 53$$

In dB,

$$N_o = 10 \log(1.38 \times 10^{-23} \times 53) = -228.6 + 17.2 = 211.4 \text{ dB}$$
$$\therefore N_o = 211.4 \text{ dB}$$

$$\left(\frac{C}{N_o}\right)_{\text{dB}} = -90 - (-211.4)$$
$$= -90 + 211.4$$
$$= 121.4 \text{ dB}$$
$$\therefore \left(\frac{C}{N_o}\right)_{\text{dB}} = 121.4$$

CARRIER-TO-NOISE RATIO (CNR, dB)

$$\text{CNR} = \frac{C}{N_o} - 10 \log(B)$$
$$= 121.4 - 10 \log(500 \times 10^6)$$
$$= 121.4 - 87$$
$$= 34.4 \text{ dB}$$

Therefore the carrier-to-noise ratio CNR = 34.4 dB.

BIT ENERGY–TO–NOISE RATIO (E_b/N_o)
The ratio of bit energy (E_b) to thermal noise density (N_o) is expressed as

$$\frac{E_b}{N_o} = \frac{C}{N_o} \times \frac{1}{f_b}$$
$$= 121.4 - 10 \log(f_b)$$
$$= 121.4 - 10 \log(720 \times 10^6)$$
$$= 121.4 - 88.6$$
$$= 32.8 \text{ dB}$$

where f_b = **transponder capacity** (60 Mb/s) or system capacity (720 Mb/s)

Therefore, $E_b/N_o \cong 32.8$ dB.

TRANSPONDER VOICE-CHANNEL CAPACITY
A satellite transponder is mainly used to carry voice-channel transmissions around the globe. For a 60 Mb/s transponder capacity for this system, the number of voice channels that can be successfully accommodated is calculated for the North American digital hierarchy, as follows:

$T_1 = 1.544$ Mb/s = 24 voice channels

$T_2 = 6.312$ Mb/s = 96 voice channels

$T_3 = 44.736$ Mb/s = 672 voice channels

60 Mb/s can accommodate one T_3, two T_2, and one T_1 group, to a total bit utilization of 58.904 Mb/s, corresponding to a maximum of 888 voice channels. Therefore, the number of voice channels = 888.

PROBLEMS

1. In a communications satellite link, the modulation scheme used is QPSK with a spectral efficiency of 1.85 b/s/Hz. Compute the maximum transponder channel capacity, assuming a 36-MHz transponder bandwidth.

2. Determine the effective isotropic radiated power (EIRP) generated by the earth station transmitter with an output power of 1.2 kW and coupled to a parabolic reflector antenna with a 15-m diameter. The system carrier frequency is 6.25 GHz and the feed and branch losses are equal to 4.5 dB.

3. Determine the up-link space losses in a communications satellite system operating at 14/12 GHz.

4. Compute the power received at the input of the earth station receiver of a communications satellite operating with the following parameters.

Earth transmitter power	1 kW
Operating frequency	6/4 GHz
Earth transmitter parabolic antenna diameter	10 m
Feed and branch losses	5 dB
Antenna efficiency factor	0.55
Parabolic reflectors on board the satellite (diameter)	1 m
On-board system gain	110 dB
Earth receiver antenna diameter	10 m

9

Fiber-Optic Communications

Objectives

- Identify the various components incorporated into fiber-optic communications systems.
- Understand the fundamental laws of light.
- Describe the operating principles of various types of optical fibers.
- Explain the structural and operating differences of various optical sources.
- Define **optical detection**.
- Describe the operating principles of various optical detector devices.

Key Terms

Optical fiber transmission	Numerical aperture	LED spectral bandwidth
Planck's constant	Optical fiber attenuation	SLEDs
Photon energy	Dispersion	Coupling efficiency
Reflection	Intermodal	Butt coupling
Refraction	Intramodal	Lens coupling
Velocity of light	Optical spectral width	DFB
Refractive index	Absorption coefficient	Laser bandwidth
Snell's Law	Intrinsic absorption	Photon density
Rayleigh scattering losses	Extrinsic absorption	PIN photodetector
Mie losses	Bending losses	Absorption layer
Core	Microbending losses	Response time
Cladding	Macrobending losses	Dark current
Single-mode fibers	LED	APD
Multimode fibers	Bandgap energy	Multiplication factor
Step-index fibers	Half-power point	Photocurrent gain
Graded-index fibers		

INTRODUCTION

The concept of electronics communications was discussed in detail earlier in this book. From that discussion, it was apparent that the volume, speed, and clarity of information processed through any communications system are the key parameters determining the performance of the system. With the advent of digital communications in the middle of 1960, relevant research and development has concentrated its efforts on an increase of volume and improved quality of information processed through the system. Since electromagnetic waves are almost always used as the primary means to transmit information electronically, the ever-increasing demand for higher volumes of processed information has generated the need for a corresponding increase of the carrier frequencies and consequently for larger bandwidths. Carrier frequencies, of course, present certain limitations in higher ranges when applied to terrestrial systems. These limitations generated the incentive to search for new ways to increase the carrier frequencies and consequently the information-carrying bandwidth.

The most suitable medium, which presents practically unlimited possibilities, is that of optical fibers. A great interest in optical communications was triggered in the 1960s with the development of optical sources. These sources were capable of generating frequencies of the order of 5×10^{34} Hz with corresponding bandwidths, having the potential to increase the information capacity of the communications system by 100,000 times. Early development of this discovery proceeded slowly due to technical limitations related to optical fiber manufacturing techniques. For example, early fibers measured transmission losses of the order of 100 dB/km; such losses were considered impractical for system implementation. Continuous effort to improve optical fiber performance resulted in a reduction of these losses from 100 dB/km down to 20 dB/km. Further improvement in semiconductor technology provided the necessary means for the development of optical fiber communication systems.

This unique transmission medium exhibits tremendous advantages over the other carriers such as coaxial cables and microwave links. The most important advantage is the large available bandwidth. Optical fiber bandwidths are capable of effectively increasing the transmission carrying capacity of the system to a level of 30 Gb/s in comparison to 0.2 Gb/s for wire pair. Furthermore, a continuous improvement in the manufacturing techniques of optical fibers resulted in a fiber attenuation of 1 dB/km. This optical fiber improvement coupled with a small size and lightweight, made it ideal for long-distance transmission.

Another important advantage of optical fibers is related to their dielectric nature. This characteristic provides resistance to conductive and radiative interference. Furthermore, because no ground loops exist between optical fibers, they can be considered secure.

Examining optical fibers from the economic point of view, it is evident that they are inexpensive because their basic manufacturing material is silicon. It is estimated that by the end of the century, less than a year away, practically all new communication systems will be optical-fiber-based.

9.1 THE NATURE OF LIGHT

Light occupies the frequency spectrum between 4.3×10^{14} Hz and 7.5×10^{14} Hz, corresponding respectively to red and ultraviolet light. As part of the electromagentic spectrum, light is used as a carrier of transfer information within the confinement of an optical fiber utilizing the frequency range between 2×10^{12} Hz to 3.7×10^{12} Hz. The utilization of this infrared spectrum as a carrier through an optical fiber is called "optical fiber transmission of information," and the general area of study is referred to as "fiber-optics communications."

The advantages of an optical fiber communications system over traditional systems were presented in the introduction of this chapter. Since light is exclusively used for the transmission in an optical fiber communications system, it is important that the laws governing its nature be examined briefly.

The formulation of the theoretical interpretation of the nature of light underwent several transformations. In the 18th century, Isaac Newton proposed that light is made up of small particles. The particle nature of light provided Newton with the required tools for exploring such phenomena as the reflection and refraction of light. Contemporary to Newton was another physicist named Christian Higgern who, in 1678, was able to demonstrate that the **reflection** and **refraction** of light can also be explained assuming light to be a wave, thus establishing the wave nature of light.

Although the wave nature of light could explain the reflection and refraction concepts, it failed to provide the theoretical background for diffraction. Early in the 18th century, Thomas Young, experimenting with several light sources, demonstrated that certain results of his experiments could only be explained when light was considered a wave and not a stream of particles as proposed by Newton. Other physicists such as Fresnel, Foucault, and Maxwell followed Young's observation. Of these three, Maxwell went furthest by suggesting that light is an electromagnetic wave traveling at a velocity of 3×10^8 m/s. In 1887, Hertz was able to generate electromagnetic waves exhibiting such properties of light waves as reflection and refraction.

Although Maxwell's wave theory of light began to gain prominence over Newton's theory among scientists, it was unable to explain certain observations made by Hertz during his experimentation, one being the photoelectric effect. When an electron is removed from the orbit of the atom of a conducting element, it becomes independent of the intensity of the light impinging upon the surface of the conducting element. This is contrary to wave theory, which states that the kinetic energy of the removed electron must be proportional to that of the impinging light intensity. It is therefore clear that both the particle nature of light (Newton) and the wave theory (Maxwell) were adequate in interpreting certain experimental observations at the time.

Early in the 20th century, the German physicist Max Planck proposed the dual nature of light. He suggested that light was composed of packets of energy called "photons" (Greek: "photonion"). Each packet carries energy given by

$$E = h \cdot f \tag{9.1}$$

where $\quad h =$ **Planck's constant**$(6.63 \cdot 10^{-34}$ J-S$)$
$\quad\quad\quad E =$ **Photon energy**
$\quad\quad\quad f =$ frequency (Hz)

Equation (9.1) establishes the dual nature of light as a particle and as a wave. The dual-nature theory of light provided a satisfactory interpretation of observations that could not be interpreted by the particle or wave theories alone. Planck's theory of light contributed significantly to the development of fiber-optic technology. For example, optical sources generating photons from electric current and optical detectors generating electric current from photons are the direct result of the implementation of Planck's theory of light.

9.2 OPTICAL LAWS

When light travels through materials, it exhibits certain behavior explained by the laws of reflection, refraction, and diffraction. The study of these laws relating to the propagation of light through a dielectric medium is very important to the understanding of optical fiber technology.

Reflection

The law of reflection states that, when a light ray is impinging upon a reflective surface at an angle of incident θ_1, it will be reflected from that surface with an angle of refraction θ_2 equal to the angle of incidence (Figure 9.1).
This law played a key role in understanding the behavior of light traveling through a dielectric medium. For example, in order for a light wave to travel through a dielectric medium, a total internal refraction must occur. Total internal refraction is relevant to the physical dimensions and other properties of the optical fiber; power loss (optical) is a direct result of the light wave traveling through the optical fiber.

Refraction

A very important parameter of the dielectric material is the **refractive index** (η). This parameter is characteristic of the dielectric material as shown in Table 9.1. Based on material density, the refractive index is expressed as the ratio of the velocity of light in free space to the velocity of light of the dielectric material:

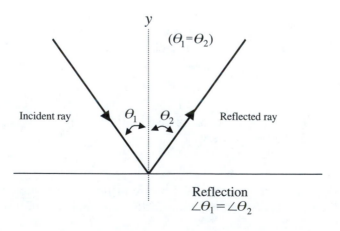

Reflection
$\angle \theta_1 = \angle \theta_2$

FIGURE 9.1
Reflection.

TABLE 9.1

Material	Refractive Index	Velocity of Light (1×10^{10} m/s)
Vacuum	1.00	3.00
Air	1.00	2.99
Water	1.33	2.25
Glass	1.46	2.05
Diamond	2.42	1.23

$$\eta = \frac{c}{V} \tag{9.2}$$

where η = refractive index
c = **velocity of light** in free space
V = velocity of light in the dielectric medium

The velocity of light in free space is given as 3×10^8 m/s. It is the maximum velocity obtained.

Snell's Law

Now let us suppose that a light ray is impinging upon the boundary of two dielectric media having refractive indexes η_1 and η_2 at an angle θ_1, where $\eta_1 > \eta_2$ (Figure 9.2). A relationship exists between the refractive index of both dielectric materials given by

$$\eta_1 \sin \theta_1 = \eta_2 \sin \theta_2 \tag{9.3}$$

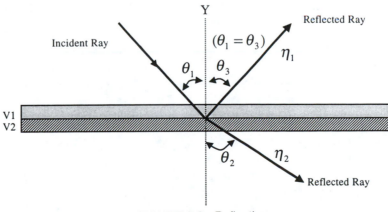

FIGURE 9.2 Reflection

or

$$\frac{\eta_1}{\eta_2} = \frac{\sin \theta_2}{\sin \theta_1} \tag{9.4}$$

Equation (9.4) is called **Snell's Law**. It shows that the ratio of the refractive index of two dielectric materials is inversely proportional to the refractive and incident angles. Substituting Eq. (9.2) into Eq. (9.4), we have

$$\eta_1 = \frac{C}{V_1} \quad \text{and} \quad \eta_2 = \frac{C}{V_2}$$

$$\frac{C/V_1}{C/V_2} = \frac{\sin \theta_2}{\sin \theta_1}$$

$$\frac{V_2}{V_1} = \frac{\sin \theta_2}{\sin \theta_1}$$

Therefore,

$$\frac{V_2}{V_1} = \frac{\sin \theta_2}{\sin \theta_1} \tag{9.5}$$

Eq. (9.5) shows that the wave velocities impinging upon the boundary of two dielectric media with two different refractive indexes are proportional to the ratio of the angle of incidence and the angle of refraction.

Let us consider now the optical ray of Figure 9.2. If the angle of incidence θ_1 is progressively increasing, there will be a progressive increase of the refractive angle, θ_2. However, there will be a point were the refractive index would be 90° at an angle θ_1. The angle of incidence θ_1 at the point at which the refractive angle θ_2 becomes 90° is called the "critical angle." A further increase beyond the critical angle will generate a total reflection of the incident ray within the denser medium with no possibility of the light escaping outside the medium.

EXAMPLE 9.1

Determine the critical angle θ of a light ray impinging upon the boundary from glass to air.

Solution:
From Table 9.1,

Refractive index of air = 1.00
Refractive index of glass = 1.50

From Eq. (9.4),

$$\frac{\eta_1}{\eta_2} = \frac{\sin\theta_2}{\sin\theta_1} \quad \text{or}$$

$$\sin\theta_1 = \frac{\sin\theta_2}{\sin\eta_1} \times \eta_2$$

$$\sin\theta_1 = \left(\frac{\eta_2}{\eta_1}\right)\sin\theta_2$$

$$\theta_1 = \sin^{-1}\left[\left(\frac{\eta_2}{\eta_1}\right)\sin\theta_2\right]$$

Since

$$\theta_2 = 90°, \ \sin 90° = 1.$$

Therefore,

$$\theta_1 = \sin^{-1}\left(\frac{1}{1.5} \times 1\right) = 41.8°$$

$$\therefore \theta_1 = 41.8°$$

Diffraction

When a light ray is intercepted by a sharp object or passes through a small hole or narrow window, it will diffract. That is, it will bend around the sharp object or the edges of the small hole or narrow window (Figure 9.3). For diffraction to occur there must be a relationship between the wavelength of the optical wave and the width of the window or diameter of the hole.

The critical parameter determining whether diffraction will take place or not is the ratio of the light wavelength to the diameter of the hole or to the width of the window.

$$\frac{\lambda}{d} \quad \text{or} \quad \frac{\lambda}{w} \tag{9.6}$$

where d = diameter of the hole
         w = width of the glass

Figure 9.3(b) shows the diameter of the hole to be comparatively large in comparison to wavelength; therefore, diffraction does not occur. Figure 9.3(c) shows diffraction to take place simply because the opening of the hole is relatively small in comparison to the wavelength.

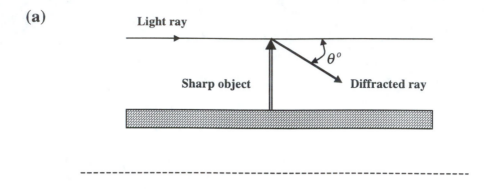

(a)

Light ray

θ^o

Sharp object

Diffracted ray

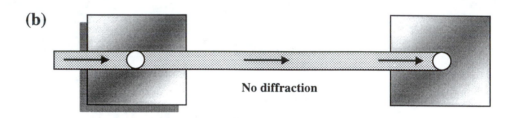

(b)

No diffraction

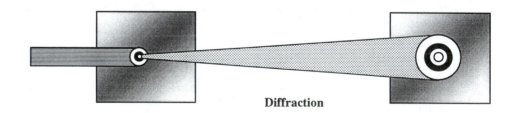

(C)

Diffraction

FIGURE 9.3 Diffraction

Light traveling through a dielectric medium suffers energy losses due to other phenomena such as scattering and absorption. Within the optical fiber, small variations in the density of the material are encountered due to fiber manufacturing imperfections. These small material density inequalities induce corresponding variations in the refractive index, resulting in small deviations of the light wave from its predetermined direction of propagation, thus causing optical energy losses. These losses, defined as "scattering losses," are divided into two categories:

1. The **Rayleigh scattering losses.** This loss occurs when the diameter of an obstacle is smaller than the wavelength of the propagating wave

$$L_{Rs} = \frac{1}{\lambda^4}$$ (9.7)

where L_{Rs} = Rayleigh scattering losses
λ = wavelength

2. Mie losses are considered when the diameter of the obstacle is larger that the wavelength. Mie-losses have been largely eliminated due to advancements in optical-fiber manufacturing techniques. A detailed examination of the propagation of a light wave in an optical fiber follows in the next section.

9.3 OPTICAL FIBERS

Industrial fibers used in electronic communications are two-layer dielectric cylindrical systems capable of conveying electromagnetic waves occupying the visible spectrum. An optical fiber consists of a dielectric cylindrical inner case surrounded by an outer dielectric layer called **cladding.** Light waves are propagated through the **core** of the optical fiber in a zigzag mode using the cladding outer layer of the fiber as the reflecting medium. For propagation of light waves to take place through the core of the fiber, the refractive index of the core's dielectric material must be higher than that of the cladding. Other plastic layers providing mechanical and environmental protection surround both the core and the cladding (Figure 9.4).

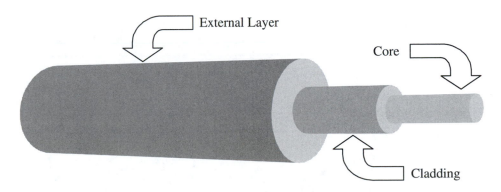

FIGURE 9.4 Three-dimensional representation of optical fiber.

Materials used in the fabrication of optical fiber are silicon dioxide (SiO_2), boric oxide–silica, and others. To achieve different refractive indexes as required by the core (larger) and the cladding (smaller), different substances are chemically diffused into the basic materials through a very complicated manufacturing process. For example, to increase the refractive index of the core, germanium oxide (GeO_2) is added to the base material during the manufacturing process, and fluorine is added in manufacturing the cladding outer layer.

Although optical fibers exhibit certain definite advantages over other transmission media, they are subject to certain operational limitations based on signal degradation suffered during the transmission process due to dispersion and attenuation phenomena.

9.3.1 Classifications of optical fibers

Based on their operational characteristics, optical fibers are classified as single-mode and multimode fibers. **Single-mode fibers** (Figure 9.5(a)), are capable of carrying only one signal of a specific wavelength. This single mode of operation has the advantage of minimizing signal distortion due to micro bending as well as minimizing overall signal attenuation and pulse expansion. **Multimode fibers** are capable of carrying signals of the same wavelength in different paths, corresponding to different arrival times at the end of the fiber (Figure 9.5(b)). Multimode fibers can operate from a very small number of modes to a large number, as high as one hundred, and can accommodate bandwidths up to 2 GHz. In

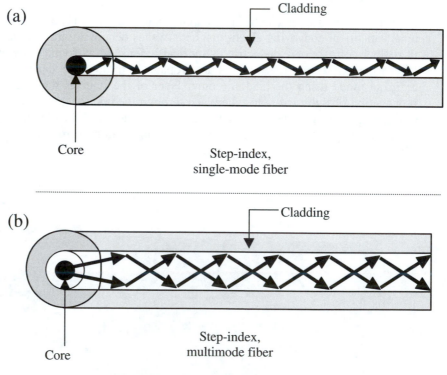

FIGURE 9.5 Classification of optical fibers.

multimode fibers, the core diameter is larger than that of a single-mode fiber. This is an advantage because it allows for better core alignment and easier splicing.

Optical fibers are also classified based on their material characteristics, such as step-index and graded-index fibers. **Step-index** fibers are the fibers in which the core and cladding materials are well defined, as is their refractive indexes. Figure 9.5(a) shows a step-index, single-mode fiber; Figure 9.5(b) shows a multimode step-index fiber. From Figure 9.5(a) and (b), notice the difference in the core diameters between the single-mode and multimode fibers. Step-index multimode fibers are used in communications systems utilizing a single fiber and higher operating bandwidths. Wavelength attenuation of step-index single-mode and multimode fibers are shown in Figure 9.6 and 9.7. Table 9.2 lists some of the physical and electrical characteristics of step-index single-mode fibers, and Table 9.3 lists some characteristics of step-index multimode fibers.

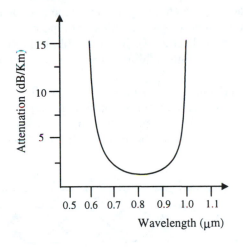

FIGURE 9.6
Wavelength versus attenuation of step-index single-mode fiber.

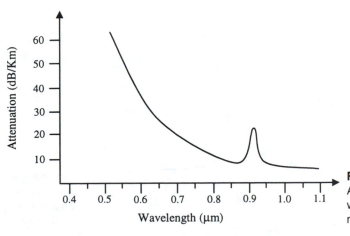

FIGURE 9.7
Attenuation versus wavelength of step-index multimode fiber.

TABLE 9.2	
Fiber core diameter	4.5 μm
Fiber outer diameter	80 μm
Numerical aperture	0.1 μm
Attenuation at 0.63 μm	20 dB/km
Attenuation at 0.85 μm	4 dB/km

TABLE 9.3	
Fiber core diameter	55 μm
Fiber outer diameter	125 μm
Numerical aperture	0.25
Attenuation at 0.8 μm	12 dB/km
Attenuation at 1.0 μm	8 dB/km
Index of refraction (core)	1.48

Graded-Index Fibers

Graded-index fibers are characterized by the gradual change of their refractive indexes in their cores and cladding materials. This type of fiber is better suited for wider bandwidth utilization because it provides a better ray-dispersion control. The graded-index optical fiber is constructed from a silicon core doped in such a way as to provide a graded refractive index and then surrounded by cladding composed of borosilicate compound. The type of core material shown in Figure 9.8 has a graded refractive index, enabling the fiber to transmit signals with larger bandwidths. Wavelength attenuation of graded-index multimode fiber is illustrated in Figure 9.9.

Based on their physical and operational characteristics, optical fibers can be utilized in communications systems as single-mode step-index, multimode step-index, and multimode graded-index fibers. A more detailed study of such modes of operation will follow the discussion of the dispersion, attenuation, and absorption of optical fibers.

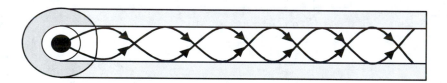

FIGURE 9.8   Graded-index, multimode fiber.

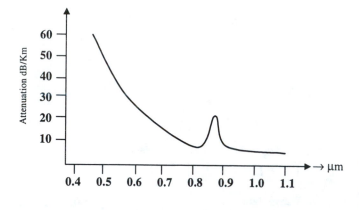

FIGURE 9.9
Wavelength versus attenuation of graded-index, multimode fiber.

Graded-Index Multimode Optical Fibers for Large-Bandwidth Applications

When the full utilization of the available bandwidth is of the outmost importance in the design of a fiber-optic communications system, a graded-index multimode optical fiber is employed. This optical fiber is composed of a specially graded inner core (graded refractive index) and a cladding outer layer composed of borosilicate material capable of handling bandwidths in the order of 900 MHz/km of optical fiber. The characteristic curve of such a fiber is shown in Figure 9.10.

9.3.2 Properties of Optical Fibers

Optical fibers have properties related to dispersion, attenuation, bandwidth, numerical aperture, and physical characteristics. All of these properties are taken into consideration when an optical fiber is to be selected for system implementation. A detailed study of these properties follows.

Dispersion

Dispersion of an optical pulse traveling through an optical fiber is the phenomenon whereby the traveling pulse appears to be spreading at the end of the fiber. This spreading of the pulse results in signal distortion, inevitably leading to system bandwidth restrictions.

Let's suppose an input pulse with a pulse width (tw_1) is traveling through an optical fiber. At the other end of the fiber the same pulse will have a pulse width tw_2, where $tw_2 > tw_1$. The gradual broadening of the input pulse while traveling through the fiber is caused by the phenomenon of dispersion. Dispersion is given by

$$\Delta t = tw_1 - tw_2 \tag{9.8}$$

where Δt = dispersion (ns)
           tw_1 = pulse width at the input of the fiber
           tw_2 = pulse width at the input of the fiber

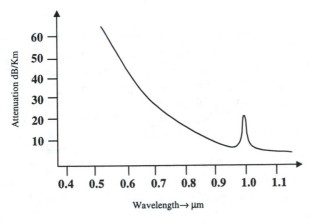

FIGURE 9.10
Attenuation of specially graded optical fiber.

The total dispersion in an optical fiber is proportional to its length and is given by

$$D = D_t \times L \qquad (9.9)$$

where $\quad D$ = total dispersion (ns)
$\quad\quad\quad\quad D_t$ = dispersion of the optical fiber given by the manufacturer (ns/K)
$\quad\quad\quad\quad L$ = length of the optical fiber (Km)

Pulse broadening will proportionally increase with the length of the fiber. Therefore, system bandwidth restrictions will determine the maximum length of the optical fiber to be considered during the design process. Optical fiber dispersion is grouped into two basic categories: intermodal, or multimode, dispersion, and intramodal, or material and waveguide, dispersion.

Intramodal or material dispersion is the broadening or spreading of the pulse traveling from the input to the output of a graded optical fiber. This broadening of the input pulse is caused by the interaction between the traveling wave and the fiber material. More specifically, three fundamental factors contribute to total material dispersion.

1. The refractive index of the fiber material (η).
2. The available optical spectral bandwidth capable of propagating through the fiber.
3. The optical spectral bandwidth generated by the input optical source.

Taking into consideration all of these factors, the total material dispersion is expressed by

$$D_m = \left(\frac{L}{C}\right)\lambda^2 \left(\frac{d_n^2}{d\lambda^2}\right)\left(\frac{\delta\lambda}{\lambda}\right) \qquad (9.10)$$

where $\quad\quad D_m$ = material dispersion
$\quad\quad\quad\quad\quad L$ = length of the optical fiber (km)
$\quad\quad\quad\quad\quad \lambda$ = optical ray wavelength
$\quad\quad\quad\quad \delta\lambda/\lambda$ = optical spectral width

A graphical expression of material dispersion as related to wavelength is shown in Figure 9.11. It is evident from Figure 9.11 that material dispersion in an optical fiber increases when the doping content of the basic material is altered to generate different refractive indexes dictated by the graded optical fiber specifications.

Another type of dispersion is caused when a light ray propagates differently through the core and cladding of the optical fiber due to their different refractive indexes.

When the two types of intramodal dispersions are plotted on a common graph (Figure 9.12), it is obvious that the material dispersion increases with the increasing wavelength while the waveguide dispersion decreases with increasing wavelength.

Figure 9.12 also shows that, at a particular wavelength, the two types of dispersion cancel each other. This cancellation point is taken into consideration when selecting the ap-

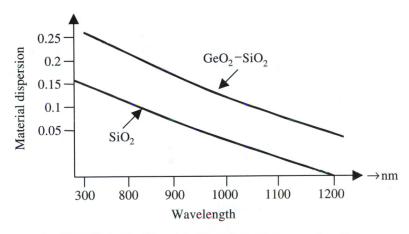

FIGURE 9.11 Material dispersion versus wavelength.

propriate operating wavelength. The cancellation point reflects the minimum optical fiber dispersion while allowing for maximum system capacity. Modification of the material properties of the optical fiber can be accomplished in such a way that zero dispersion, or cancellation of the material and waveguide dispersion, can be accomplished at a desired wavelength as required by the system design specifications.

The qualitative relationship between dispersion and optical fiber bandwidth limitations is given as follows: As we have mentioned previously, the pulse traveling though the fiber is subject to multimode and chromatic dispersions. **Intermodal,** or multimode dispersion is the result of a monochromatic pulse traveling through the fiber at different velocities; chromatic or intramodal dispersion is related to the limited optical source spectral width (single mode). Chromatic dispersion is also subdivided into waveguide and material dispersion.

It is evident from this that the available bandwidth of an optical fiber is directly related to the mode of operation and consequently to the total fiber dispersion. If an electric

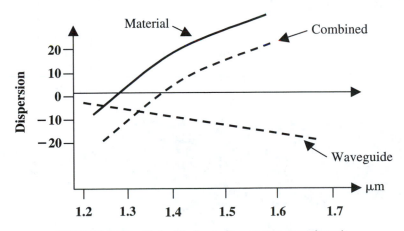

FIGURE 9.12 Material dispersion versus wavelength.

pulse with a finite width (W_{t_1}) is converted into optical domain, then coupled to an optical fiber, and at the end of the optical fiber the optical power is converted back to electric power, then the detected electric pulse will have a pulse width $W_{t_2} > W_{t_1}$. The optical fiber bandwidth can be established as follows.

$$W_{\text{total}} = \sqrt{(W_{t_2})^2 - (W_{t_1})^2} \qquad (9.11)$$

where $\quad W_{t_1}$ = pulse width at the input of the fiber
$\qquad\qquad W_{t_2}$ = pulse width at the output of the fiber

Numerical Aperture (NA)

One of the most difficult problems encountered in the early stages of optical fiber communications development was the coupling of the light ray generated by the light source into the fiber. This was due mainly to very small fiber core diameter, typically of the order of 50 to 100 μm.

For a light ray to propagate through an optical fiber, it must be coupled into the fiber at a specific angle determined by the material characteristics of the fiber (Figure 9.13). The sign of the angle θ_c between the axis of the core of the fiber and the impinging light ray is called the **numerical aperture** of that fiber and is given by

$$N_A = \sin \theta_c = \sqrt{(\eta_1)^2 - (\eta_2)^2} \qquad (9.12)$$

where $\quad \eta_1$ = refractive index of the core
$\qquad\qquad \eta_2$ = refractive index of the cladding

The numerical aperture of an optical fiber can take typical values between 0.2 to 0.6.

Attenuation

Optical fiber attenuation is the progressive amplitude reduction of the light ray traveling through the fiber. This amplitude reduction or signal attenuation is expressed in terms of

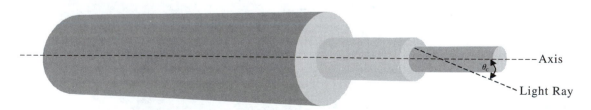

FIGURE 9.13   Coupling into the fiber through a specific angle.

power loss in dB/klm of the optical fiber:

$$L_{dB} = 10 \log \frac{P_o}{P_i}$$

(9.13)

where L = optical fiber loss (dB)
         P_o = power output (W)
         P_i = power input (W)

The attenuation of the light wave traveling through the fiber is an important parameter that must be taken into account when determining the number of repeaters required in an optical fiber communication system.

The maximum power received at the other end of the fiber is related to its length and the **absorption coefficient** of the material:

$$P_r = P_o e^{\alpha L}$$

(9.14)

where P_r = maximum power received
         P_o = incident power, or input power
         α = material absorption coefficient
         L = length of the optical fiber

Power attenuation through an optical fiber has to two main causes: scattering and absorption.

Scattering   Scattering is one of the principal causes of optical fiber signal attenuation. It is the result of the interaction of the traveling light wave through the fiber and the small variations in the material density composing the fiber. Scattering is also divided into two main categories: Rayleigh scattering and Mie scattering.

Rayleigh scattering is caused when the material anomalies within the fiber material are of the order of one tenth or smaller in diameter in comparison to the wavelength of the traveling light ray. Light scattering occurs when the light wavelength propagating through the fiber encounters density irregularities with correspondingly different refractive indexes. Rayleigh scattering causes a small part of the optical ray to escape from its predetermined path through the fiber, thus causing the forward-propagated ray to be slightly attenuated.

The signal attenuation caused by Rayleigh scattering is given by

$$L_{sc} = \frac{C_0}{\lambda^4}$$

(9.15)

where L_{sc} = losses in dB (scattering losses)
         C_0 = 0.7 dB/km $\cdot$ μm^4
         λ = wavelength

Equation (9.15) illustrates that Rayleigh scattering is inversely proportional to the fourth power of the wavelength of the propagating ray. It also indicates that the smaller the wavelength compared to the size of material irregularities, the higher the scattering loss. Because of the importance of scattering phenomenon in optical fiber system design, manufacturers have successfully managed to reduce the effect of Rayleigh scattering by substituting phosphorus pentoxide (P_2O_5) for germanium dioxide (GeO_2) as the principal dopant. Figure 9.14 shows optical fiber attenuation due to Rayleigh scattering.

Mie-scattering occurs when the material anomalies are larger than one tenth of the operating wavelength and closer to the operating wavelength. These material anomalies are the result of the improper mixing of the fabrication materials during the manufacturing process of the fiber. Today, Mie scattering has been almost completely eliminated due to sophisticated manufacturing techniques.

Absorption   The second but not less important property of optical fiber attenuation is absorption. Absorption is also classified into two basic categories: intrinsic and extrinsic. **Intrinsic absorption** is the result of the interaction of the free electrons within the fiber material and the light wavelength.

The propagating light waves cover a wide wavelength spectrum, from ultraviolet to infrared, and this spectrum interacts differently with the atoms of the fiber material. For example, the ultraviolet region (shorter wavelength) interacts with the outer-shell electrons of the atoms, whereas the infrared region (longer wavelength) interacts with the lattice structure of the atoms. Figure 9.15 shows the absorption of light energy as related to the wavelength for a single-mode fiber.

Figure 9.14 shows a peak absorption occurring in the ultraviolet region and extending very close to the infrared region. The absorption occurring in the UV region is given by

$$A_{\mathrm{UV}} = Ce^{\frac{\lambda_{\mathrm{UV}}}{\lambda}} \tag{9.15}$$

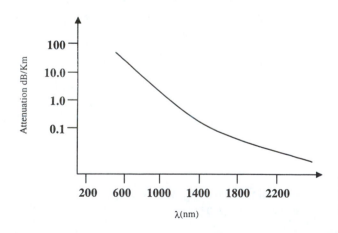

FIGURE 9.14 Optical-fiber attenuation due to Rayleigh scattering.

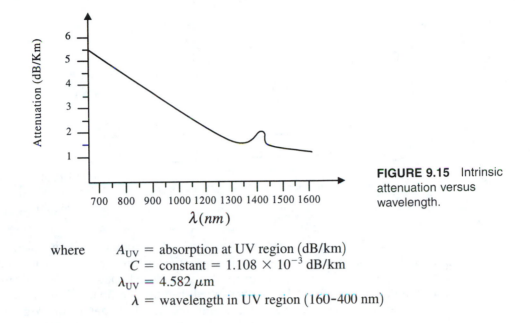

FIGURE 9.15 Intrinsic attenuation versus wavelength.

where A_{UV} = absorption at UV region (dB/km)
           C = constant = 1.108×10^{-3} dB/km
          λ_{UV} = 4.582 μm
           λ = wavelength in UV region (160–400 nm)

The absorption of the operating wavelength due to its interaction with the materials lattice structure is given by

$$A_l = A_{lo}e^{-\lambda_l/\lambda} \tag{9.16}$$

where A_l = absorption due to lattice (dB/km)
         A_{lo} = constant = 4×10^{-11} dB/km
         λ_l = constant (reference wavelength 48 μm)
         λ = operating wavelength

Extrinsic absorption is attributed to impurities unintentionally injected into the optical fiber mix during the fabrication process. The most undesirable impurity in an optical fiber mix is metal ions. The presence of metal ions in an optical fiber influences and alters the characteristic transmission properties of that fiber, resulting in a loss of light power. The metal-ion contaminant in a fiber mix is so critical that it must not exceed the level of one part per billion (ppb). Metal contaminants such as iron (Fe), chromium (Cr), and nickel (Ni) are significant contributors to light energy absorption. Figure 9.16 shows absorption levels in dB/km in relation to wavelength. Modern fabrication techniques have all but eliminated the metal-ion contamination of the optical fibers as well as other contaminants, such as OH^-.

Bending-Losses   Another element contributing to the loss of light power traveling through the fiber is that of bending. The very fact that optical fibers are physically placed in a communication network can induce losses in the system. These **bending losses** are due

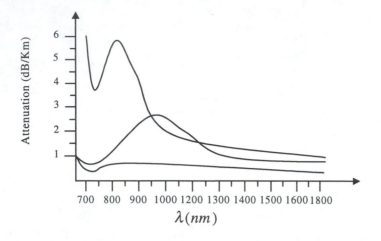

FIGURE 9.16 Intrinsic attenuation versus wavelength.

to the bending of the fibers in order to fit the physical environment within which they must operate and are classified into two basic categories: microbending and macrobending. **Microbending losses** are related to various stresses induced upon the fiber and stresses related to coating abnormalities. Even a very small bend in the fiber will alter the length of the optical path, thus contributing to the overall optical fiber attenuation. **Macrobending losses** occur when optical fibers are packaged for transportation to the field of installation and during the installation process. Figure 9.17 shows graphs of optical attenuation related to microbending and macrobending, given in dB/km versus wavelength.

Cutoff Wavelength

In order for the cutoff wavelength of an optical fiber to be determined, additional information related to the propagation of light through that fiber must be given. Since optical fibers are no more then dielectric cylindrical waveguides characterized by the refractive index and the fiber radius, knowledge of the behavior of the light propagating through such waveguides is essential. The light wave traveling through an optical guide is an electromagnetic wave; hence an electromagnetic wave is a transverse wave. It is composed of an electric and a magnetic field perpendicular to each other and both perpendicular to the direction of propagation.

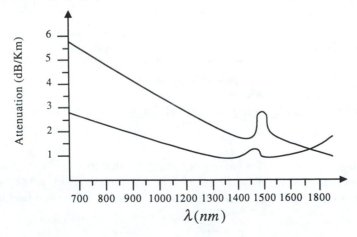

FIGURE 9.17 Bending losses.

The analysis of the electric field component of the electromagnetic wave through the implementation of the wave equations led to two different conclusions. First, through the utilization of the polar component of the electric field, the precise number of the modal component was established. This approach of establishing the modal components of an optical fiber exhibits certain disadvantages because it requires broad approximations to be incorporated into the mathematical formulas. The second method, utilizing the Cartesian component of the electric field, led to the establishment of the linear-polarized (LP) field. This method, although providing only an approximate representation of the field distribution, was fairly accurate and easy to calculate.

One of the most important parameters determining the guiding mechanism of an optical fiber is its normalized frequency (V). This normalized frequency is related to the physical and operating characteristics of the optical fiber, represented by

$$V = \frac{2\pi\alpha\text{NA}}{\lambda} \tag{9.17}$$

where
$$A = \text{optical fiber radius}$$
$$\text{NA} = \text{numerical aperture}$$
$$\lambda = \text{wavelength of free space}$$
$$\alpha = \text{material absorption coefficient}$$

From Eq. (9.17), an increase in λ will decrease the normalized frequency and an increase of the aperture or core diameter will increase the normalized frequency. There exists a value of the normalized frequency (V) below which a particular mode cannot propagate through the fiber. This value is called the "cutoff normalized frequency" (V_c) and is related to linear propagation. For a specific value of normalized frequency, a specific number of modes will propagate. As the normalized frequency decreases, the number of propagating modes also decreases.

A progressive decrease of V will reach a point at which only one mode will be able to propagate. This mode of propagation is referred as "fundamental-mode" (LP_{01}) propagation. Optical fibers operating at the fundamental mode are referred to as "single-mode fibers." Beyond this value of V, the number of modes increases. Therefore, optional fibers are operating at normalized frequencies above V_c. Solving for λ in Eq. (9.17) we have

$$\lambda_{c_{th}} = \frac{2\pi\alpha N}{V_c} \tag{9.18}$$

where
$$V_c = \text{cutoff normalized frequency}$$
$$\lambda_{c_{th}} = \text{theoretical value of } \lambda$$

Equation (9.18) shows that the theoretical cutoff wavelength is proportional to the optical fiber aperture and core radius, and inversely proportional to the cutoff normalized frequency (V_c). This theoretical cutoff wavelength $\lambda_{c_{th}}$ is also subject to various external stresses such as bending and microbending and optical fiber length. Therefore, considering $\lambda_{c_{th}}$ in practical applications is unrealistic.

Let's suppose that a value of λ is being selected slightly higher than $\lambda_{c_{th}}$ for a practical application. In such a case, the external stresses in addition to the total length of the wire will lower the selected value of λ to the point where $\lambda \leq \lambda_{c_{th}}$. Therefore, a practical wavelength λ_c is considered in real applications. The value of λ_c is determined from the graph in Figure 9.18. Figure 9.18 shows that the value of the practical wavelength is subject to the total fiber light and external stresses. An increase in length or increase in external stresses decreases the value of the practical wavelength.

9.4 OPTICAL SOURCES

9.4.1 Introduction

The primary optical source of a fiber optics transmission system is the semiconductor light-emitting diode (**LED**). The selection of this device as the primary optical source was based on its ability to provide optical power ranging from 0.05 mW to 2 mW over optical fibers several kilometers in length. Semiconductor optical sources are very reliable devices with

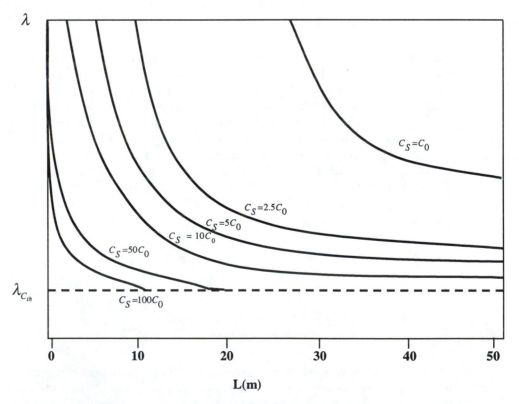

FIGURE 9.18
Behavior of λ_c as function of the fiber length *(L)* for different values of stress parameter C$_S$.

a long operating life. Another advantage of the devices is that, for ordinary system applications, they require no modulating circuitry because the optical power of the device can be altered in accordance with the input current variations. In the early 1970s the basic device structure was composed of gallium arsenide (GaAs) and aluminum–gallium arsenide (AlGaAs), materials capable of generating wavelengths of the order of 800 nm to 900 nm.

These LED devices were limited in their application to short-transmission optical links because of the optical fibers' large signal attenuation at high data rates. Continuous improvement in optical fiber manufacturing technology, especially in the 1500-nm to 1700-nm wavelengths, significantly improved the overall data transmission capability of the LED devices. The continuous drive to satisfy an ever-increasing system performance demand generated the need for the development of improved LED devices such as those composed of indium–gallium arsenide–phosphorus (InGaAsP). These devices were capable of generating light power covering the 920-nm to 1650-nm spectrum.

Today, there are two basic types of semiconductor optical sources: the LEDs we have just briefly described and laser diodes. Laser diodes are used for long-distance high-data-rate transmissions; LED devices are used for shorter-distance, lower-data-rate transmission. The progressive evolution of LED technology led to the development of two types of LED devices: edge-emitting LEDs (ELEDs), and surface-emitting LEDs (**SLEDs**).

Edge-emitting LEDs are used for both single-mode and multimode operations, with bit rates in excess of 400 Mb/s. Surface-emitting LEDs are better suited to multimode medium-range optical fiber transmission. The fundamental theory for the design and construction of LED and laser optical sources is based on electron–hole recombination within a semiconductor material. This electron–hole recombination results in the generation of photons with frequencies determined by the physical parameters of the semiconductor material. The two phenomena derived from the interaction between matter and light are emission and absorption. Emission can also be divided into two basic categories: stimulation emission and spontaneous emission. Both stimulation emission and spontaneous emission can be better explained with the assistance of the following oversimplified atomic models (Figure 9.19).

An electron of energy E_1 can be elevated to level E_2 by absorbing a photon of energy $E = hf$; or an electron already at energy level E_2 can decay to energy level E_1 by releasing a photon (spontaneous emission). The third phenomenon is observed when an electron already at energy level E_2 absorbs a photon and decays to energy level E_1. Through this process, two photons having the exact same phase direction and energy levels are simultaneously released. If continuous radiation is maintained, the generation of photons will also be continued (stimulation emission). Light-emitting diode operation is based on sponta-

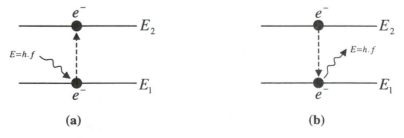

FIGURE 9.19 (a) Spontaneous (b) Stimulated

neous emission, and LASER operation is based on stimulated emission. In general, for both devices, the principle of operation is based on the interaction of light and matter within a semiconductor material.

9.4.2 Light-Emitting Diode (LED)

Perhaps the best representative of the optical source device family is the light-emitting diode (LED). LEDs are classified as electroluminescent devices composed of semiconductor materials capable of generating light when forward biased by a current source. The basic semiconductor materials composing LED structures are GaAsP, GaAlAs, GaAs, and GaP. Gallium arsenide–phosphate (GaAsP) LEDs generate light between 640 nm and 700 nm, with peak optical power at 660 nm. Gallium–aluminum arsenide (GaAlAs) LEDs produce light between 650 nm and 700 nm, with peak optical power at 670 nm. Gallium phosphate (GaP) LEDs generate light between 520 nm and 570 nm, with a peak optical power at 550 nm. Indium–gallium arsenide–phosphorus devices are a recent development presenting advantages over the GaAsP semiconductor devices. These devices can generate optical wavelengths between 930 nm and 1650 nm. The selection of semiconductor materials described here and their proportional contribution to optical device fabrication is indicative of the wavelengths required. The optical energy obtained from these semiconductor material combinations is measured in electron volts (eV) as follows:

Ge = 0.7 eV
Si = 1.1 eV
CdS = 2.4 eV
GaP = 2.2 eV
GaAs = 1.4 eV
GaAlP = 0.8–2.0 eV

The center wavelength (λ_c) of an LED device is determined by the **bandgap** energy (E_g) in eV at the active layer given by

$$\lambda = \frac{h_c}{E_g} \qquad \textbf{(9.19)}$$

where
h = Planck's constant $(6.63 \times 10^{-34}$ J-S$)$
c = velocity of light in vacuum $(3 \times 10^8$ m/s$)$
E_g = bandgap in eV

Edge-Emitting LEDs

Edge-emitting LEDs were first introduced in the mid-1970s. The basic structure of this device closely resembles the laser diode, with one fundamental difference. For laser diodes, positive feedback is promoted in order to enhance stimulated emission, whereas with edge-

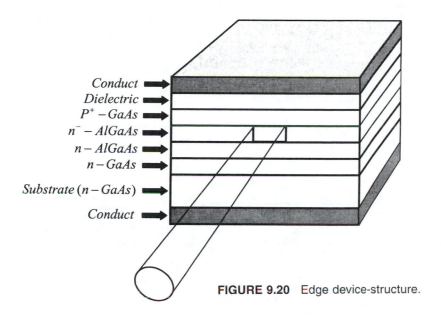

FIGURE 9.20 Edge device-structure.

emitting LEDs, the feedback mechanism is suppressed for the exact opposite reason; that is, to prevent the device from going to a saturated emission mode of operation. The active layer of n-AlGaAs with a thickness of 0.05 μm is confined by two layers of p and n semiconductor materials such as p-AlGaAs and n-AlGaAs with a thickness of 0.115 μm. External to these two optical guiding layers are another two layers of p$^+$GaAs and n$^+$GaAs with a corresponding thickness of 3.5 μm. The optical-guide layers confine the generated light into the active layer. This represents an advantage when coupling the source to the optical fiber. The basic device structure is shown in Figure 9.20.

An external dc source connected across the device will provide the necessary forward biasing. The ejected electrons from the n-AlGaAs and holes from the p-AlGaAs layers will recombine at the thin n-AlGaAs layer. During this recombination process, a certain number of photons will escape through the edge of the active layer. By reducing the active layer, self-absorption is reduced to a minimum, thus eliminating the possibility of stimulation emission.

A typical voltage/current characteristic curve of an edge-emitting LED is shown in Figure 9.21. Figure 9.21 shows the current increasing exponentially beyond the threshold biasing voltage that is characteristic of the device. The current (I) generated by the forward-biasing voltage of the ELED device is used to determine the optical power generated by that device (Figure 9.22).

LED Characteristics

One of the most important characteristics of an LED source is that of power efficiency. This power efficiency is subdivided into two categories: internal power efficiency, also called "quantum efficiency" (η_g), and external power efficiency (η).

Internal power efficiency is defined as the ratio of photons generated to the number of electrons induced into the active layer of the device:

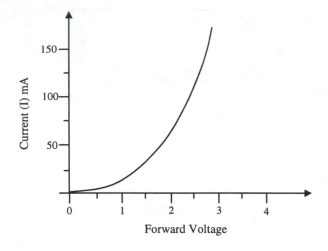

FIGURE 9.21
Voltage/Current charac-
teristic curve.

$$\%\eta_g = \frac{N_{\text{ph}}}{N_{e^-}} \times 100 \tag{9.20}$$

where η_g = quantum efficiency (%)
N_{ph} = number of photons
N_{e^-} = number of electrons

External power efficiency (η_c) is defined as the ratio of the optical power coupled into the fiber to the electrical power applied by the optical device

$$\%\eta_c = \frac{P_f}{P_{\text{in}}} \times 100 \tag{9.21}$$

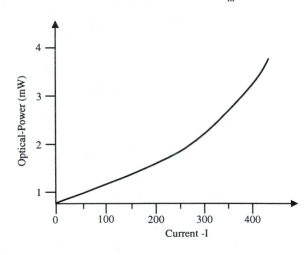

FIGURE 9.22
Voltage/Current charact-
eristic curve.

where η_c = external power efficiency
                P_f = optical fiber power
                P_{in} = input power

Optical fiber power (P_f) is only a fraction of the power generated internally by the optical device. This optical power loss is relevant to the device–optical fiber coupling efficiency, expressed by

$$\eta_c = (\text{NA})^2 \qquad\qquad (9.22)$$

where   NA = numerical aperture of the optical fiber

EXAMPLE 9.2

A light source generating an optical power output equal to 1 μW is coupled into an optical fiber with a cross-sectional area larger than that of the active area of the light source. Determine the power coupled into the fiber with a fiber θ equal to 15°.

Solution

$$\theta = 15°$$
$$\sin \theta = \text{NA}$$
$$\sin (15°) \cong 0.26 = \text{NA}$$

Coupling Efficiency $\eta_c = (\text{NA})^2 = (0.26)^2 = 0.0676$

Power Coupled into the Fiber $\eta_c = \dfrac{P_f}{P_{in}}$

$$P_f = \eta_c \times P_{in} = 0.067 \times 1 \ \mu\text{W} = 667.6 \ \mu\text{W}$$

It is evident from these examples that only a small fraction of the optical power generated in the active region of the optic device will be coupled into the optical fiber.

EXAMPLE 9.3

Calculate the optical power coupled into the fiber generated by an optic device with a bias current of 20 mA and forward voltage of 1.5 V. The device's internal efficiency is 2% and the fiber $\theta = 20°$.

Solution
Power Input to the Optical Device $P_{in} = I_f \times V_f = 1.5 \ \text{V} \times 20 \ \text{mA} = 30 \ \text{mW}$

$$\therefore P_{in} = 30 \ \text{mW}$$

The power output of the optical device is given by

$$\eta_{\text{in}} = \frac{P_o}{P_{\text{in}}}$$

$$0.02 = \frac{P_o}{30 \text{ mW}}$$

Therefore, $P_o = 6$ nW of the optical source

   This power output (P_o) of the optical source becomes the power input to the optical fiber, $P_{\text{in}} = 6$ nW.

$$\%\eta_c = \frac{P_f}{P_{\text{in}}}$$

Since $\eta_c = \text{NA}^2$ and $\text{NA} = \sin \theta,$

$$\eta_c = (\sin 20°)^2 = 0.116$$
$$P_F = \eta_c \times P_{\text{in}} = 0.1169 \times 6 \text{ nW}$$

Therefore, the optical power coupled into the optical fiber is equal to 0.7 nW.

A more detailed relationship establishing the maximum power coupled into the optical fiber is given by

$$P_{\text{max}} = I_o A \pi \, (\text{NA})^2$$

$$I_o = \left(\frac{P_o}{\text{Active area}} \right)$$

where P_o = power output of the optical device to its active area
      P_{max} = maximum power coupled into the fiber
       A = selection of the smaller of the two areas, that of the active
          region of the optical devices or that of the cross-sectional area
          of the optical fiber
     NA = numerical aperture of the fiber

LED Spectral Bandwidth at the Half-Power Point ($\Delta\lambda$)

The **LED spectral bandwidth** is determined at the **half-power point** (50%) of the spectral density in reference to wavelength (Figure 9.23). Light-emitting diode optical devices exhibit a spectral bandwidth of 20 nm to 200 nm at the half-power point. This spectral bandwidth is translated to the pulse, broadening as it travels through the optical fiber. Light-

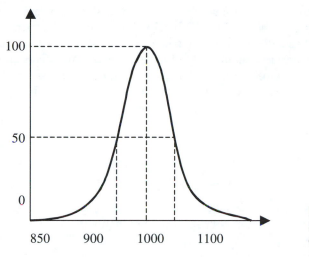

FIGURE 9.23 LED Spectral Bandwidth Curve λ (nm).

emitting diodes emitting at a peak wavelength of 800 nm exhibit a pulse broadening of 5 ns/km. This disadvantage of the LED device can be controlled by shifting the peak wavelength from 800 nm to 1300 nm. At this wavelength, a smaller dispersion is encountered.

LED Bandwidth

Light-emitting diodes are intensity-modulated devices. That is, the input current can directly affect the output intensity of the device. In digital transmission, the LED device is turned ON and OFF in accordance with the input binary data. Ideally, turning the device ON and OFF must occur simultaneously with the input binary data. In reality there is a time delay between the bias current changes and the turning ON and OFF of the LED. This delay is caused by the rise time (t_r) and fall time (t_f) of the LED device (Figure 9.24).

Rise time (t_r) is measured at between 10% and 90% of the power output, that is, the time it takes the output power to increase from 10% to 90%. Fall time (t_f) is the time it

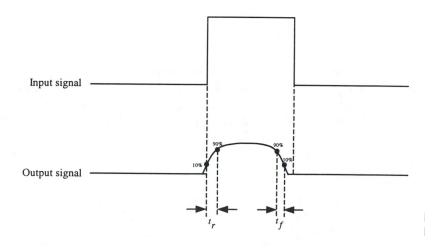

FIGURE 9.24 Time Delay.

takes the output to decrease from 90% to 10%. Rise time and fall time are the two most significant factors contributing to LED bandwidth limitations, which in turn establish the maximum data bit rate the device is capable of handling. This total time delay is the result of such factors as carrier recombination, time and space change, capacitance inherent to the LED device, physical dimensions, and semiconductor properties.

The relationship establishing total optical bandwidth at the half-power point is given by

$$BW = \frac{1}{2\pi r} \tag{9.23}$$

where   BW = bandwidth
         r = lifetime of carrier recombination

A more practical formula establishing optical bandwidth is given by

$$BW = \frac{0.35}{t_r} \tag{9.24}$$

where t_r = rise time of the device

EXAMPLE 9.4

Determine the bandwidth of an LED device with a rise time equal to 10 ns.

Solution

$$BW = \frac{0.35}{t_c} = \frac{0.35}{10 \times 10^{-9}\text{s}} = 35 \text{ MHz}$$

Therefore, 35 MHz is the maximum operating bandwidth of this LED device.

The modulation bandwidth of an LED device can be increased by increasing the carrier concentration in the active region, with a simultaneous decrease of the carrier lifetime. This of course has a negative effect on the LED overall optical output power. Therefore, a compromise between modulation bandwidth and power output must be reached during the design process.

Surface-Emitting LEDs (SLEDs)

The design of a surface-emitting LED was based on a massive electron injection into a thin optically transparent layer of p-material. This thin layer, confined between two other layers with larger bandgaps, secures the confinement of the injected carriers, thus promoting a higher degree of recombination and ultimately generation of a larger number of photons (Figure 9.25).

In contrast to edge-emitting LEDs, the optical radiation of the surface LED takes place from the surface of the active layer. An examination of the SLED cross section reveals the following: The ScO_2 layer acts as an insulator between the GaAs p-layer and the metallic conductor. The other two AlGaAs and GaAs p-materials are performing dual functions. One is light confinement and the other is the minimization of the recombination process close to the p-GaAs and n-AlGaAs junction.

The active region substrate of the SLED device is etched away in a well-type manner. This reduction of the active region dramatically reduces the recombination process, and the well-shaped area enhances the focusing of the emitted light into the optical fiber. Another structure of a SLED semiconductor device is shown in Figure 9.26. This SLED optical source is designed to generate optical power in the 1.3-μm range. It uses an InP substrate because InP is transparent to this wavelength. The other four layers are grown epitaxially on this substrate with varying doping levels and thicknesses in order to facilitate the device's design objectives, such as maximization of optical power and modulation speed.

The first layer grown on the wafer substrate is a buffer composed of n-InP with an average thickness of 3 μm and a doping density of $2 \times 10^{18}/cm^3$. The second is the much thinner active layer of p-InPGaAsP with an average thickness of 1 μm doped with Zn to an average density of $2 \times 10^{18}/cm^3$. The third is the cladding layer, which is somewhat

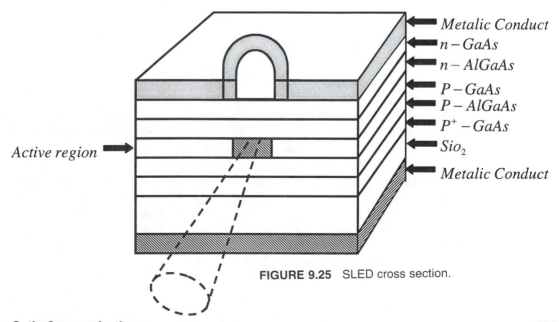

FIGURE 9.25 SLED cross section.

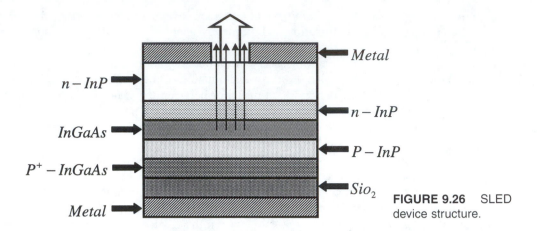

FIGURE 9.26 SLED device structure.

thicker than the first layer, having an average thickness of 2.5 μm and similarly doped with Zn to an average density of $2.75 \times 10^{18}/cm^3$. The fourth and final layer is composed of a very thin (0.25 nm) layer of p-ZnGaAsP, which is heavily p-doped and is intended to minimize conductor resistivity.

Coupling of an SLED into Optical Fiber   One of the major problems facing optical fiber communications systems is that of the coupling of optical power generated by the optical source into the optical fiber. The ratio of the optical power coupled into the fiber to that of the power generated by the optical source is called the coupling efficiency and is given by

$$\eta_c = \frac{P_f}{P_s} \qquad\qquad (9.25)$$

where η_c = **coupling efficiency**
P_f = optical power coupled into the fiber
P_s = optical power generated by the source

This inability to transfer all the generated optical power of the source into the fiber is caused by the different physical characteristics of the source and the fiber. Either the optical area of the source is larger than the cross-sectional area of the core of the fiber or the numerical aperture (NA) of the light beam is larger than the NA of the optical fiber. In the first case, when the optical area of the source is larger than the cross-sectional area of the optical fiber, only a fraction of the generated optical power of the source can be coupled into the fiber. Similarly, when the NA of the beam is larger than the NA of the core of the fiber, only some of the optical power will be coupled into the fiber. To maximize the coupling efficiency between the optical source and the optical fiber, two methods have been implemented: the Butt method and the lens method.

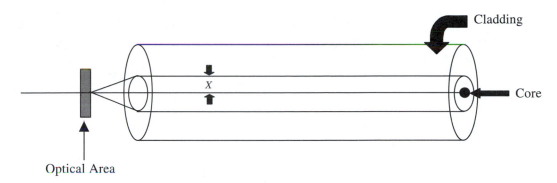

Cladding

Core

Optical Area

FIGURE 9.27 Butt-coupling method.

The Butt Method   The simplest method of coupling optical power generated by the SLED device into the fiber is the Butt direct coupling method. The fundamental requirement for an efficient **Butt coupling** is that the core cross-sectional area of the fiber must be at least equal to the optical emission area of the optical source. If the source area is larger than the core area, only a fraction of the generated optical power will be coupled into the cladding of the fiber, resulting in a quick attenuation of that optical power. This attenuation of the power coupled into the cladding section of the optical fiber can be considered as optical power loss. Figure 9.27 illustrates the Butt-coupling method.

The coupling efficiency obtained by the Butt method for a uniformly exited guiding source (lambertian) such as that of the SLED devices varies with step-index and graded-index fibers. For $y > x$, the coupling efficiency is given by

$$\eta_c = T\left(\frac{\text{NA}}{\eta_0}\right)^2 \cdot \left(\frac{x}{y}\right)^2 \cdot \left(\frac{\alpha}{\alpha + 2}\right) \tag{9.26}$$

where T = transmissivity of the medium
NA = numerical aperture of the optical fiber
η_0 = refractive index of the medium between the source of the fiber
x = radius of the fiber core
y = half-length of the optical source
α = fiber core refractive-index parameter (for step-index, $\alpha = \infty$; for graded-index, $\alpha = 2$)

For $y < x$, the coupling efficiency is given by

$$\eta_c = \frac{T(\text{NA}/\eta_0)^2[\alpha + 1 - (y/x)^2]}{\alpha + 2} \tag{9.27}$$

EXAMPLE 9.5

Determine the coupling efficiency of an optical source coupled into the fiber, given the following data:

$$T = 1 \text{ (air)}$$
$$\eta_0 = 1$$
$$NA = 0.3$$
$$x = y$$

Solution

For Step-Index Fiber

$$\alpha = \infty$$

$$\eta_c = T\left(\frac{NA}{\eta_0}\right)^2 \cdot \left(\frac{x}{y}\right)^2 \cdot \left(\frac{\alpha}{\alpha + 2}\right)$$

$$= 1\left(\frac{0.3}{1}\right)^2 \cdot \left(\frac{1}{1}\right)^2 \cdot (1) = 0.09$$

Therefore, $\eta_c = 9\%$.

For Graded-Index Fiber

$$\alpha = 2$$

$$\eta_c = \frac{T(NA/\eta_0)^2[\alpha + 1 - (y/x)^2]}{\alpha + 2}$$

$$= \frac{1(0.3/1)^2[2 + 1 - (1)^2]}{2 + 2}$$

$$= \frac{0.09(2 + 0)}{4} = \frac{0.18}{4} = 4.5\%$$

Therefore, $\eta_c = 4.5\%$.

EXAMPLE 9.6

Determine the coupling efficiency η_c when an optical source is coupled into a fiber given the following data:

$$T = 1 \text{ (air)}$$
$$\eta_0 = 1$$
$$NA = 0.3$$
$$y = 0.75x$$
$$x > y$$

Solution

For Step-Index Fiber

$$\eta_c = T\left(\frac{NA}{\eta_0}\right)^2 \cdot \left(\frac{x}{y}\right)^2 \cdot \left(\frac{\alpha}{\alpha + 2}\right) = 1(0.3/1)^2 \cdot (1.3)^2 \cdot (1) = 0.15$$

Therefore, $\eta_c = 15\%$.

For Graded-Index Fiber

$$\alpha = 2$$

$$\eta_c = \frac{T(NA/\eta_0)^2[\alpha + 1 - (y/x)^2]}{\alpha + 2} = \frac{1(0.3/1)^2[2 + 1 - (1.33)^2]}{2 + 2}$$

$$= \frac{0.09(2 + 1 - 1.77)}{4} = \frac{0.09(2 + -0.77)}{4} = \frac{0.09 \times 1.23}{4} = 2.75\%$$

Therefore, $\eta_c = 2.76\%$.

It is evident from these examples that, for both step-index and graded-index fibers the Butt-coupling efficiency is very small. It is also evident that the coupling efficiency is higher when the optical area of the source is smaller than the cross-sectional area of the optical fiber.

The Lens Method   It is evident from the foregoing examples that only a small fraction of the generated optical power from an optical source can be coupled into the narrow fiber angle of acceptance. To improve the coupling efficiency of optical power into the fiber, a lens is inserted between the radiating area of the source and the fiber core cross-sectional area. The objective of such an insertion is to equalize the optical area of the source to that of the fiber core cross-sectional area. This lens scheme allows for a maximum coupling efficiency when the ratio of the radius of the power optical sources to the radius of the fiber core (x/y) becomes equal to the magnification factor of the inserted lens (Figure 9.28).

Figure 9.28 shows that when the optical source radiating area is smaller than the core cross-sectional area of the fiber, the solid optical angle of the source is larger than the solid angle of acceptance of the core. The physical properties of the lens and its precise location between the source and the core achieve an equalization of these two solid angles, $\theta_1 = \theta_2$, resulting in an optimum coupling efficiency.

There are two methods of **lens coupling.** One requires that the radiating surface of the optical source be placed behind the focal point of the lens (Figure 9.29); the second requires that the source be placed in front of the focal point (Figure 9.30). Figure 9.29 illustrates the image of the optical source concentrated at the fiber core surface area, resulting in an improved coupling efficiency. In Figure 9.30, the optical source is placed between the lens and the focal point. In this arrangement, the optical solid angle can be made equal or smaller to the numerical aperture of the fiber core, thus enhancing coupling efficiency.

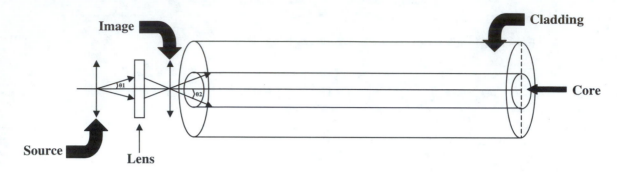

FIGURE 9.28   The lens coupling method.

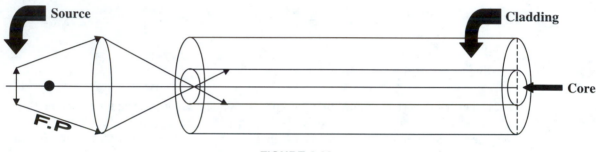

FIGURE 9.29

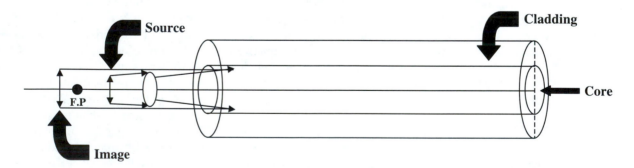

FIGURE 9.30

There are several methods of incorporating a lens into the source-to-fiber assembly. The three most commonly used methods are shown in Figures 9.31(a) through (c). In Figure 9.31(a), the lens is embedded at the input of the fiber core. In Figure 9.31(b), the lens is an integral part of the optical area of the source, and in Figure 9.31(c), the lens is between the optical source and the fiber core. These methods of lens coupling result in a significant improvement in efficiency over Butt coupling.

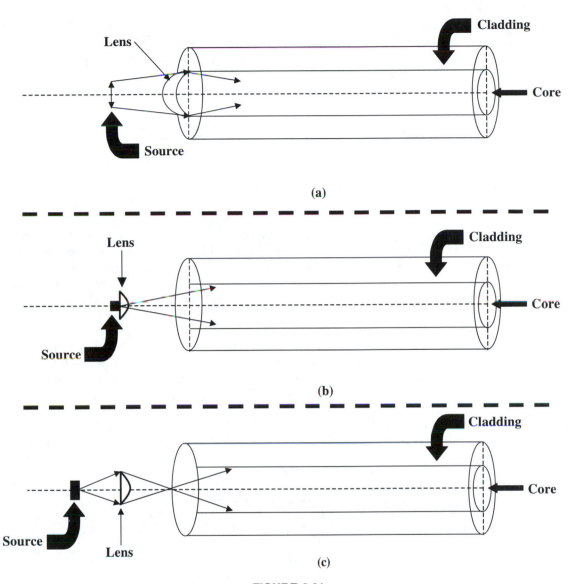

(a)

(b)

(c)

FIGURE 9.31

9.4.3 Laser Diodes

Laser (*l*ight *a*mplification by *s*timulated *e*mission of *r*adiation) devices were first introduced in 1961, and their operation was based on stimulated emission instead of spontaneous emission of radiation. Stimulated emission of radiation is the process whereby photons are used to generate other photons having the exact phase and wavelength as the parent photons (Figure 9.32).

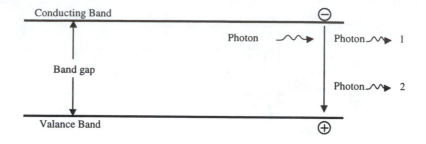

FIGURE 9.32 Stimulated emission of radiation model.

Early in their development stage, laser devices exhibited a substantial increase in optical power output confined to a very narrow beamwidth in comparison with standard LED devices. Some of these early structures were based on a heavily pumped (2×10^{18} e$^-$/cm^3), thin (0.2-μm) GaAs alloy forming the laser cavity and sandwiched between four epitaxially grown layers of AlGaAs, providing carrier and optical power confinement. (Figure 9.33).

For stimulation emission to occur, the number of electrons in the conducting band must exceed the number of electrons in the valence band. This increase can be accomplished through the process of "pumping." Pumping, or the elevation of an electron from the valence band to the conducting band, is achieved by passing a sufficient amount of current through the active region of the laser device. Stimulated emission of radiation will start to take place at a minimum current level, called the "threshold current," induced to the active region of the laser structure. Below this threshold current, stimulated emission does not take place, because radiation and absorption losses occurring inside the active region

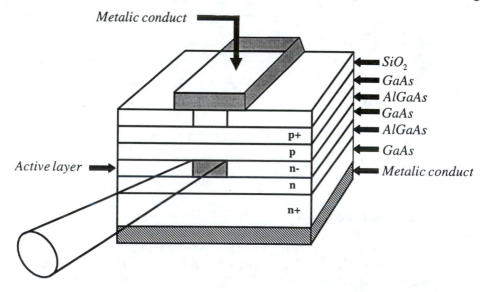

FIGURE 9.33 Early structure of Laser devices.

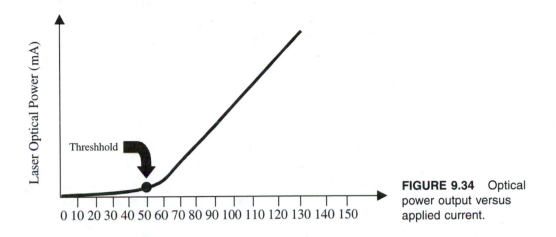

FIGURE 9.34 Optical power output versus applied current.

offset the additional photon generation of this below-threshold current. The relationship between threshold current and optical power output of a laser device is illustrated in Figure 9.34.

Figure 9.34 shows that at approximately 50 mA of current injected into the active region of the laser device, stimulated emission of radiation begins. Below 50 mA, the device operates by spontaneous emission of radiation. The early laser structures suffered from a lack of efficiency. That is, the electron-to-photon conversion ratio was low. In addition, they required very high currents, on the order of 50 kA/cm^2 of active area, for stimulation emission to take place. These deficiencies of early laser diodes were addressed with the introduction of the double heterostructure laser device. A very thin layer of low-bandgap, high-refractive-index AlGaAs, the active region, is confined between n-AlGaAs and p-AlGaAs layers with higher bandgaps and lower refractive indices (Figure 9.35).

In 1975, a threshold current of 0.5 kA/cm^2 was achieved in a laser device with an active layer 0.1 μm thick. This device was capable of generating optical power in the range

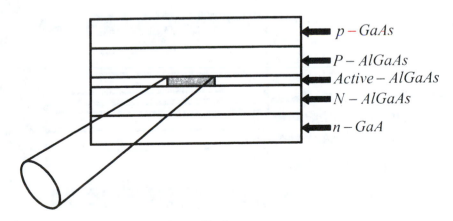

FIGURE 9.35 Double Heterostructure Laser Diode.

of 800 nm to 900 nm. The optical wavelength generated by the laser device is related to the alloy composing the structure and is expressed by

$$\lambda = \frac{h \cdot c}{E_g}$$

(9.28)

where
λ = wavelength in (nm)
h = Planck's constant = 6.63×10^{-34} J-s
c = velocity of light = 3×10^8 m/s
E_g = bandgap energy of the active region (eV) (for InGaAsP alloy,
E_g = 0.74 to 1.13 eV; for AlGaAs, E_g = 1.42 to 1.61 eV

For operating wavelengths in the spectral region between 1300 nm and 1500 nm, substances such as InGaAsP were introduced for the fabrication of laser semiconductor structures requiring a threshold current on the order of 0.7 kA/cm² to 1.8 kA/cm². The schematic diagram of such a device is shown in Figure 9.36.

If a forward-biasing voltage of 1.5 to 2.0 V is applied across the laser diode, a carrier concentration will gradually be built up into the active region. This carrier concentration is denser at the center of the active region than in the lateral regions. This high carrier density at the center of the active region far exceeds the transparency concentration, resulting in an overall optical gain. The lesser lateral carrier concentration is far below the transparency level; thus, an overall optical power loss is encountered in the lateral regions of the active layer. The double-heterostructure laser is a gain-guided multimode device. The spectral density of such a device is shown in Figure 9.37.

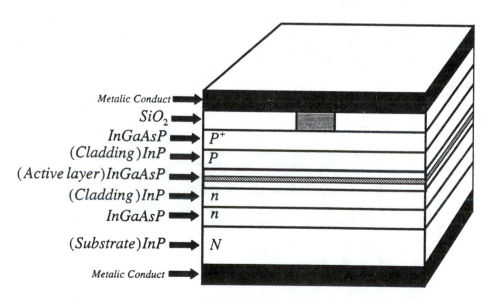

FIGURE 9.36 In.Ga.As.P laser Device

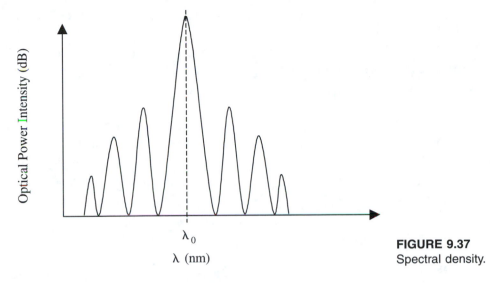

FIGURE 9.37
Spectral density.

The multimode behavior of a gain-guided laser device can be remedied by the reduction of the active region's width. This width reduction permits the formation of a substantial number of discrete dielectric levels, which allow for the development of a fundamental mode instead of a multimode spectral output and result in a guided-index laser. Both multimode and single-mode devices have been utilized in optical fiber communications systems, taking advantage of their fundamental properties. That is, for multimode transmission systems, the multimode laser device is used as the optical power source, and for single-mode transmission systems, the single-mode optical source is utilized. The implementation of the single-mode optical source in such a transmission system complies with fundamental system design requirements such as low optical modal noise. Laser diodes exhibit a single spectral line with side lines reduced by at least 25 dB (Figure 9.38).

In optical fiber communications systems, direct optical modulation of the optical source is an established practice. At very high rates of modulation, the behavior of a single-mode optical source begins to change, resulting in an output spectrum showing more than one spectral line. To maintain a single-mode spectral output, a modified version of the

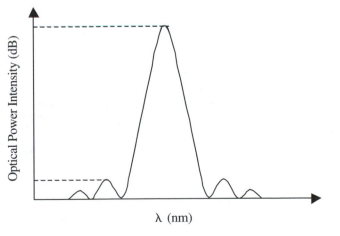

FIGURE 9.38 Laser diode spectral line.

single-step-index laser device is fabricated; that is, a distributed-feedback laser device (DFB).

Distributed-Feedback (DFB) Laser Structures

In a **DFB** device structure (Figure 9.39) the main design objective is to generate a single-line spectrum at the output under high data rates of modulation. This objective is achieved by incorporating a corrugated layer below the active layer of the DFB device. These hills and valleys generate a constant change of the refractive index, which contributes to the device's feedback mechanism so that a single mode is produced and undesirable modes are suppressed. For the structure to operate as a DFB device, the grating period must satisfy

$$GP = \frac{\lambda_{mode}}{\eta} \qquad (9.29)$$

where: GP = grating period
λ_{mode} = operating wavelength
η = refractive index of the effective mode

Typical operating characteristics of a DFB laser diode are shown in Table 9.4.

TABLE 9.4

Operating wavelength	1300 nm
Output power (maximum)	5 × 10 mW
Threshold current	40 mA
Temperature coefficient (for threshold current)	1.3 mA/K
Modulation bandwidth	800 MHz
Temperature coefficient	0.07 nm/K
Spectral bandwidth	20 MHz

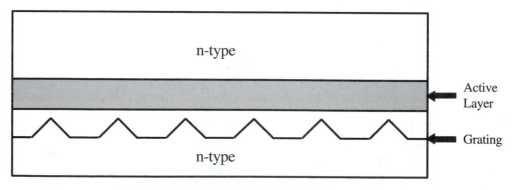

FIGURE 9.39 Distributed-Feedback Laser (DFB)

Distributed Bragg Reflectors (DBRs)

Another single-mode feedback laser is the distributed Bragg feedback device (Figure 9.40). The fundamental structural difference between the DFB and DBR laser devices lies in their grating mechanisms. In the DFB device, the grating is at the bottom of the active layer; in the DBR device, the grating is at both ends of the active region. Here, it can act as a perfect optical mirror because of the difference between the constant refractive index of the active layer and the continuously changing refractive index of the grating layer. This structural arrangement provides the required feedback mechanism for optical power generation and spectral purity. DBR devices require a higher threshold current than DFB structures. They also exhibit a higher degree of susceptibility to temperature variations, and their line width is more sensitive to the rate of modulation of DFB laser devices.

Laser Diode Characteristics

There is a spectral difference between gain-guided and index-guided laser devices. As we mentioned previously, gain-guided lasers are multimode devices and index-guided lasers are single-mode devices. This is attributed to the fact that the spontaneous emission component within the index-guided structures is smaller than that of the gain-guided structures. It is also observed that under modulated conditions, the single-mode index-guided laser can become multimode, with additional broadening of the modal line width. This results from a small change of the modal frequency attributed to the carrier density variations due to pulse modulation. Maintaining a narrow modal line width is crucial when laser devices are used as optical sources for long-distance communications systems designed to process high data rates while operating under the most stringent system noise restrictions.

Temperature versus Optical Power   Experimental results have shown that lasers are temperature-dependent devices. The relationship of the threshold current and optical power output is subject to operating temperature conditions. This relationship between threshold current and temperature is given by

$$I_{th} = I_0 \cdot e^{T/T_0} \qquad\qquad (9.30)$$

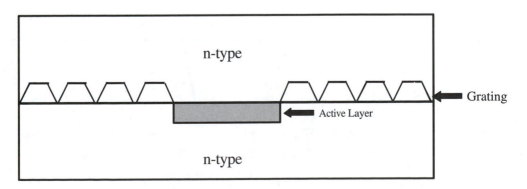

FIGURE 9.40   Distributed Bragg reflector.

where I_0 = threshold current at room temperature (characteristic of the laser)
                 I_{th} = current of operating temperature
                 T_0 = room temperature (characteristic of the laser device)
                 T = operating temperature

T_0 is an intrinsic value, and it is different for GaAs/AlGaAs and InGaAsP/InP laser structures. For example, for InGaAsP devices, T_0 has an average value of 65 K, whereas for AlGaAs devices, the average T_0 is observed to have a value of approximately 125°K. The lower T_0 value for InGaAsP devices is attributed to various phenomena occurring inside the device, such as heterobarrier carrier leakage and bond absorption.

When laser diodes are used as optical sources in optical fiber communications systems, the generated output optical power is different for the two basic modes of operation: carrier wave and pulsating. This difference is shown in Figure 9.41. Figure 9.41 shows that, under pulsating modulation, the laser diode requires less threshold current for laser action to take place and its optical power output is considerably higher than in the carrier-wave mode of operation. The threshold current variations relationship to different operating temperatures are shown in Figure 9.42.

Figure 9.42 illustrates a proportional increase of the threshold current in relation to a corresponding increase of the operating temperature. These temperature variations also have an effect on the operating wavelength (λ) of the laser device. It is observed that an increase in the operating temperature of 1°C shifts the operating wavelength by 0.3 nm. This wavelength change is attributed to device cavity expansion due to the increase of the operating temperature.

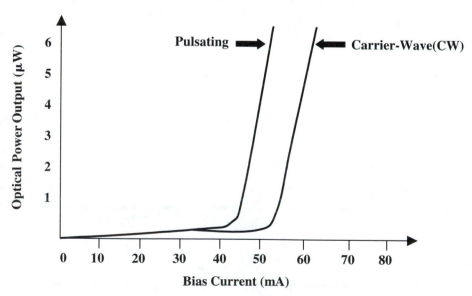

FIGURE 9.41   Optical power versus biasing current (mA)

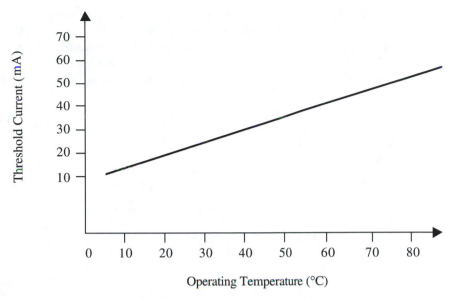

FIGURE 9.42   Threshold current versus temperature.

Laser Bandwidth

One of the fundamental advantages of optical fiber communications systems are their ability to directly modulate the optical source at a very high data rate. Although these modulating rates are very high (approx. 20 GHz), there are limits beyond which the laser diode cannot respond. Figure 9.43 shows the output frequency response of a typical laser diode under modulating conditions.

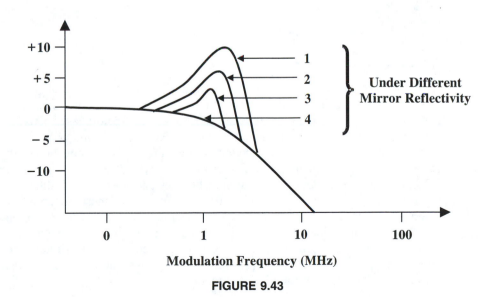

FIGURE 9.43

Figure 9.43 illustrates that under relatively low modulating frequencies, the spectral output intensity is constant. When the modulating frequency reaches a certain level, the carriers injected into the device cavity interact with the generated photons, enhancing a self-oscillatory process and thus sharply reducing the output spectral density. Another serious problem is that of noise intensity. There is a direct relationship between the noise spectral density and the carriers injected into the active region. The result is an increase of the threshold current beyond the level required for laser action to take place. This spectral noise density is observed to take a maximum value at maximum modulating frequencies. The self-oscillating frequency of a laser device is expressed by

$$f_{so} = \frac{1}{2\pi}\sqrt{\frac{A \cdot P_0}{\tau}} \tag{9.31}$$

where
f_{so} = self-oscillating frequency
A = gain
τ = average lifetime of the photon, inside the cavity
P_0 = **photon density**

It is evident from Eq. (9.31) that in order to increase the range of the oscillating frequency, and consequently in order to increase the modulation rate and to enhance output stability, both photon density and gain must be increased with a simultaneous decrease of the photon (cavity) lifetime. Stabilizing the device operating temperature can be achieved through a thermoelectric aperture (heat sinking); output power stabilization can be achieved by incorporating a feedback mechanism that can increase the differential gain through a PIN diode. Since the lifetime of photons inside the cavity depends on the physical characteristics of the laser structure, a decrease of the length of the cavity will result in a corresponding decrease of the (τ). This of course has its drawbacks. For example, a decrease in the length of the cavity will increase the carrier density and increase temperature beyond a certain level.

Laser Device Reliability

Laser devices present a number of challenging problems as a result of their fabrication processes. Two of the problems are control of the physical dimensions and control of the heterobarrier lattice growth. Various techniques such as liquid-phase epitaxial (LPE) growth have been utilized in the fabrication of laser devices. This technique represents certain difficulties related to uniform reproduction of larger areas. The introduction of the vapor-phase epitaxial (VPE) growth method has eliminated some of the problems encountered by the LPE fabrication technique. Fabrication techniques and operating conditions are key factors in determining a device's reliability and life span. Both of these factors are very important when the optical device is considered for utilization in an optical fiber communications system. The operating life span of an optical source is the period for which the device is capable of delivering specified optical power at a predetermined maximum threshold current. Over time, internal degradation based on operating temperatures, crystal defects, facets, and conducting damages limit a device's life span.

9.5 PHOTODETECTION

Photodetection is the process whereby optical power is detected and converted to electrical power. Photodetector devices or optical detectors perform photodetection. Optical detectors perform the exact opposite function to that of the optical sources; that is, they convert electric power into optical power.

In any optical fiber communications system, the optical source is part of the transmitter section and optical detectors are part of the receiver section. The performance of an optical detector incorporated into the receiver section of an optical fiber communications system will be determined by its ability to detect the smallest optical power possible (detector sensitivity) and will generate a maximum electric power at its output with an absolute minimum degree of distortion (low noise). Optical detectors must also exhibit a comparatively wide bandwidth and sharp response to accommodate the high bit rate required by such a system. Other criteria for selecting a particular photodiode for implementation into an optical fiber communications system are ability to interface with optical cables, a long operating life, and cost.

Although there are several types of photodetectors, not all of them are suitable for use in optical fiber communications systems. In such systems the optical detector device that is almost exclusively utilized is the semiconductor photodiode. Photodiode design criteria are set forth by system parameters such as size, sensitivity, bandwidth, and degree of tolerance to temperature variations. The two photodetector devices most commonly used in optical fiber communications systems are the PIN and APD devices.

9.5.1 PIN Photodetector

The abbreviation for the *P*-region, *I*-intrinsic *N*-region semiconductor diode is PIN. The principal theory on which a **PIN photodetector** device is based is illustrated in Figure 9.44. When a photon is incident upon a semiconductor photodetector device with energy larger than the bandgap energy of that device, the energy of the photon is absorbed by the band gap and an electron–hole pair is generated across the band gap. The energy of the incident photon is given by

$$E_{ph} = \frac{hc}{\lambda} \qquad (9.32)$$

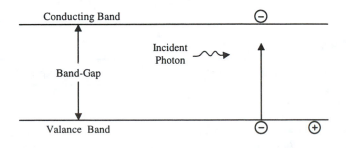

FIGURE 9.44
Photo-detector device
Model.

where E_p = energy of the photon
         h = Planck's constant (6.62×10^{-34} W-s^2)
         c = velocity of light (3×10^8 m/s)
         λ = wavelength (m)

It is evident from Eq. (9.32) that the photon energy (E_{ph}) is inversely proportional to the wavelength (λ). Therefore, there exists a wavelength at which the photon energy becomes equal to the bandgap energy. At this photon energy level, electron–hole generation will occur. The wavelength at which the photon energy becomes equal to the bandgap energy is called the "cutoff wavelength" (λ_c). From Eq. (9.32), solving for λ we have

$$E_{ph} = \frac{hc}{\lambda}$$

$$\lambda = \frac{hc}{E_{ph}}$$

Since $E_{ph} = E_g$,

$$\lambda = \frac{hc}{E_g}$$

Substituting $h = 6.62 \times 10^{-34}$ W-s^2 and $c = 3.0 \times 10^8$ m/s,
Therefore,

$$\lambda_c = \frac{1.24}{E_g} \, \mu m \qquad \qquad \textbf{(9.33)}$$

Semiconductor materials employed in the fabrication of photodetectors are the same as the materials employed in the fabrication of optical sources. Each individual element or substance is classified by a bandgap energy level (E_g) characteristic of that element or substance. Therefore, different materials exhibit different cutoff wavelengths. Some materials with their corresponding bandgap energy levels are listed in Table 9.5.

TABLE 9.5

Elements/Substances	Bandgap Energy (eV)
Germanium, Ge	0.67
Silicon, Si	1.11
Indium–Gallium Arsenide, InGaAs	0.77
Indium–Gallium Arsenide–Phosphorus, InGaAsP	0.89

Applying these energy gap levels to Eq. (9.33) yields the following cutoff wavelengths.

For Ga

$$\lambda_c = \frac{1.24}{E_g} = \frac{1.24}{0.67} = 1.85 \; \mu m$$

Therefore, the cutoff wavelength for Ge $\lambda_c = 1.85 \; \mu$m.

For Si

$$\lambda_c = \frac{1.24}{0.77} = 1.6 \; \mu m$$

Therefore, the cutoff wavelength for Si $\lambda_c = 1.12 \; \mu$m.

For In Ga As

$$\lambda_c = \frac{1.24}{0.77} = 1.6 \; \mu m$$

Therefore, the cutoff wavelength for InGaAs $\lambda_c = 1.6 \; \mu$m.

For InGaAsP

$$\lambda_c = \frac{1.24}{0.89} = 1.4 \; \mu m$$

Therefore, the cutoff wavelength for InGaAsP is $\lambda_c = 1.4 \; \mu$m.

The cross-sectional area of a silicon PIN diode is shown in Figure 9.45. When a photon impedes upon the photodetector, the photon is absorbed by the low-bandgap **absorption layer** and an electron–hole is generated. This electron–hole pair is called a "photocarrier." These photocarriers, under the influence of a strong electric field generated by a reverse-bias potential difference across the device, are separated, thus forming a photo current intensity proportional to the number of the incident photons. The dc biasing of a PIN-diode photodetector is shown in Figure 9.46.

The photocurrent generated from the PIN photodetector device develops a potential difference across the load resistance R_L with a frequency given by

$$f = \frac{E_{ph}}{h} \quad \text{or} \quad E_{ph} = fh \tag{9.34}$$

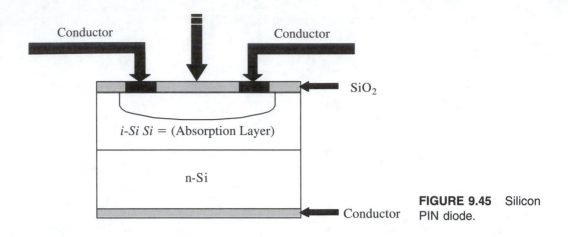

FIGURE 9.45 Silicon PIN diode.

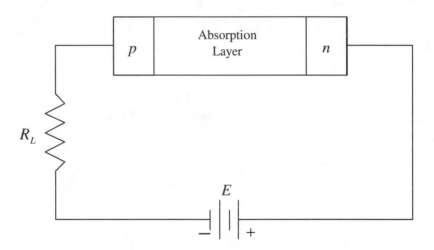

FIGURE 9.46 PIN-diode biasing.

where
E_{ph} = photon energy (eV)
h = Planck's constant $(6.62 \times 10^{-34} \text{ W} \cdot \text{s}^2)$
f = frequency

Since there exists a cutoff wavelength for each substance used in the fabrication of a PIN photodetector, there also exists a cutoff frequency for each element or substance. This cutoff frequency can be calculated as follows:

For Ge

$$f_c = \frac{E_{ph}}{h}$$

First the E_{ph} must be converted from eV to joules:

$$1 \text{ eV} = 1.6 \times 10^{-19} \text{ J}$$

For Germanium, $E_{ph} = 0.67 \text{ eV}$, or in joules, $1.6 \times 0.67 \times 10^{-19} \text{ J} = 1.07 \times 10^{-19} \text{ J}$. Thus,

$$f_c = \frac{1.07 \times 10^{-19}}{6.62 \times 10^{-34} \text{ J-s}} = 160 \times 10^{12} \text{ Hz}$$

Therefore, the cutoff frequency for Ge $f_c = 160$ Te Hz (Terra).

For Si

$$f_c = \frac{1.776 \times 10^{-19} \text{ J}}{6.62 \times 10^{-34} \text{ J-s}} = 268 \times 10^{12} \text{ Hz}$$

Therefore, the cutoff frequency for Si $f_c = 286$ Te Hz.

For InGaAs

$$f_c = \frac{1.232 \times 10^{-19} \text{ J}}{6.62 \times 10^{-34} \text{ J-s}} = 186 \times 10^{12} \text{ Hz}$$

Therefore, the cutoff frequency for InGaAs $f_c = 186$ Te Hz.

For InGaAsP

$$f_c = \frac{1.42 \times 10^{-19} \text{ J}}{6.62 \times 10^{-34} \text{ W-s}^2} = 215 \times 10^{12} \text{ Hz}$$

Therefore, the cutoff frequency for InGaAsF $f_c = 215$ Te Hz.

The combination of different semiconductor alloys operating at different wavelengths allows the selection of material capable of responding to the desired operating wavelength. For example: GaAs/AlGaAs substances operate at wavelengths between 800 and 900 nm, and photodetector devices composed of InGaAs/InP alloys operate at wavelengths between 1000 and 1600 nm.

9.5.2 PIN Photodetector Characteristics

The fundamental PIN photodiode operational characteristics are quantum efficiency (η), responsivity (R), speed, and linearity. Quantum efficiency (η) is defined by the number of electron–hole pairs generated per impeding photon, expressed by

$$\eta = \frac{N(e^-, p^+)}{N_{ph}}$$ (9.35)

where $N(e^-, p^+)$ = number of electron-hole pairs generated
            N_{ph} = number of photons
            η = quantum efficiency

The number of electron–hole pairs generated is translated to current by

$$I_p = q \times N(e^-)$$ (9.36)

$$N(e^-) = \frac{I_p}{q}$$ (9.37)

where I_p = photocurrent (mA)
            q = electron charge = 1.5×10^{-19} C
            $N(e^-)$ = number of electrons

Consequently, the number of incident photons is translated to light power by

$$P_o = N_{ph} \times hv$$ (9.38)

$$N_{ph} = \frac{P_o}{hv}$$ (9.39)

where P_o = light power
            N_{ph} = number of photons
            h = Planck's constant (6.62×10^{-34} m-s)
            $v = c$ (c = velocity of light in space, 3×10^8 m/s)
            λ = wavelength

Substituting Eqs. (9.37) and (9.38) into the efficiency Eq. (9.35), we have

$$\eta = \frac{N_{e-}}{N_{ph}} = \frac{I_p/q}{P_{o/hv}} = \frac{I_p hv}{qP_o}$$

Substituting $v = c/\lambda$,

$$\eta = \frac{I_p hc}{q p_o \lambda} \quad \left(\text{since } E_{\text{ph}} = \frac{hc}{\lambda} \right)$$

Therefore,

$$\eta = \frac{I_p E_{\text{ph}}}{q p_o} \tag{9.40}$$

From the quantum efficiency equation, it is evident that the efficiency of a PIN photodetector is proportional to the photon energy absorbed by the absorption layer of the device. Larger photon energy requires a thicker absorption layer, allowing a longer time for electron–hole pair generation to take place.

Response Time (speed)

The **response time,** or speed, of a photodetector is the time required by the carriers generated within the absorption region to travel through that region under reverse-bias conditions. The main factor determining this time is the thickness of the absorption region. The thicker the absorption region, the longer the time. Here, there is a conflict between photodetector efficiency and response time. For higher efficiency, a thicker absorption region is needed; for higher speed, a thinner absorption region is required. In practice, trade-offs between efficiency and speed are made to accommodate design objectives. The response time (t_r) of a photodetector is given by

$$t_r = \frac{\text{Thickness of the absorption layer}}{\text{Saturation velocity}}$$

The saturation velocity (V) for a typical InGaAs alloy is 10^7 m/s. Given the device absorption layer thickness, the response time can be calculated.

EXAMPLE 9.7

Compute the response time of a PIN photodetector composed of InGaAs with 5 μm of absorption layer thickness.

Solution

$$t_r = \frac{5 \times 10^{-6} \text{ m}}{1 \times 10^7 \text{ m/s}} = 5 \times 10^{-13} \text{ s} = 0.5 \times 10^{-12} \text{ s}$$

Therefore, $t_r = 0.5$ ps.

The key parameter for determining photodetector device performance is responsivity, the current generated in the absorption region per unit optical power incident to the region. Responsivity is closely related to quantum efficiency and is expressed by

$$R = \eta \frac{q}{E_{ph}} \qquad (9.41)$$

where
R = responsivity
η = quantum efficiency
q = electron charge (1.59×10^{-19} C)
E_{ph} = energy of the photon ($h\nu$)

Substituting Eq. (9.35) (quantum efficiency) into Eq. (9.40), we have

$$R = \frac{I_p E_{ph}}{q p_o} \times \frac{q}{E_{ph}} = \frac{I_p}{p_0}$$

Therefore, responsivity

$$R = \frac{I_p \; \mu A}{p_o \mu W} \qquad (9.42)$$

The responsivity of a PIN photodiode is the ratio of the generated photocurrent per unit of incident-light power.

A graphical representation of quantum efficiency (η) and responsivity is shown in Figure 9.47, which illustrates the fundamental difference between responsivity and quantum efficiency. For different semiconductor materials, the responsivity is linear up to a particular wavelength, then drops quickly. Beyond this point, the photon energy becomes smaller than the energy required for electron–hole generation.

Dark current (I_d)

Dark current is the reverse leakage current of the photodetector device in the absence of optical power impinging upon the photodetector device. Dark current is an unwanted element caused by such factors as current recombination within the depletion region and surface leakage current. The negative effects of such unwanted currents contribute to shot noise. Shot noise power is expressed by

$$P_n = 2I_d q \qquad (9.43)$$

where
P_n = shot noise power (W)
I_d = dark current (A)
q = electron charge (1.59×10^{-19} C)

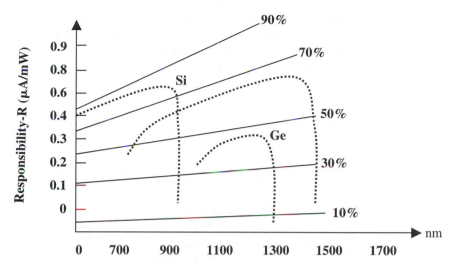

FIGURE 9.47 Graphical representation of quantum efficiency (η) and responsivity (R) in reference to operating wavelengths. (Key: quantum efficiency, η; responsivity, R)

In reference to shot-noise voltage (V_n)

$$V_n = 2I_d \, \text{BW} \qquad\qquad (9.44)$$

where V_n = noise voltage
              BW = receiver operating bandwidth

Equation (9.44) indicates that an increase of dark current will decrease the overall receiver operation bandwidth by maintaining the noise-voltage level constant. Although it can be considerably reduced by proper material selection and controlled fabrication techniques, the dark current cannot be totally eliminated. This unwanted current is also temperature dependent. That is, it increases with an increase in the operating temperature. Therefore, the proper control of material fabrication techniques and operating temperatures is a key factor in reducing dark current and, consequently, in enhancing the operating performance of the PIN photodetector device.

9.5.3 Avalanche Photodetectors (APDs)

Avalanche photodetectors are very similar to PIN diodes with one exception: the addition to the **APD** device of a high-intensity electric field region. In this region the primary electron–hole pairs generated by the incident photons are able to absorb enough kinetic energy from the strong electric field to collide with atoms present at this region, thus generating more electron–hole pairs. This process of generating more than one electron–hole pair from one incident photon through the ionization process is referred to as the "avalanche effect."

It is apparent that the photocurrent generated by an APD photodetector device exceeds the current generated by a PIN device by a factor referred to as the "multiplication factor" (M). Since the current generated by a PIN device is

$$I = qN(\text{e}^-)$$

the generated photocurrent is

$$I_p = qN(\text{e}^-)M \qquad\qquad \textbf{(9.45)}$$

where
$$\begin{aligned}
I_p &= \text{generated photocurrent} \\
q &= \text{electron charge } (1.59 \times 10^{-19}\,\text{C}) \\
N(\text{e}^-) &= \text{carrier number} \\
M &= \text{multiplication factor}
\end{aligned}$$

The **multiplication factor** depends on the physical and operational characteristics of the photodetector device, such as the width of the avalanche region, the strength of the electric field, and the type of semiconductor material employed. The cross-sectional area of a short-wavelength silicon APD device is shown in Figure 9.48. This structure is composed of a $\text{p}^+\text{p}^-\text{pn}^+$ semiconductor material. A lightly doped p^- region is epitaxially grown on a heavily doped p^+-type substrate.

When a reverse-biased voltage is gradually applied across the diode, an electric field develops across the avalanche region with its strongest intensity measured at the pn^+ junction. As the reverse-biasing voltage gradually increases, the corresponding electric-field across the region also increases; as a consequence, there is an expansion of the depletion

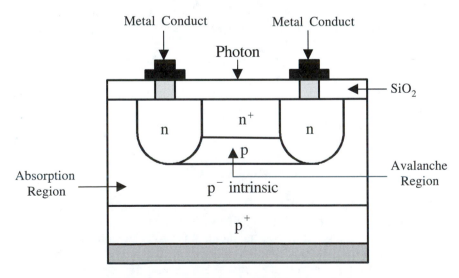

FIGURE 9.48 APD silicon photodetector device.

region. If the electric field intensity increases just below the avalanche breakdown point, the depletion region has almost reached the total width of the p⁻ layer. Under these circumstances, if a photon is incident upon the device, it will be absorbed by the p⁻ layer. The energy released by the incident photon causes the generation of the first electron–hole pair. Under the influence of the strong electron field, the generated electron is guided from the p⁻ intrinsic region closer to the pn⁺ junction. At this point, the electric field intensity is at its maximum. Under the influence of the strong electric field, the generated electron collides with other atoms, thus generating a new electron pair. The secondary electron, still under the influence of the strong electric field, generates another electron–hole pair, and so on (avalanche effect).

The number of secondary electron–hole pair generations is proportional to the carrier distance traveled through the avalanche region and to the semiconductor materials used for the fabrication of the photodetector devices. Figure 9.49 shows the relationship between electric field strength and ionization rates for different alloys employed in the fabrication of photodetector diodes.

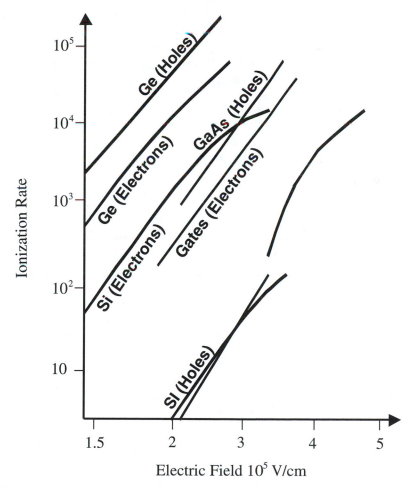

FIGURE 9.49 Ionization rate versus electric field strength for various alloys.

The equivalent circuit of an avalanche-photodetector diode is shown in Figure 9.50. When it is reverse biased, an electric field develops across the depletion region with its maximum field strength across the pn junction. The avalanche effect will occur only when the depletion region has reached its maximum or is fully developed.

The cross-sectional area of an InGaAs photodetector device is shown in Figure 9.51. This is a double heterostructure device incorporating an InGaAs low-bandgap absorption layer and using the InP as a confinement layer with a band gap higher than that of the absorption layer. This higher band gap of InP ensures that no photon absorption will occur in the confinement layer.

One of the fundamental operating characteristics of a photodetector device is its available bandwidth. This bandwidth is limited by the number of holes present in the device. The InGaAP layer is epitaxially grown on the absorption layer, with a bandgap somewhere between the absorption and confinement layers. This arrangement removes the bandwidth-limiting holes and thus enhances the overall bandwidth characteristics of the photodetector device. The absorption layer is elevated to a higher-bandgap region, resulting in the first electron–hole pair generated by the incident photon. The generated electron under the influence of the very strong electric field is guided to the p⁻InP ring. At the pn⁺ junction, the electric field is maximum, with a maximum depletion region. Through this region, impact ionization takes place, resulting in photocurrent multiplication. Another basic performance characteristic of an APD photodetector device is the ratio of the electron to hole ionization rates, which are indicative of different semiconductor materials used during the fabrication process. This electron–hole ionization ratio is a key factor determining the gain–bandwidth product and noise-performance characteristics. For example, let us suppose that electrons are the primary carriers in an APD structure. A higher electron-to-hole ratio reflects low noise and a higher gain–bandwidth product; a low electron-to-hole ratio reflects a higher noise and lower gain–bandwidth product.

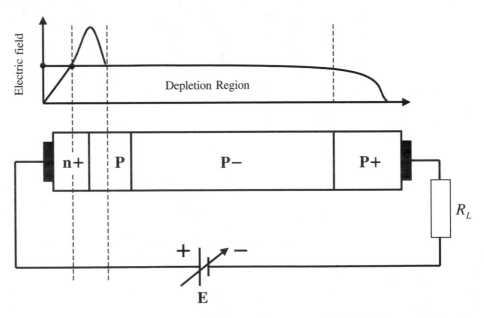

FIGURE 9.50 Equivalent circuit for an APD (minimum electric field required for avalanche effect).

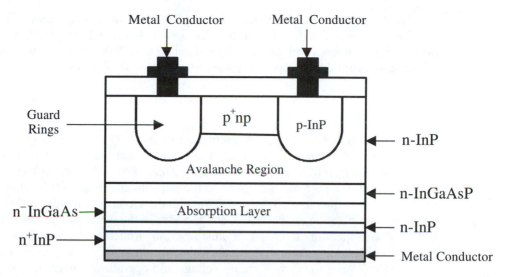

FIGURE 9.51 InGaAs photodetector device cross section.

The **photocurrent gain** in an APD device is a function of several elements, such as (a) the wavelength of the incident photons, (b) the electric field strength as a result of the reverse-bias voltage, (c) the width of the depletion region, and (d) the types of semiconductor materials used for the fabrication of the APD device. The relationship of the photocurrent gain to biasing voltage for different wavelengths is shown in Figure 9.52. It is evident that higher photocurrent gain is observed at higher wavelengths at specific reverse-biasing voltages.

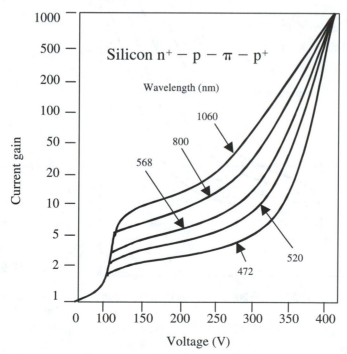

FIGURE 9.52
Photocurrent gain versus reverse-biasing voltage for different wavelengths.

Fiber-Optic Communications

337

The function of the guard rings in an APD structure is to prevent edge breakdown around the avalanche region. When InGaAsP materials are used in the fabrication of APD devices, these devices exhibit operating wavelengths of 900–1600 nm; when silicon materials are used, they exhibit operating wavelengths of 400–900 nm.

Gain

The important parameter of photocurrent gain of an APD device is also temperature dependent. Figure 9.53 shows an increase of the multiplication factor with a corresponding increase of the operation temperature at a constant reverse-biasing voltage. This current fluctuation with temperature is an undesirable phenomenon and must be confined to a tolerable level or completely eliminated. To maintain a constant multiplication factor for a wide range of operating temperatures, any increase in the generated photocurrent due to an increase in the operating temperature must be compensated for by a proportional decrease of the reverse-biasing voltage. The equation determining the current gain multiplication factor (M), in reference to reverse-biasing voltage and operating temperature, is given by

$$M = \cfrac{1}{1 - \left(\cfrac{V_\alpha - I_p R_l}{V_{bk}[1 + \alpha(T_1 - T_0)]} \right)} \tag{9.46}$$

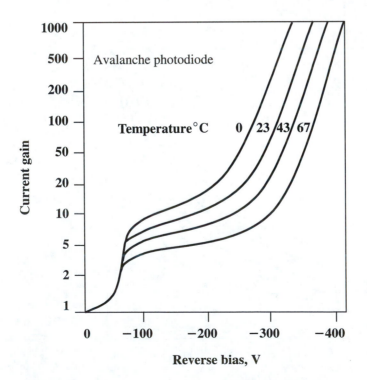

FIGURE 9.53 Photo current gain versus reverse-biasing voltage at different operating temperature.

where
V_a = reverse biasing voltage (V)
I_p = multiplied photocurrent (mA)
R_l = device resistance plus load resistance
V_{bk} = breakdown voltage at room temperature
T_0 = room temperature
T = operating temperature
α = factor determined from gain versus temperature graphs

Photodetector Noise

Avalanche photodetectors exhibit higher noise levels than PIN devices as a result of the ionization and photocurrent multiplication process that takes place within the APD device. The random nature of the incident photons on the APD device results in a random photocurrent generation at the output of the device. This current fluctuation is classified as shot noise, expressed by

$$\frac{d(j_P)^2}{df} = 2qI(M)^2 \tag{9.47}$$

where
(J_p) = mean square spectral density
f = frequency (Hz)
q = electron charge (1.6×10^{-19} C)
$*I$ = primary photocurrent
$(M)^2$ = mean square of the avalanche gain

$$I = I_p + I_{br} + I_{dk}$$

where
I_p = photocurrent
I_{br} = background current
I_{dk} = dark current

Equation (9.46) is further modified to incorporate the gain nonlinearities resulting from the statistical nature of the ionization and avalanche processes:

$$\frac{d(j_p)^2}{df} = 2qI(M^2)F(M) \tag{9.48}$$

Where $F(M)$ is the excess noise factor, it is related to the semiconductor material used in the fabrication of the APD device, the electron-to-hole carrier ratio and the electric field profile across the depletion region. An empirical equation determining the excess noise factor $F(M)$ is given by

$$F(M) = 2(1 - K) + K(M) \tag{9.49}$$

Fiber-Optic Communications

where M = multiplication factor
K = ratio of the smallest to the largest ionization coefficients.
(for silicon, $K \cong 0.02$–0.1; for germanium, $K \cong 0.5$;
for InGaAsP, $K \cong 0.3$–1)

Dark Current

Dark current is the current present at the photodetector output with an absence of incident light. For an APD device the dark current is also multiplied by the device multiplication factor (M), resulting in an overall reduction in device sensitivity. The dark current is a non-linear function of the reverse-biased voltage at avalanche breakdown levels and is referred to as "tunneling current." Different semiconductor materials exhibit different levels of tunneling current, resulting from different bandgap sizes. For example, devices with small band gaps measure small tunneling currents and large-bandgap devices measure larger tunneling currents. A practical solution for a substantial reduction of the tunneling current is the fabrication of structures with a separation between the absorption (low-bandgap) region and the avalanche (high-bandgap) region.

Response Time

The response time of a photodetector device is the time a carrier takes to cross the depletion region. For APD devices, the response time is almost double that of PIN devices. Since the APD structure incorporates a large-bandgap region produced by the large electric field, the generated photocarriers must travel twice the distance from the low-bandgap to the higher-bandgap region and back after the multiplication process has taken place. It is therefore evident that response time is directly related to depletion region width. The larger the width, the larger the response time. If a reduction of the depletion region is attempted in order to reduce response time, inevitably a substantial quantum efficiency reduction will result. Therefore, in APD photodetector devices, a trade-off between quantum efficiency and response time is necessary. A typical response time of 0.5 ns at 800–900 nm has been achieved.

Capacitance

In a photodetector device, internal capacitance is a parasitic component affecting the overall response time of the detector. As with any other capacitance, the junction capacitance of an APD device is determined by the cross-sectional area and width of its depletion region and is expressed by

$$C = \frac{\varepsilon q A N}{2(V + V_j)} \tag{9.50}$$

where C = junction capacitance (F)
ε = dielectric constant
A = depletion area
N = doping density (depletion region)
V = reverse bias voltage (V)
V_j = junction voltage (V)

QUESTIONS

1. What are the advantages of utilizing the visible light spectrum as a transmission medium?
2. Describe the contributions of Isaac Newton, Christian Higgern, Thomas Young, and James Maxwell to the formulation of the related theories governing the behavior of light.
3. Briefly explain the fundamental optical laws.
4. List the basic classifications of optical fibers and briefly explain their operating characteristics.
5. What are the properties of optical fibers?
6. Define *numerical aperture*.
7. What are the main components contributing to optical fiber attenuation?
8. Describe the difference between Rayleigh scattering and Mie scattering.
9. What are the main components contributing to optical fiber absorption?
10. Describe the difference between stimulation and spontaneous emission.
11. Define the principle theory of an LED (structure, operation).
12. What are the characteristics of an edge-emitting diode?
13. What is the critical element limiting LED bandwidth?
14. Briefly describe the structure and operating characteristics of a surface-emitting diode.
15. List the various methods of coupling LEDs into the fiber and briefly describe the advantages and disadvantages of each method.
16. Describe the operating principle of a laser diode.
17. What are the operating characteristic of a distributed-feedback laser structure?
18. List the components determining laser diode bandwidth and define their relationship.
19. What is photodetection?
20. Briefly describe the operating principle of a PIN photodetector.
21. Describe the basic PIN photodetector characteristics.
22. What is dark current and how is it related to device noise power and noise voltage?
23. Describe the operating principle of an avalanche photodetector (APD).
24. With the assistance of a cross-sectional schematic diagram, briefly describe the operation of an InGaAs photodetector device.
25. Name the components contributing to APD gain.
26. What is photodetector noise and what are the contributing components?

PROBLEMS

1. If a light ray is impinding upon the boundary from a diamond to air, compute the angle (θ).

2. An optical source generates 2.1 μW of optical power. This power is coupled into an optical fiber with an angle of 12°. Assuming that the active area of the optical source is larger than the optical fiber cross-sectional area, compute the optical power coupled into the fiber.

3. Compute the optical power coupled into the optical fiber at a 25° angle. The source bias current is 10 mA, the forward voltage is 2.0 V, and the source internal efficiency is 3.5%.

4. A light-emitting diode exhibits an 8-ns rise time. Compute its operational bandwidth.

10

Optical Fiber Communications Systems Analysis and Design

Objectives

- Describe the various components of an optical communication link.
- Identify the system parameters.
- Compute all the fundamental components of an analog optical link.
- Describe the essential components constituting an FM/FDM video signal transmission system.
- Design an optical fiber link operating in FM/FDM mode capable of transmitting eight video channels.
- Explain the differences between analog and digital optical fiber systems.
- Design a digital optical fiber link.

Key Terms

Optical source
Optical fiber
Optical detector
Optical index of
    modulation
Peak optical power
Average optical power
Wavelength-division
    multiplexing
Fusion splices

Bandwidth–length product
Fiber attenuation
Dark current
Coupling loss
Optical receiver sensitivity
System bandwidth
Multimode width
Time and temperature
    degradation
Link power margin

VSB/AM/FDM
Sideband power
Filter weighting factor
Insertion loss
Responsivity
Detector noise power
Power margin

INTRODUCTION

The basic function of an optical fiber communications link is to convey analog or digital information from one point to another via an optical fiber. The main components of such an optical link are the transmitter, the **optical source,** the **optical fiber**, the **optical detector,** and the receiver (Figure 10.1). When analog or digital information is to be transmitted through an optical fiber communication link, the information is fed into the transmitter, whose function is to condition the signal (multiplex and amplify) to the level required by the optical source.

The transmitter output current becomes the driving current for the optical source, which in turn converts it to optical power. This optical power generated by the optical source is coupled into the optical fiber and through its core; it travels in a zigzag mode to the other end of the fiber. This end is also coupled into the input of the optical detector. The function of the optical detector is to convert the impinging light into an electric signal. This electric signal generated by the optical detector is fed into the receiver, whose main function is to convert it back to its original form.

For digital transmission, the voice or video analog signals are digitally coded to a specified format and in compliance with pre-established information standards. The conversion of voice analog signals to digital information is discussed extensively in Chapter 2. The selection of a particular transmission format, either analog or digital, for an optical communications link is based on various system considerations. For example, for short- or medium-distance video transmission, the analog form is utilized almost exclusively. The main difficulty with long-distance analog transmission through an optical fiber link lies in the fact that such a transmission requires a considerably higher signal-to-noise (SNR) ratio than a digital transmission. Another problem with such a transmission is the maintenance of stringent operating characteristics for the optical semiconductor devices (optical detector and optical source) as well as for the optical fiber. Despite these difficulties, analog transmission is in great demand today by some telephone utility and video cable providers because of the already existing analog infrastructure. Recent developments in laser semiconductor device technology operating in the 1300-nm to 1500-nm range and in single-mode optical fiber have made possible the transmission of video signals to a distance of 60 km with an impressive SNR of better than 60 dB. For long-distance optical fiber transmission of voice, data, or video, the digital mode is almost universally accepted as the most efficient method of transmission. This preference is based on the fact that digital receivers exhibit higher sensitivity than their analog counterparts, with smaller optical fiber losses.

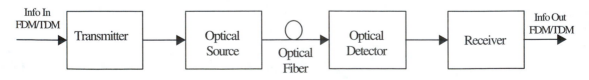

FIGURE 10.1   Block diagram of a basic optical fiber communications link.

10.1 SYSTEM PARAMETERS

One of the most critical performance parameters of an optical fiber communication link is the **bandwidth–length product** (BL), where B is the system bit rate and L is the length of the link between repeaters. Optical links can vary from a relatively small to a very large bandwidth–length product. For example, a single-mode optical fiber with a very small dispersion, operating at 1300 nm to 1500 nm and coupled to a laser source device operating at the same wavelength, can provide an impressive BL product of several GHz. Other system parameters include system power requirements, baseband bandwidth, in MHz if analog and Mb/s if digital, length of the link, receiver SNR or BER for either analog transmission or digital transmission, system rise and fall times, and optical fiber attenuation in dB/km. Other considerations to be taken into account are the number of connectors, number of splices and their corresponding connector and splice losses, detector coupling losses, and time and temperature degradation losses.

Selection of the Optical Source   Since the selection of the optical source must come from the two existing types, either LED or laser diode, the system requirements will determine the type of the device to be selected. These requirements are operating wavelength, optical power, spectral width, frequency response, rise time, drive current, and power coupled into fiber. For optical links requiring higher optical power generated by the optical source, faster rise time, longer spectral width, and higher frequency response, the laser diode is preferable to the LED. If cost is a factor influencing the design criteria, then for short-distance links, LEDs can be considered in conjunction with single-mode fibers.

Selection of the Optical Fiber   It was mentioned previously that the principal parameter determining the selection of a particular type of optical fiber to be utilized in an optical fiber link is the bandwidth–length product. For maximum system capacity, a very low-loss optical fiber is required. Another consideration that must be taken into account is that of system flexibility, which must be capable of accommodating modifications pertaining to future demands for system capacity increases. For example, the only type of fiber that can carry a large volume of binary data to a distance of up to 10 km while using a LED as the optical source is the single-mode fiber. Single-mode fibers are also ideal for future system upgrading. With such fibers, system modifications can be accomplished by changing only the optical source from LED to laser type. This kind of modification can increase the system capacity increase by tenfold. Other optical fiber characteristics to be considered by the system designer are fiber material, operating modes, attenuation, dispersion, temperature coefficients, microbending, splicing, macrobending losses, upgradability, and cost.

Selection of the Optical Detector   The primary function of an optical detector is to convert optical power to electric power. In the optical detector section, we examine two basic types of optical detectors: the avalanche photodetector (APD) and the PIN diode. Both types of devices exhibit reasonable operating bandwidth characteristics and high rise time, but the APD diode is more sensitive and consequently more expensive than the PIN diode. Other detector device characteristics to be considered are power output, threshold current, pulse forward current, continuous forward current, rise and fall times, radiation patterns, and so on.

Selection of the Receiver   The receiver's primary function in an optical fiber communication link is to amplify the weak electrical signals generated by the optical detector to a level required by the encoder circuit. In practice, the receiver is a highly sensitive, high-gain amplifier designed to maintain a pre-established BER, usually 10^{-9} for digital transmission, and a very high SNR, usually 60 dB or better for analog transmission. Other receiver specifications to be considered are output impedance, dynamic range, digital output rise time, output signal level, and power supply requirements.

Selection of Modulation   There are two basic types of modulation for optical fiber communications systems: analog and digital. The analog method is used for relatively short distances, from 10 km to 20 km. Voice or video channels are frequency-division multiplexed (FDM). The output current of such multiplexers modulates the optical source. Analog modulation presents certain disadvantages compared to the digital method. For example, the high SNR required by the analog system limits the usable system bandwidth, thus reducing the link information capacity as well as the path length between repeaters. Another disadvantage of analog transmission is increased chromatic distortion due to optical source nonlinearities generated by the intensity modulation.

Digital modulation is much preferable to analog, especially for long-distance transmission, because of its increased information carrying capacity and resistance to various link impairments. Although the binary signal traveling through the core of the optical fiber is subject to a considerable loss due to signal dispersion, a BER of 10^{-9} can easily be maintained. Furthermore, the optical sources, whether LEDs or laser diodes, exhibit a very small rise time, leading to switching frequencies in the order of several MHz for LED and of several GHz for laser diodes. These optical-source-switching capabilities dramatically increase the system's available bandwidth and consequently the overall system capacity.

10.2 ANALOG OPTICAL FIBER COMMUNICATION LINKS

Voice or video analog signals transmitted through an optical fiber communications link are first modulated, then frequency-division multiplexed, before the optical source is modulated. Figure 10.2 shows the block diagram of an analog optical fiber communications link. The modulation techniques employed in an analog optical fiber communications system are amplitude modulation and frequency modulation. The use of subcarrier frequencies for voice or video signal transmission through an optical fiber link is made possible through the utilization of single-mode optical fibers and laser diodes.

When amplitude modulation (AM) is implemented, the overall performance of such a system employing subcarrier modulation is a direct function of the **optical index of modulation** (m_o):

$$m_o = \frac{P_{\max} - P_o}{P}$$

(10.1)

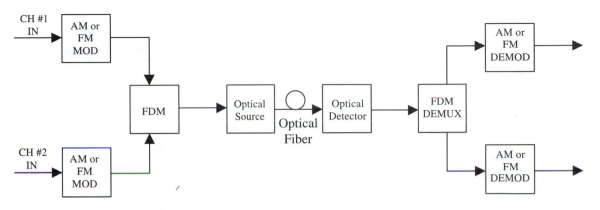

FIGURE 10.2 Block diagram of an analog optical fiber communication link.

where m_o = optical index of modulation
P_{max} = **peak optical power** at the input of the optical detector
P_o = **average optical power** at the input of the optical detector

In such a system, the received optical power is a direct function of the optical index of modulation (m_o). The higher the index of modulation, the higher the receiver optical power and consequently the higher the SNR. Of course, this increase of receiver optical power due to the increase of the modulation index results in a proportional increase of the inter-modulation distortion. The intermodulation distortion occurs because the optical source is forced to operate in the nonlinear region of its operating characteristic curve. This distortion, coupled with cochannel interference, substantially contributes to the overall system performance degradation. It is therefore essential that a design compromise between maximum optical power received at the input of the optical detector, index of optical modulation, and distortion levels must be reached.

AM subcarrier-modulation schemes employed in optical fiber communications systems exhibit certain advantages such as low cost, simplicity, and high system capacity. On the negative side, they require higher SNR, are shorter in length between repeaters, and are more susceptible to noise and cochannel interference. The alternative to an AM subcarrier modulation scheme and a much preferred method in optical fiber link designs is the FM/FDM subcarrier-modulation method. In this method the input voice or video analog signals are frequency modulated and then frequency-division multiplexed. The composite FDM signal is then used to modulate the optical source.

The implementation of the FM/FDM method of modulation in optical links has certain advantages over the AM/FDM method. For example the SNR requirements are more relaxed with FM/FDM than with AM/FDM. Such systems are less susceptible to optical source nonlinearities and optical-fiber-induced distortions. Furthermore, for IF frequencies of 70 MHz or 140 MHz, the optical link is compatible with satellite FM/FDM transmission links, also operating at 70 MHz or 140 MHz IF frequencies. The disadvantages arising from the implementation of such systems are the higher costs and reduced system capacities.

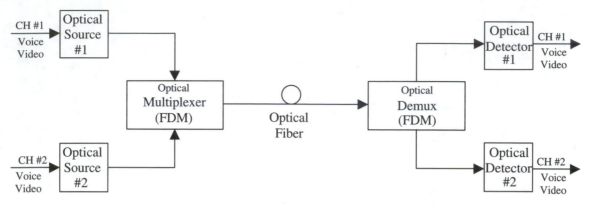

FIGURE 10.3   Block diagram of WDM optical fiber link.

Another scheme of modulation and multiplexing used in optical fiber communication links is **wavelength-division multiplexing** (WDM). In such a scheme, the analog input signals are used to modulate separately optical sources operating at different wavelengths. The outputs of these sources are combined through an optical multiplexer, then coupled into a common fiber for transmission. At the receiver end the combined signal is fed to an optical demultiplexer, whose function is to separate the combined optical signal to single channels and then to feed them to separate optical detectors. Figure 10.3 shows the block diagram of a wavelength-division multiplexed optical fiber link.

The implementation of a wavelength-division multiplexing scheme in an optical fiber communications system results in specific advantages over the implementation of AM/FDM and FM/FDM schemes. The most important advantage is channel independence. Since each input signal employs its own optical source operating at different wavelengths, failure of one or even two of these channels does not result in complete system failure. Therefore, system reliability is higher with WDM than with the other two schemes. Another important advantage is that such a system can easily be upgraded. However, WDM schemes are incompatible with other existing communications systems, are more complex in their implementation, and are more expensive.

10.2.1 Design Procedure

A successful optical fiber communications link, like any other communication system, must satisfy all the system performance requirements while remaining within the preset budget restrictions. A number of important design considerations must be taken into account by the system designer, and a preliminary list of operating characteristics and fundamental system components must be compiled prior to the design work. Such a list of important components is shown in Table 10.1.

Optical Wavelength   The two commercially available optional sources are the light-emitting diode (LED) and the laser diode. Both devices operate at short wavelengths (780–900 nm), with typical operating wavelength at 830 nm, and at long wavelengths

TABLE 10.1

Optical wavelength
Optical source
Type of optical fiber
Optical detector
Number of **fusion splices**

(1200–1600 nm), with typical operating wavelengths at 1300 nm and 1550 nm. These three optical-source operating wavelengths—namely, 830 nm, 1300 nm, and 1550 nm—are in absolute correlation with the silicon-based optical fiber operating windows.

In the early stages of their developments, optical links were designed to operate in the 850-nm range. At this wavelength, the optical fiber was exhibiting an attenuation of 3.0 dB/km and a modal dispersion in the order of 100 ps/km. Advances in optical fiber fabrication techniques led to the development of optical fibers with operating wavelength windows of 1300 nm and 1550 nm, and attenuation of 0.2 dB/km. Single-mode-fibers exhibit an almost nonexistent multimode dispersion while dramatically increasing their operating bandwidth. These types of fibers are now available to the system designer. For long-distance transmission, the 1300-nm-wavelength window is preferable because of its low achromatic dispersion and very low attenuation.

Optical Sources   Since cost is a very important factor, it must be considered in every step of the design process. Here too the selection of the optical source must be based on performance as well as economics. LED devices operating in the 850-nm region are quite inexpensive, but their cost significantly increases in the1300-nm region. They also exhibit low coupling efficiency, a typical spectral width of 40 nm, a relatively large chromatic dispersion, low modal noise, and a relatively high degree of reliability.

In comparison to LEDs, LASER devices are very expensive and have almost the exact opposite performance characteristics of the LED devices. Laser sources exhibit high coupling efficiency and low spectral width (3 nm), small chromatic dispersion, and high operating bandwidth. On the negative side, they exhibit larger modal noise and lower reliability. Laser devices are available in monochromatic and polychromatic modes of operation. Monochromatic lasers exhibit spectral widths in the order of 0.01 nm, but they are very susceptible to reflections resulting from feedback noise. Monochromatic laser diodes are best suited to operate with single-mode fibers. For optical systems requiring multimode fibers, polychromatic lasers are utilized almost exclusively.

Optical Fibers   Cost, system performance, and upgrading capabilities must be the basic criteria for optical fiber selection. If the system parameters require a very large operating bandwidth, a single-mode fiber must be selected. Single-mode fibers exhibit a maximum of 200 GHz/km bandwidth, while the step-index fiber bandwidth is approximately 20 MHz/km. Single-mode fibers reflect a very low attenuation and low scattering losses. On the negative side, their coupling efficiency is as low as 20%. From this, it is evident that the selection of the optical fiber to be utilized in an analog optical fiber system must be based on three main considerations: cost, coupling efficiency, and operating bandwidth.

Optical Detector The proper selection of the detector device in an optical fiber system is very critical because its operating characteristics must correlate with the pre-established optical receiver system specifications. Two optical detector devices are readily available for immediate use: APD and PIN-diode detectors. The structure and operating characteristics of these devices were dealt with extensively in Chapter 9. Nevertheless, a brief summary of their operating characteristics is essential. Avalanche photodiodes exhibit good sensitivity, a very important characteristic for achieving the high SNR required by an analog system and the BER of 10^{-9} required by a digital system. The APD's excellent sensitivity overrides certain disadvantages such as poor linearity and a high dc biasing-voltage requirement. In contrast to APD, PIN photodetectors provide better linearity and higher operating bandwidth but poorer sensitivity. The third photodetector device is the PIN–FET diode. This device combines the characteristics of both APD and PIN diodes by providing reasonably good sensitivity and very good stability while eliminating the need for high dc biasing-voltage requirements.

The system design process will start with very simple optical fiber communications links and then progress to very complex systems, including synchronous optical fiber transmission.

DESIGN EXAMPLE 10.1

Design a 3.5-km single-channel analog optical fiber communications link with a usable bandwidth of 40 MHz and required signal-to-noise ratio of 60 dB.

Solution

Component Selection

The basic components to be utilized in this link are the optical source, the fiber, and the optical detector.

1. Optical Sources

   The selection of the optical source and consequently the system operation wavelength is mainly based on the length of the system. Since 3.5 km is considered to be a short distance, an LED is preferable to a laser diode source because it generates enough optical power to satisfy the system requirement and is also inexpensive, particularly at the 830-nm wavelength.

2. Optical Fiber

   Optical fiber selection is based on two basic factors: First, the fiber must maintain an operating bandwidth of 40 MHz along the 3.5-km length; second, it must have a relatively large core size in order to maintain a satisfactory coupling efficiency between the optical source and the fiber.

3. Optical Detector

   The main performance characteristic of the optical detector to be used in this link is its sensitivity. Since both the APD and the PIN detectors can be used with equal success, the remaining important parameter that will determine the selection is the cost. In this case, the PIN device is finally selected because it is less expensive than the APD detector.

Compute Bandwidth–Length Product (*BL*)

$$BL = 40\ \text{MHz} \times 3.5\ \text{km} = 140\ \text{MHz/km}$$

From manufacturers' product lists, select a fiber with a bandwidth–length product larger than 140 MHz/km. Also determine **fiber attenuation** specifications. Assume for this system a 1 dB/km fiber attenuation.

Compute Total Fiber Attenuation (*L_f*)
Since 1 dB/km is the specified attenuation for the selected fiber, then

$$L_f = 3.5\ \text{dB}$$

Select Optical Source
An LED device is selected, capable of generating 50 μW of optical power (P_o). Assuming a 50% coupling efficiency, the optical power coupled into the fiber is 25 μW or -46 dB.

$$P_o = -46\ \text{dB}$$

Determine the Number of Connectors Required by the Link (*L_c*)
For short-distance links, usually two connectors are required. Since the average optical connector loss is approximately 1 dB, the total connector loss is 2 dB.

$$L_c = 2\ \text{dB}$$

Establish Temperature and Line Degradations (*L_{T-t}*)
The average loss is 1 dB.

$$L_{T-t} = 1\ \text{dB}$$

Establish Fiber–to–Optical Detector Coupling Loss (*L_{f-d}*)
The average loss is 1 dB.

$$L_{f-d} = 1\ \text{dB}$$

Calculate Total Losses

$$
\begin{aligned}
L_t &= L_f + L_c + L_{T-t} + L_{f-d} \\
&= 3.5 + 2 + 1 + 1 \\
&= 7.5\ \text{dB}
\end{aligned}
$$

Calculate Receiver Sensitivity (P_r)

Optical receiver sensitivity is established by algebraically adding the link total loss and the optical power coupled into the fiber.

$$P_r = L_t + P_o$$
$$= -7.5 + (-46)\,\text{dB}$$
$$= -53.5\,\text{dB}$$
$$P_r = -53.5\,\text{dB} \quad \text{or} \quad P_r = -23.5\,\text{dBm}$$

Link Design Summary List

Length of the optical list	3.5 km
System bandwidth	40 MHz

Optical Source

Operating wavelength	830 nm
Optical power output	−46 dB (50 μW)
Power coupled into the fiber	25 μW

Optical Fiber

Multimode width	40 μm
Fiber attenuation	1 dB/Km
Total fiber loss	3.5 dB
Bandwidth–length product	140 MHz/km
Number of connectors	2
Total connector loss	2 dB
Time and temperature degradation	1 dB
Fiber-to-detector loss	1 dB
Link power margin	7.5 dB
Receiver sensitivity	−29.5 dBm.

10.2.2 Multichannel Analog Systems

An analog multichannel optical fiber link must be designed to process four video channels in a **VSB/AM/FDM** mode of operation as shown in Figure 10.4. In the VSB/AM/FDM optical fiber link of Figure 10.4, each individual video channel is band limited by the low-pass filter, amplified, and then modulated by a separate carrier signal while a corresponding local oscillator generates each carrier signal. The modulated video signals are then frequency-division multiplexed, and the combined signal is fed to the input of the optical

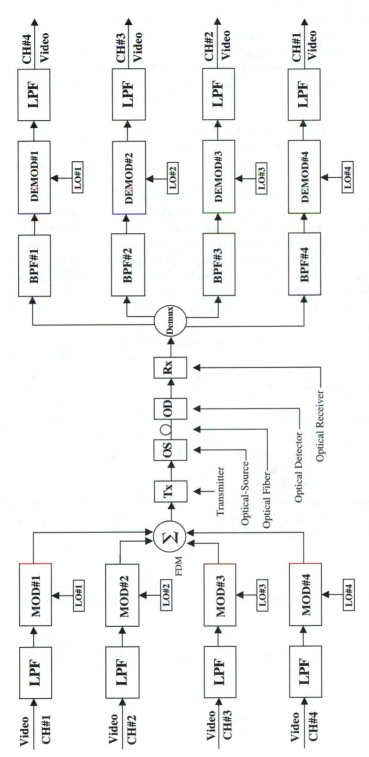

FIGURE 10.4 Block diagram of an analog multichannel system.

transmitter. The transmitter in an optical link performs several functions such as filtering of the modulated video channels in order to maintain bandwidth restrictions and adjusting the signal level to establish compatibility with the dynamic range of the optical source. It also converts the FDM output voltage to current in order to drive the optical source. This current is adjusted through a feedback mechanism to a level appropriate to that required by the source. Any current variations attributed to temperature changes will also be offset by this feedback mechanism.

At the other end of the link the detector converts the optical power to electric power while the receiver amplifies the generated electric signal to the level required by the demultiplexer circuit. The demultiplexer separates the combined FDM signal into single carriers. The bandpass filter makes sure that bandwidth restrictions are observed while the demodulator circuits demodulate each individual video channel to its original form. Frequency-division-multiplexing schemes can also accommodate frequency demodulation video or audio signals as well as amplitude-, frequency-, and phase-shift-keying signals. Time-division multiplexing (TDM) is best suited for pulse-code-modulation (PCM) signals.

Each of these multiplexed formats reflects different optical power and SNR system requirements. The selection of a particular modulation technique is directly related to the multiplexing scheme employed into the system, the baseband bandwidth, and the required signal-to-noise ratio. The modulation of the baseband signals with different subcarriers is highly recommended for systems in which SNR must strictly be maintained to a pre-established level. Since the available optical power from the source is comparatively very small and the system available bandwidth is very large, the SNR can be maintained at a desirable level though trade-offs between optical power and system bandwidth.

In a VSB/AM system, the FDM total bandwidth is determined by the type and number of baseband signals and the guard bands required. Guard bands between the various modulated signals are absolutely necessary in order to prevent the possibility of cross-talk, a very important source of system degradation. The FDM bandwidth will also determine the required optical receiver bandwidth, usually selected to be longer than the FDM bandwidth. This higher receiver bandwidth is necessary in order to relax the otherwise strict receiver filter requirements. In reference to optical power requirements, the available optical power from the source is unevenly distributed among all the channels occupying the system bandwidth. The larger portion of the optical power is distributed at the higher-subcarrier-frequency band and the smaller portion to the lower. Since the noise power density of the receiver is higher at higher frequencies, lower optical power at these higher frequencies results in overall signal-to-noise channel degradation. Therefore, one of the most important parameters to be established in a VSB/AM/FDM multichannel video system transmitting over an optical fiber link is the carrier-to-noise ratio (CNR). This ratio is given by

$$\text{CNR} = \text{SNR}_{\text{dB}} + 10\log\left(\frac{P_c}{P_{\text{sb}}}\right) + 20\log\left(\frac{1}{m}\right) + 20\log(C_{\text{pp/rms}}) + 20\log(C_{\text{pc}}) - W_{f_{\text{dB}}} \quad \textbf{(10.2)}$$

where SNR_{dB} = signal-to-noise ratio over the channel bandwidth in dB
               P_c = carrier power (W)
             P_{sb} = **sideband power** (W)

$$m = \text{modulation index (\%)}$$
$$C_{\text{pp/rms}} = \text{peak-to-peak–to–rms VSB signal conversion (dB)}$$
$$C_{\text{pc}} = \text{picture-to-composite (signal + sync) VSB signal}$$
$$\text{conversion (dB)}$$
$$W_f = \text{\textbf{filter weighting factor} used (dB)}$$

It is evident from the CNR expression that the CNR is subject to various system components such as modulation index (m), carrier power (P_c) harmonic power (P_{sb}), conversion from C_{pp} to C_{rms}, and from picture to composite signal, and finally the filter weighing factor (W_f). It follows that each individual component in the formula must be calculated separately and then incorporated into the final expression:

$$P_c = \frac{V^2}{R} \tag{10.3}$$

and

$$V = V_c + V_{\text{sb}} \tag{10.4}$$

Substituting Eq. (10.3) and (10.4), we have

$$P_c = \frac{(V_c + V_{\text{sb}})^2}{R}$$

Assuming

$$V_c = 2V_{\text{sb}},$$
$$P_c = \frac{(2V_{\text{sb}} + V_{\text{sb}})^2}{R} = \frac{(3V_{\text{sb}})^2}{R} = \frac{9(V_{\text{sb}})^2}{R} \tag{10.5}$$

because $P_{\text{sb}} = \dfrac{(V_{\text{sb}})^2}{R}$.

Substituting Eq. (10.5) into (10.4), we have

$$P_c = 9P_{\text{sb}} \tag{10.6}$$
$$\frac{P_c}{P_{\text{sb}}} = 9$$

In dB,

$$\frac{P_c}{P_{sb_{dB}}} = 10 \log(9)$$

$$\frac{P_c}{P_{sb_{dB}}} = 9.54$$

For the maximum modulation index of 87% (0.87), the second term in the relationship of CNR is

$$20 \log\left(\frac{1}{m}\right) = 20 \log\left(\frac{1}{0.87}\right) = 1.2 \text{ dB}$$

The conversion factor from peak-to-peak to rms is given by

$$C_{pp/rms} = 20 \log\left(\frac{2 + 2\sqrt{2}}{2}\right) = 7.66 \text{ dB}$$

The picture-to-composite signal conversion factor is expressed by

$$C_{pc} = 20 \log\left(\frac{140}{100}\right) = 2.92 \text{ dB}$$

The filter-measuring factor due to white noise is established at 6.2 dB (CCIR rec.),

$$\therefore W_f = 6.2 \text{ dB}$$

Substituting the terms into the CNR equation gives

$$CNR_{dB} = SNR + 10 \log\left(\frac{P_c}{P_{sb}}\right) + 20 \log\left(\frac{1}{m}\right) - 20 \log(C_{pp/rms}) + 20 \log(C_{pc})$$

$$= SNR_{dB} + 9.54 + 1.2 - 7.66 + 2.92 - 6.2 = SNR_{dB} - 0.2$$

From this expression, it is evident that the signal-to-noise (SNR) ratio of a VSB/AM/FDM video optical system must be set at 0.2 dB above the CNR level. The level of CNR is also determined by the composite power of the second and third harmonics. The calculation of these harmonics is given as follows:

$$P_{th_{dB}} = P_{2h_{dB}} + P_{3h_{dB}} \qquad (10.7)$$

where $\quad P_{th}$ = total harmonic power (dB)
$\quad\quad\quad P_{2h}$ = power of the second harmonic (dB)
$\quad\quad\quad P_{3h}$ = power of the third harmonic (dB)

Harmonic distortions as a result of the optical source nonlinearity are evaluated by the following expressions:

Second Harmonic (P₂ₕ)

$$P_{2h_{dB}} = 10 \log(T_1) + l_{dB} + 20 \log\left(\frac{K_2}{4}\right) - (B_{cf})_{2dB} \quad\quad (10.8)$$

where $\quad T_1$ = number of sideband terms.
$\quad\quad\quad l$ = level of side band terms above the second harmonic set at 6 dB
$\quad\quad\quad \dfrac{K_2}{4}$ = magnitude of the second harmonic below the carrier

B_{cf} = bandwidth conversion factor for the second-order-term distortion power spread, related to carrier modulation, set at 1.5 dB.

Third Harmonic (P₃ₕ)

$$P_{3h} = 10 \log\left(T_2 + \frac{T_3}{4}\right) + l_2 + 20 \log\left(\frac{K_3}{4}\right) - (B_{cf})_{3dB} \quad\quad (10.9)$$

where $\quad T_2$ = number of $f_a \pm f_b \pm f_c$ terms in the band
$\quad\quad\quad T_3$ = number of $2f_a \pm f_b$, and $f_a - 2f_b$ terms of the band.
$\quad\quad\quad l_2$ = level of T_2 terms above the third harmonic, set at 15.6 dB.
$\quad\quad\quad \dfrac{K_3}{4}$ = magnitude of the third harmonic below the unmodulated carrier
$\quad\quad\quad (B_{cf})_3$ = bandwidth conversion factor for the third-order distortion term power spread, related to carrier modulation, set at 2.4 dB

The values of T_1, T_2, and T_3 can be calculated based on the frequency used in a specific design problem.

DESIGN EXAMPLE 10.2: MULTICHANNEL ANALOG FIBER

Four video channels are to be transmitted through an optical fiber line in a VSB/AM/FDM mode of operation.

System Components Required

Optical source
Fiber
Number of connectors
Number of splices
Optical detector
Receiver

Optical Source

The selection of the optical source will be either laser or LED.
Operating wavelength (nm)
Spectral width (nm)
Coupled power into fiber (dBm)
Rise/fall time (ns)
Source linearity versus index of modulation

Optical Fiber

Single mode or multimode
Operating wavelength (nm)
Attenuation (dB/Km)
Dispersion (P_s/nm/Km): material
Dispersion (P_s/nm/Km): multimode
Temperature variations (dB/km)

Connectors

Insertion loss (dB)
Deviation due to temperature variations

Splices

Splice losses (dB)

Optical Detector

The optical detector device can be an APD or PIN diode. In either case, device characteristics are as follows:
Responsivity (A/W)
Gain
Dark current (A)
Rise/fall time (ns)
Capacitance (F)

The Receiver and Its Characteristics

Input impedance (Ω)
Input capacitance (F)
Gain
Noise factor (r)

Optical Source Selected

Laser diode
Optical power output, 0.5 dBm
Second harmonic distortion, -55 dB
Third harmonic distortion, -75 dB

System characteristics

Type and number of baseband signals to be processed: four, video
Modulation scheme: VSB/AM/FDM

FDM Bandwidth

Assume the following frequency scheme:

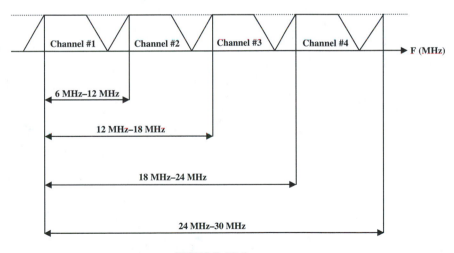

FIGURE 10.5

Channel Positions

X = 1: For channel #1
Y = 4: For channel #2

Channel Frequencies

CH #1: 6–12 MHz
CH #2: 12–18 MHz
CH #3: 18–24 MHz
CH #4: 24–30 MHz

Channel Number (n)

$n = 1, 2, 3, 4$

Compute Carrier-to-Noise Ratio (CNR_{dB})
The CNR of the optical link is limited by the total harmonic distortion. Therefore the second and third harmonic distortion power must be calculated first.

POWER OF THE SECOND HARMONIC ($P_{2h_{dB}}$).

$$P_{2h} = 10 \log(T_1) + l_1 + 20 \log\left(\frac{K_2}{4}\right) - (B_{cf})_2 \qquad \textbf{(10.10)}$$

where

l_1 = level of each term above the second harmonic, 6 dB

$\dfrac{K_2}{4}$ = magnitude of the second harmonic below the carrier, 55 dB

$(B_{cf})_2$ = bandwidth conversion factor = 1.5 dB.

T_1 = number of sideband terms

T_1 is calculated as follows:

$$T_1 = \frac{n+1}{2} - x + (y - n - x + 1) \qquad \textbf{(10.11)}$$

where

n = number of the video channel in the frequency scheme

x = position 1

y = position 4

Assume $n = 3$; then

$$T_1 = \frac{3+1}{2} - 1 + (4 - 3 - 1 + 1) = 2 - 1 + 1 = 2$$

$$\therefore T_1 = 2$$

$$P_{2h} = 10 \log(2) + l_1 + 20 \log\left(\frac{K_2}{4}\right) - (B_{cf})_2$$

$$= 3 + 6 + (-55) - 1.5 = -47.5$$

$$\therefore P_{2h} = -47.5 \text{ dB}$$

POWER OF THE THIRD HARMONIC $(P_{3h_{dB}})$.

$$P_{3h_{dB}} = 10 \log\left(T_2 + \frac{T_3}{4}\right) + l_2 + 20 \log \frac{K_3}{4} - (B_{cf})_3 \qquad (10.12)$$

where
$$l_2 = 15.6 \text{ dB}$$
$$\frac{K_3}{4} = -75 \text{ dB}$$
$$B_{cf} = 2.4 \text{ dB}$$

The values of T_2 and T_3 are calculated as follows:
For $x = 1, y = 4, n = 3$,

$$T_2 = \left(\frac{y - n - 2x}{2}\right)^2 + \left(\frac{n - 1}{2}\right)\left(\frac{n - 2}{2}\right) + (n - x)(y - n) + \left(\frac{y - n}{2}\right)\left(\frac{y - n - 1}{2}\right)$$

$$= \left(\frac{4 - 3 - 2(1)}{2}\right)^2 + \left(\frac{3 - 1}{2}\right)\left(\frac{3 - 2}{2}\right) + (3 - 2)(4 - 3) + \left(\frac{4 - 3}{2}\right)\left(\frac{4 - 3 - 1}{2}\right)$$

$$= \left(\frac{-1}{2}\right)^2 + \left(\frac{2}{2} \times \frac{1}{2}\right) + (2 \times 1) + \left(\frac{1}{2}\right) \times 0 = \frac{1}{4} + \frac{2}{4} + \frac{2}{1} = 2.75 \qquad (10.13)$$

$$\therefore T_2 = 2.75$$

$$T_3 = \frac{n - x}{2} - \frac{n}{3} + \frac{n - 1}{3} - x + 1 + \frac{y - n}{2} + \frac{n - x}{2} + \frac{y - n}{2} - x + 1$$

$$= \frac{3 - 1}{2} - \frac{3}{3} + \frac{3 - 1}{3} - \cancel{x} + \cancel{x} + \frac{4 - 3}{2} + \frac{3 - 1}{2} + \frac{4 - 3}{2} - \cancel{x} + \cancel{x}$$

$$= \frac{2}{2} - \cancel{x} + \frac{2}{3} + \frac{1}{2} + \frac{2}{2} + \frac{1}{2} \qquad (10.14)$$

$$\therefore T_3 = 2.7$$

$$P_{3h_{dB}} = 10 \log\left(T_2 + \frac{T_3}{4}\right) + l_2 + 20 \log\left(\frac{K_3}{4}\right) - (B_{cf})_3$$

$$= 10 \log\left(2.7 + \frac{2.7}{4}\right) + 15.6 + (-75) - (2.4)$$

$$= 5.3 + 15.6 - 75 - 2.4 = -56.5 \text{ dB}$$

Therefore: $P_{3h} = -56.5 \text{ dB}$

TOTAL HARMONIC DISTORTION $(P_{th_{dB}})$

$$P_{th_{dB}} = P_{2h_{dB}} + P_{3h_{dB}} = -46.98$$
$$\therefore P_{th} = -47 \text{ dB}$$

Since CNR is limited by the level of the total harmonic distortion, then

$$CNR = 47 \text{ dB}$$

Previously the relationship between CNR and SNR was established

$$CNR = SNR - 0.2 = 47.2 \text{ dB}$$
$$\therefore \text{Optical fiber system SNR} = 47.2 \text{ dB}.$$

Compute Receiver Sensitivity ($P_{or_{dBm}}$)

The receiver sensitivity is limited by the noise powers attributed to (a) the receiver amplifier $(\iota_{nr})^2$ and (b) the optical detector **dark current** $(\iota_{nd})^2$.

RECEIVER NOISE POWER (ι_{nr}^2)

If a PIN–FET detector–amplifier combination is employed at the receiver end, then the receiver noise power $(\iota_{nr})^2$ is given by

$$(\iota_{nr})^2 = \frac{4\,K\,T\,(\text{BW})}{R_i} + \frac{16\pi^2 K\,T\tau C^2(\text{BW}_3)^3}{g_m} \qquad \textbf{(10.15)}$$

where
$$(\iota_{nr})^2 = \text{detector–receiver amplifier noise power (dB)}$$
$$K = \text{Boltzmann's constant } (1.38 \times 10^{-23}\text{J-K})$$
$$T = \text{operating temperature } 273 \text{ K} + 17 = 300 \text{ K})$$
$$\text{BW} = \text{channel bandwidth (6 MHz)}$$
$$R_i = \text{FET input impedance (1 M)}$$
$$g_m = \text{FET transconductance (0.03)}$$
$$C = \text{amplifier input capacitance (1.5 pF)}$$
$$\tau = \text{FET channel noise factor (1.6)}$$

$$\left.\begin{array}{l} f_h = 30 \text{ MHz} \\ f_l = 24 \text{ MHz} \end{array}\right\} \text{Channel 4}$$

$$(\text{BW}_3)^3 = \frac{(f_h)^3 - (f_l)^3}{3} = \frac{(30 \times 10^6)^3 - (24 \times 10^6)^3}{3}$$

$$= \frac{(27 \times 10^3) \times 10^{18} - (14 \times 10^3) \times 10^{18}}{3}$$

$$\therefore (\text{BW}_3)^3 = 4.3 \times 10^{21} \text{ Hz}$$

Substituting these values into the receiver noise power relationship, we have

$$(\iota_{nr})^2 = \frac{4(1.38 \times 10^{-23})(300)(6 \times 10^6)}{1 \times 10^6} + \frac{16(\pi)^2(1.38 \times 10^{-23})(300)(1.6)(1.5 \times 10^{-12}) \cdot (4.3 \times 10^{21})}{0.03}$$

Therefore,

$$(\iota_{nr})^2 \cong 43 \times 10^{-20}\,\text{W}.$$

DETECTOR NOISE POWER $(\iota_{nd})^2$

The detector noise power due to dark current is given by

$$(\iota_{nd})^2 = 2eI_d(\text{BW}) \tag{10.16}$$

where
e = electron charge $(1.59 \times 10^{-19}\,\text{C})$
I_d = Dark current (12 nA)
BW = bandwidth (6 MHz)

Therefore,

$$(\iota_{nd})^2 = 2(1.59 \times 10^{-19})(12 \times 10^{-9})(6 \times 10^6) = 229 \times 10^{-22}$$

or

$$(\iota_{nd})^2 = 2.3 \times 10^{-20}\,\text{W}$$

The relationship establishing optical receiver sensitivity (P_{or}) is expressed by

$$P_{or} = \frac{2 \cdot e \cdot \text{CNR} \cdot \text{BW} \cdot N^2}{r \cdot m^2} \left(1 + \sqrt{1 + \frac{m^2(\iota_{nr})^2 + (\iota_{nd})^2}{N^2 \cdot 2\,(e \cdot \text{BW})^2 \cdot \text{CNR}}} \right) \tag{10.17}$$

where
e = electron charge $(1.59 \times 10^{-19}\text{C})$
CNR = 47 dB or 5×10^4
BW = channel bandwidth $(6 \times 10^6\,\text{GHz})$
N = number of channels (4)
m = modulation index (0.87)
$(\iota_{nr})^2$ = receiver noise power $(43 \times 10^{-20}\,\text{W})$
$(\iota_{nd})^2$ = detector noise power due to dark current $(2.3 \times 10^{-20}\,\text{W})$
r = **responsivity** (0.6)

Substituting these values into the receiver sensitivity equation, we have

$$P_{or} = 7 \times 10^{-6}\,\text{W}$$

or, in dBm,

$$P_{or} = -21.5\,\text{dBm}$$

Compute the Optical Link Operational Distance (d_{km})
The distance between repeaters in an optical fiber link is given by

$$d_{km} = \frac{P_{or} - P_{or} - L_c - M_{ar}}{L_s + L_f} \tag{10.18}$$

where

P_{ot} = optical transmitted power (dBm)
P_{or} = optical receiver sensitivity (dBm)
L_c = **coupling loss** (dB)
L_s = splice loss (dB)
L_f = optical fiber attenuation (dB/km)
M_{ar} = margin power considered as loss, incorporating various system degradations.

Assuming a laser source with an optical power output (P_{ot}) of -2 dBm, a coupling loss (L_c) of 2 dB, a splice loss (L_s) of 0.5 dB, using an optical fiber with attenuation (L_f) of 0.5 dB/km, and allowing a power margin of 4 dB, the maximum link distance d_{km} is calculated by

$$d_{km} = \frac{P_{ot} - P_{or} - L_c - M_{ar}}{L_s + L_f} = \frac{-2 - (-21.5) - 2 - 4}{0.5 + 0.5}$$

$$= \frac{-2 + 21.5 - 6}{1} = 21.5 - 8 = 13.5 \text{ km}$$

Therefore, link distance $d = 13.5$ km.

10.2.3 FM/FDM Video Signal Transmission

The transmission of a video signal through an FM/FDM mode of operation exhibits two significant advantages over the VSB/AM/FDM system: larger achievable link distance and substantial improvement in the signal-to-noise ratio. The main disadvantage of this mode of operation is the limited number of video channels that can be processed. This limitation is based on the fact that a larger number of video channels in the system increases the harmonic distortion and consequently decreases the link distance. The FM/FDM block diagram is shown in Figure 10.5.

The relationship between the baseband SNR and CNR in the FM/FDM system is given by

$$\text{CNR}_{\text{rms}_{dB}} = \text{SNR}_{\text{rms}_{dB}} = 20 \log \frac{\Delta f}{\text{BW}_b} - 10 \log \frac{\text{BW}_c}{\text{BW}_b} - 1.76 \text{ dB} \tag{10.19}$$

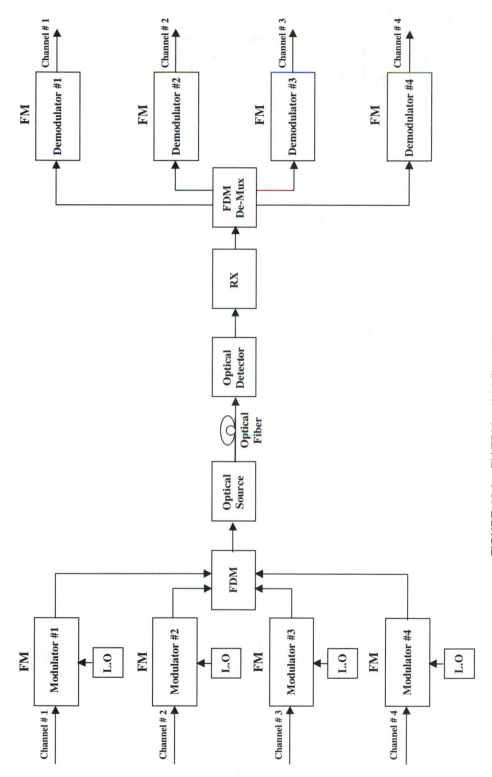

FIGURE 10.6 FM/FDM optical fiber video transmission link.

where $SNR_{rms_{dB}}$ = baseband bandwidth SNR_{rms} in dB
Δf = frequency deviation (set at 8 MHz)
BW_b = baseband bandwidth of the video channel (6 MHz)
BW_c = carrier bandwidth (MHz)

The specified SNR_{pp} is converted to the required SNR_{rms} for Eq. (10.19) as follows:

$$SNR_{rms_{dB}} = SNR_{pp_{dB}} - C_{pp/rms_{dB}} - D_{f_{dB}} - W_{f_{dB}} \qquad \textbf{(10.20)}$$

where $C_{pp/rms}$ = conversion factor from peak-to-peak to rms of a video signal, given by

$$C_{pp/rms} = 20 \log\left(\frac{2 + 2\sqrt{2}}{2}\right) = 7.66 \text{ dB} \qquad \textbf{(10.21)}$$

D_f = FM pre-emphasis/de-emphasis improvement noise factor (set at 10 dB)
W_f = filter weighting factor, set at 6.2 dB (CCIR Rec.)
BW_c = carrier bandwidth in MHz, given by

$$BW_c = 2(\Delta f + BW_b) \qquad \textbf{(10.22)}$$

Therefore,

$$\begin{aligned}
SNR_{RMS} &= SNR_{pp_{(dB)}} - C_{pp/rms_{(dB)}} - D_{f_{dB}} - W_{f_{dB}} \\
&= SNR_{pp_{dB}} - 7.66 \text{ dB} - 10 \text{ dB} - 6.2 \text{ dB} \\
&= SNR_{pp_{dB}} - 23.86 \text{ dB} \qquad \textbf{(10.23)}
\end{aligned}$$

Substituting Eq. (10.23) into (10.19), CNR can be established:

$$CNR = SNR_{pp} - 23.86 - 20 \log\frac{\Delta f}{BW_b} - 10 \log\frac{BW_c}{BW_b} - 1.76 \text{ dB} \qquad \textbf{(10.24)}$$

DESIGN EXAMPLE 10.3

Design an optical fiber link capable of transmitting eight Video channels to a maximum allowable distance through an FM/FDM mode of operation. The required SNR_{pp} is to be maintained at 58 dB.

Solution

Compute CNR_{dB}

$$CNR_{dB} = SNR_{pp} - 23.86 - 20 \log\frac{\Delta f}{BW_b} - 10 \log\frac{BW_c}{BW_b} - 1.76 \text{ dB}$$

where:
$$\Delta f = 8 \text{ MHz}$$
$$BW_b = 6 \text{ MHz}$$
$$BW_c = (2 \Delta f + BW_b) = 2(8 \text{ MHz} + 6 \text{ MHz}) = 28 \text{ MHz}$$

$$CNR = (58 - 23.86) - 20 \log\left(\frac{8}{6}\right) - 10 \log\left(\frac{28}{6}\right) - 1.76$$
$$= 34.14 - 2.5 - 6.7 - 1.76 = 34.14 - 10.96$$
$$\therefore CNR = 23.18 \text{ dB}$$

Compute FDM System Bandwidth (BW$_{FDM}$)

The required FDM system bandwidth is given by

$$BW_{FDM} = N(BW_b + BW_g)$$

where

N = number of video channels (8)
BW_b = baseband bandwidth (6 MHz)
BW_g = guard band (16 MHz) (required in order to minimize adjacent-channel interference)

Therefore,

$$BW_{FDM} = 8(28 + 16) = 352 \text{ MHz}$$

Establish Frequency Plan

TABLE 10.2 Frequency Plan

Channel No.	$f_{l_{MHz}}$	$f_{c_{MHz}}$	$f_{h_{MHz}}$
1	46	60	74
2	86	100	114
3	126	140	154
4	166	180	194
5	206	220	234
6	246	260	274
7	286	300	314
8	326	340	354

Compute Receiver Sensitivity P_{or}

Receiver sensitivity is limited by the sum of the detector and the receiver noise power, and varies for each individual video channel. Therefore, in order for receiver sensitivity to be established, both detector and receiver noise power must be determined for each individual channel.

RECEIVER NOISE POWER $(\iota_{nr})^2$

$$(\iota_{nr})^2 = \frac{4KTBW}{R_i} + \frac{16\pi^2 K\tau C^2 (BW_3)^3}{g_m}$$

where
K = Boltzmann's constant $(1.38 \times 10^{-23}$ J-K$)$
T = operating temperature (assumed 300 K)
BW = FM bandwidth (28 MHz)
R_i = FET input impedance (1 MΩ)
g_m = FET transconductance (0.03 siemens)
C = amplifier input capacitance (1.5 pF)
τ = channel noise factor (1.6)

It is evident from the formula establishing $(\iota_{nr})^2$ that only the $(BW_3)^3$ factor changes, while all the other parameters remain the same. Therefore, the factor $(BW_3)^3$ for each video channel is calculated based on frequency plan values of Table 10.2.

Channel 1:

$$(BW_3)^3 = \frac{(f_h)^3 - (f_l)^3}{3} = \frac{(74 \times 10^6) - (46 \times 10^6)^3}{3} = 0.1 \times 10^{24}$$

Repeating the process for the rest of the channels, $(BW_3)^3$ is given in Table 10.3.

Substituting the values into equation for $(\iota_{nr})^2$, we have the following values for channel 1.

RECEIVER NOISE POWER $(\iota_{nr})^2$:

$$(\iota_{nr})^2 = \frac{4(1.38 \times 10^{-23}) \times 300 \times 28 \times 10^6}{1 \times 10^6}$$
$$+ \frac{16(\pi)^2 \times 300 \times 1.38 \times 10^{-23} \times 1.6 \times (1.5 \times 10^{-12})^2 (0.1 \times 10^{24})}{0.03}$$

$(\iota_{nr})^2 = 82.6 \times 10^{-19}$

DETECTOR NOISE POWER $(\iota_{nd})^2$

$$(\iota_{nd})^2 = 2e\, I_d\, (BW)$$

where
e = electron charge $(1.59 \times 10^{-19}\, C)$
I_d = detector device dark current (2 nA)
BW = FM bandwidth (28 MHz)

Substituting these values into the equation for $(\iota_{nr})^2$, we have

$$(\iota_{nd})^2 = 2 \times 1.59 \times 10^{-19} \times 12 \times 10^{-9} \times 28 \times 10^6$$
$$(\iota_{nd})^2 = 1.06 \times 10^{-19}$$

From this, it is evident that all video channels have the same detector noise power.

The process is repeated for each individual video channel, and the obtained data is listed in Table 10.4.

Receiver sensitivity differs for each video channel because each is based on different values of $(\iota_{nr})^2 + (\iota_{nd})^2$ listed in Table 10.4. The general expression determining receiver sensitivity is the same as that of a VSB/AM/FDM system:

$$P_{or} = \frac{2e \cdot CNR \cdot BW \cdot N^2}{rm^2} \left(1 + \sqrt{1 + \frac{m^2[(\iota_{nr})^2 + (\iota_{nd})^2]}{N^2 2[e \cdot BW]^2 CNR}} \right)$$

where

$$
\begin{aligned}
e &= \text{electron charge } (1.59 \times 10^{-19}\text{C}) \\
CNR &= 23.18 \text{ dB or } 208 \\
BW &= 28 \text{ MHz} \\
N &= 8 \text{ channels} \\
m &= 0.5 \text{ (modulation index assumed)} \\
(\iota_{nr})^2 + (\iota_{nd})^2 &= \text{from Table 10.4} \\
r &= \text{responsivity } (0.6)
\end{aligned}
$$

Channel 1

$$P_{or} = \frac{2(1.59^{-19})(208)(28.10^6)(8^2)}{(0.6)(0.5)^2}$$

$$\left(1 + \sqrt{1 + \frac{(0.5)^2(83.69 \cdot 10^{-19})}{8^2 \cdot 2[1.59 \cdot 10^{-19}(28.10^6)^2 \cdot 208]}} \right) = 2.5 \times 10^{-6}$$

TABLE 10.3

Channel	$(BW_3)^3$
1	0.1×10^{24}
2	0.29×10^{24}
3	0.55×10^{24}
4	0.9×10^{24}
5	1.4×10^{24}
6	1.9×10^{24}
7	2.2×10^{24}
8	3.2×10^{24}

TABLE 10.4

Channel	$(\iota_{nr})^2$	$(\iota_{nd})^2$	$(\iota_{nr})^2 + (\iota_{nd})^2$
1	82.6×10^{-19}	1.09×10^{-19}	83.69×10^{-19}
2	230×10^{-19}	1.09×10^{-19}	321×10^{-19}
3	430×10^{-19}	1.09×10^{-19}	431×10^{-19}
4	706.6×10^{-19}	1.09×10^{-19}	707.6×10^{-19}
5	1092×10^{-19}	1.09×10^{-19}	1093×10^{-19}
6	1482×10^{-19}	1.09×10^{-19}	1483×10^{-19}
7	1716×10^{-19}	1.09×10^{-19}	1717×10^{-19}
8	2496×10^{-19}	1.09×10^{-19}	2497×10^{-19}

In dBm,

$$P_{or} = -26.02 \text{ dBm}$$

Channel 2

The preceding process is repeated with the only variable factor being $(\iota_{nr})^2 + (\iota_{nd})^2$. Therefore,

$$P_{or} = -24.56 \text{ dBm.}$$

Channel 3

$$P_{or} = -23.52 \text{ dBm}$$

Channel 4

$$P_{or} = -23.3 \text{ dBm}$$

Channel 5

$$P_{or} = -21.87 \text{ dBm}$$

Channel 6

$$P_{or} = -21.3 \text{ dBm}$$

Channel 7

$$P_{or} = -20.99 \text{ dBm}$$

Channel 8

$$P_{or} = -20.26 \text{ dBm}$$

From this it is evident that there are different optical power requirements for different video channels. For example, channel 1 requires 2.5 μW of optical power whereas channel 8 requires 9.4 μW of optical power. This problem can be solved by selecting the receiver with the minimum sensitivity equal to that of the higher video channel or by adjusting the transmitter power proportionally to satisfy the individual channel requirements.

Compute optical fiber link distance (d_{km})
The maximum distance of an optical fiber link is given by

$$d_{km} = \frac{P_{ot} - P_{or} - L_c - P_m}{L_f + L_s}$$

where P_{ot} = optical source power (dBm)
$\quad\quad\quad\quad P_{or}$ = receiver sensitivity
$\quad\quad\quad\quad L_c$ = coupling loss
$\quad\quad\quad\quad L_s$ = splice loss per kilometer
$\quad\quad\quad\quad L_f$ = fiber loss per kilometer
$\quad\quad\quad\quad P_m$ = power margin (between 5 and 7 dB)

Assuming,

P_{ot} = 0 dBm
P_{or} = −20.26 (selected from the highest channel)
L_c = coupling loss $\cong$ 2 dB
L_s = 0.05 dB/km
L_f = 0.5 db/km

and substituting into (1), we have

$$d_{(km)} = \frac{0 - (-20.26) - 2 - 6}{0.5 + 0.05} = \frac{20.26 - 8}{0.55}$$

$$\therefore d_{(km)} \cong 22$$

10.3 DIGITAL OPTICAL FIBER SYSTEMS

In the introduction of this chapter, the advantages of digital transmission over analog transmission were clearly stated. A brief summary of these advantages is listed as follows:

Increase of information capacity
Resistance to link impairments
BER 10^{-9} easily achievable

Digital optical fiber systems incorporate the same basic components in their design as analog systems. For example, the system designer must select the following:

Optical source: LED or laser diode
Optical fiber: single-mode or multimode
Operating wavelength: 800–900 nm or 1300–1600 nm
Optical detector: APD or PIN–FET
Modulation: digital (two-level or multilevel)
Multiplexing: time division multiplexing (TDM) or wavelength-division multiplexing (WDM)

Design Procedures

1. *Establish system performance:* In a digital optical fiber communications system the performance of the system is established by setting the BER to a level commonly accepted as the normal level, that is, BER $= 10^{-9}$.

2. *Establish system bandwidth–length product (BL):* This important parameter specifies the maximum system carrying capability over the required distance while maintaining the pre-established BER at 10^{-9} for most systems.

3. *Select optical fiber:* The selection of optical fiber is mainly based on the required maximum system capacity and operating wavelength. It can be a

Single-mode fiber with a maximum bandwidth–length product of 1Ghz-km

Step-index fiber with bandwidth–length product of 100 MHz-km

Graded-index fiber with bandwidth–length product of 1.5 GHz-km

4. *Select optical source:* The optical source can be either a LED or a LASER diode. One of the main device operating characteristics is its optical power output. LED devices exhibit optical power ranging between -20 dBm to -10 dBm; laser devices exhibit power ranging between -10 dBm to $+6$ dBm.

5. *Select optical detector:* Optical detectors are either APD or PIN–FET devices. The selection of the detector device to be utilized in a particular optical system is based on the system's operating wavelength and receiver sensitivity. The sensitivity of an APD device ranges between -60 dBm and -42 dBm, and the PIN–FET device sensitivity ranges between -60 dBm and -30 dBm. This applies to a system capacity of between 10 Mb/s and 1 Gb/s respectively.

6. *Establish maximum link losses* $(L_{\text{max}_{\text{dB}}})$: Maximum link loss is established as the difference between the optical source power output P_o and the receiver sensitivity P_{or}. Therefore,

$$L_{\text{max}_{\text{dB}}} = P_{o_{\text{dB}}} - P_{\text{or}_{\text{dB}}}$$

7. Determine fiber loss $(L_{f_{\text{dB}}})$: Optical link fiber loss is given by

$$L_{f_{\text{dB}}} = L_{\text{max}_{\text{dB}}} - L_{c_{\text{dB}}} - L_{s_{\text{dB}}} - P_{m_{\text{dB}}}$$

where
L_c = connector loss (dB)
L_s = splice loss (dB)
P_m = power margin (dB)

The **power margin** (P_m) is set to compensate for various system losses related to temperature variations, component aging, and so on. This power margin varies from design to design between 6 and 10 dB.

8. *Establish link maximum distance* $(d_{\text{max}_{\text{km}}})$: d_{max} is established as the ratio of optical fiber losses (L_f) to manufacturer's specified optical fiber losses per km (L_{f_m}) in dB/km.

$$d_{\text{max}_{(\text{km})}} = \frac{L_f}{L_{f_m}}$$

For the optical link to be functional,

$$d_{\text{max}} \geq d$$

Otherwise the system power budget must be reassessed.

DESIGN EXAMPLE 10.4

Design an optical fiber digital communications link capable of transmitting 15 Mb/s of data to a desired distance of 4 km.

Solution

Determine (BL)

$$(BL) = 15 \text{ Mb/s} \times 4 \text{ km} = 60 \text{ (Mb/s) km}.$$

Select Optical Source
Since the link is not very long, a relatively inexpensive LED device operating at 820 nm and generating -10 dBm of optical power will be used.

Select Optical Detector
Reliability and low cost dictate the selection of a PIN–FET device exhibiting a -50 dBm sensitivity.

Select Optical Fiber
A step-index multimode fiber is selected. This fiber exhibits a 100 (Mb/s) km *BL* product at relatively low cost; therefore, it will be appropriate for this application.

System Analysis

COMPUTE MAXIMUM ALLOWABLE SYSTEM LOSS $(L_{\text{max}_{\text{dB}}})$

In dB,
$$L_{\text{max}_{\text{dB}}} = P_{\text{o}_{\text{dB}}} - P_{\text{or}_{\text{dB}}}$$

where P_{o} = optical source output power (-10 dBm)
P_{or} = optical receiver sensitivity (-50 dBm)

Therefore,

$$L_{\text{max}_{\text{dB}}} = -10\,\text{dB} - (-50\,\text{dBm}) = -10 + 50 = +40\,\text{dBm}$$

COMPUTE SPLICE LOSSES $(L_{s_{\text{dB}}})$, CONNECTOR LOSSES $(L_{c_{\text{dB}}})$ AND POWER MARGIN $(P_{m_{\text{dB}}})$

Assume:

$L_s = 0.5$ dB/slice
$L_c = 1.5$ dB
$P_m = 8$ dB

Therefore, $L_s + L_c + P_m = 0.5 + 1.5 + 8 = 10$ dB.

COMPUTE FIBER LOSSES (L_f)

$$L_f = L_{\text{max}_{\text{dB}}} - (L_s + L_c + P_m) = 40 - 10 = 30\,\text{dB}$$

The maximum fiber loss in the optical link must not exceed 30 dB in order for the link to operate reliably.

Assuming a selected step-index multimode fiber with a specified 6 dB/km attenuation, then,

$$d_{\text{max}} = \frac{L_f}{L_{f_m}} = \frac{30}{6} = 5\,\text{km}$$

Since the maximum distance is greater than the link required distance, the system is functional.

DESIGN EXAMPLE 10.5

Design an optical fiber digital communication link capable of transmitting 150 Mb/s of digital information over a distance of 8 km.

Solution

Determine bandwidth–length product (*BL*)

$$BL = 150\,\text{Mb/s} \times 8\,\text{km} = 1.2\,(\text{Gb/s})\text{km}$$

Select Optical Source
For this design, an LED source operating at 820 nm and a power output of -10 dBm is selected.

Select Optical Fiber
The basic difference between the design in Design Example 10.5 and this design is the fact that in this case, the BL product is very high [1.2 (Gb/s)km], so a graded-index optical fiber is highly desirable.

Select Optical Detector
For this design example the PIN–FET detector, which exhibits good reliability, low cost, and low noise power, is selected.

$$P_{or} = -50 \text{ dBm}$$

System Analysis

COMPUTE MAXIMUM SYSTEM LOSSES (L_{max})

$$L_{max_{dB}} = P_{o_{dB}} - P_{or_{dB}}$$

where P_o = optical source power output (-10 dBm)
          P_{or} = optical receiver sensitivity (-50 dBm)

Therefore, $L_{max} = P_{o_{dB}} - P_{or_{dB}} = -10 - (-50) = -10 + 50 = 40 \text{ dB}.$

ESTABLISH SPLICE AND CONNECTOR LOSSES AND POWER MARGIN

Splice losses $(L_{s_{dB}})$: $0.5 \times 2 = 1.0$ dB (4 km per splice, 2 splices required)
Connector losses (L_c): 1.5 dB
Power margin (P_m): 10 dB

$$L_s + L_c + P_m = 1 + 1.5 + 10 = 12.5 \text{ dB}$$

COMPUTE FIBER LOSS $(L_{f_{dB}})$

$$L_f = L_{max} - (L_s + L_c + P_m) = 40 - 12.5 = 27.5 \text{ dB}$$

DETERMINE MAXIMUM LINK DISTANCE (d_{max})

$$d_{max} = \frac{L_f}{L_{f_m}}$$

Assuming the selected graded-index multimode fiber with attenuation specified at 2 dB/km, then,

$$d_{\max} = \frac{L_f}{L_{f_m}} = \frac{27.5}{2} = 13.75$$

Since $d_{\max} \gg d$, $(d = 8 \text{ km})$ the link is functional.

DESIGN EXAMPLE 10.6

Design an optical fiber digital communications link with a system capacity of 100 Mb/s and a length of 120 km.

Solution

Determine the bandwidth–length product (*BL*)

$$BL = 100 \text{ Mb/s} \times 120 \text{ km} = 12 \text{ (Gb/s)km}$$

Select Optical Source

Since the link distance is quite long and the corresponding bandwidth–length product is in the Gb/s range, a laser diode with an optical power output of 0 dBm and operating wavelength of 1500 nm is appropriate.

Select Optical Fiber

The optical fiber to be employed for such a long distance must have the minimum possible attenuation characteristics. Only a single-mode optical fiber is capable of satisfying such system requirements. Assume a fiber attenuation of 0.40 dB/km operating at the optical source wavelength window.

Select Optical Detector

A PIN–FET detector preamplifier is selected having the same operating characteristics as in Design Example 10.5: $P_{\text{or}} = -50 \text{ dBm}$.

System Analysis

COMPUTE MAXIMUM SYSTEM LOSSES ($L_{\max_{dB}}$)

$$L_{\max} = P_{o_{(dB)}} - P_{\text{or}_{(dB)}} = 0 - (-50 \text{ dB}) = 50 \text{ dB}.$$

ESTABLISH SPLICE AND CONNECTOR LOSSES AND POWER MARGIN

Splice losses ($L_{s_{dB}}$): Assuming one splice for every 4 km of optical link, then

$$\frac{120}{4} = 30$$

The maximum number of splices required is 30 for a 0.4-dB per splice attenuation. The sum of splice attenuation is $30 \times 0.4 = 12$ dB.

Connector losses $(L_{c_{dB}})$: Assume $L_c = 2.5$ dB.
Set power margin: $P_m = 15$ dB.

The total losses are

$$L_s + L_c + P_m = 12 + 2.5 + 15 = 29.5 \text{ dB}.$$

COMPUTE FIBER LOSS $(L_{f_{dB}})$

$$L_{f_{(dB)}} = L_{max_{(dB)}} - (L_s + L_c + P_m)_{dB}$$
$$= 50 - 29.5 = 20.5 \text{ dB}$$

Optical fiber losses must *not* exceed 20.5 dB.

DETERMINE OPTICAL LINK FUNCTIONALITY
The link will be considered functional if $d_{max} \geq 120$ km.

$$L_f = \frac{L_f}{L_{f_m}} = \frac{20.5 \text{ dB}}{0.4 \text{ dB/km}} = 51.25 \text{ km}$$

where L_{f_m} = manufacturer's optical fiber specifications related to fiber attenuation (0.4 dB/km)
It is evident that this link is nonfunctional because $d_{max} < 120$ km (required link distance).

Modifications
It is obvious that the selected components do not satisfy the system requirements, so new components must be selected.

1. Select an optical source (laser diode) with a higher optical output power, ranging from 0 dBm to 0.5 dBm.
2. Reduce splice losses from 0.4 dB per splice to 0.3 dB per splice.
3. Reduce system power margin from 15 dB to 12 dB.
4. Select an optical fiber with a 0.2 dB/km attenuation instead of 0.4 dB/km.
5. Reduce connector losses to 2.0 dB.
6. Perform a new system analysis.

System Analysis

COMPUTE MAXIMUM ALLOWABLE SYSTEM LOSS ($L_{max_{dB}}$)

$$L_{max} = P_o - P_{or} = 0.5 - (-50) = 0.5 + 50 = 50.5 \text{ dB.}$$

ESTABLISH SPLICE AND CONNECTOR LOSSES AND POWER MARGIN

Splice losses ($L_{s_{dB}}$): $L_s = 30 \times 0.3 = 9$ dB.

Connector losses are 2.0 dB.

Power margin is 12 dB.

The total losses are

$$L_s + L_c + P_m = 9 + 2 + 12 = 23 \text{ dB}$$

COMPUTE FIBER LOSSES ($L_{f_{dB}}$)

$$L_{f_{dB}} = L_{max_{dB}} - (L_s + L_c + P_m)_{dB} = 50.5 - 23 = 27.5 \text{ dB.}$$

Optical fiber losses must not exceed 27.5 dB.

DETERMINE LINK FUNCITONALITY

$$d_{max} \geq d$$
$$d_{max} \geq 120 \text{ km}$$
$$d_{max} = \frac{L_f}{L_{f_m}} = \frac{27.5 \text{ dB}}{0.2 \text{ dB/km}} = 137.5 \text{ km}$$

Since $d_{max} > d$ (137.5 km > 120 km), this design is successful.

DESIGN EXAMPLE 10.7: PCM/TDM VIDEO TRANSMISSION OVER OPTICAL FIBER

Design an optical fiber communications link capable of transmitting eight video channels in a PCM/TDM mode at a distance of 60 km.

System Parameters

Required: $SNR_{in_{pp}} = 70$ dB (per video channel); $SNR_{out_{pp}} = 60$ dB.

Video bandwidth: $(BW)_v = 4.2$ MHz.

Audio bandwidth: $(BW)_a = 20$ KHz.

Audio: $SNR_{rmf} = 60$ dB.

System performance: BER $= 10^{-9}$.

System Analysis

$$\text{SNR}_{\text{pp(sys)}} = \text{SNR}_{\text{out}_{\text{pp}}} - \text{SNR}_{\text{in}_{\text{pp}}} = -10 \log\left[\frac{1}{10^6} - \frac{1}{10^7}\right] = 60.46 \text{ dB}$$

CONVERT SNR_{pp} TO SNR_{rms}

$$\text{SNR}_{\text{rms}} = \text{SNR}_{\text{pp}} - C_{\text{pp/rms}} - W_f$$

where $\quad C_{\text{pp/rms}} =$ peak-to-peak–to–rms video signal conversion $= 7.66 \text{ dB}$
$W_f =$ filter weighting factor $= 6.2 \text{ dB}$.

Therefore, $\text{SNR}_{\text{rms}} = 60.46 - 7.66 - 6.2 = 46.6 \text{ dB}$.

COMPUTE NUMBER OF BITS REQUIRED FOR VIDEO SIGNAL ENCODING (n)
Linear PCM systems were discussed in Chapter 2. From these discussions, it was determined for the quantization noise (N_q) that

$$N_q = q^2 \quad \text{or} \quad q = \frac{Aq}{12}$$

If the maximum peak-to-peak signal measurement range of a linear quantizer is A_q and is given by

$$A_q = q \cdot 2^n \quad \text{or} \quad q = \frac{A_q}{2^n}$$

where $\quad q =$ step size
$n =$ number of quantization bits
$2^n =$ number of steps

Then,

$$\text{SNR}_{\text{rms}} = \frac{\text{Signal power}}{\text{Quantization noise power}} \frac{(V_{\text{rms}})^2}{\dfrac{q^2}{12}} = \frac{12 \, (V_{\text{rms}})^2}{\dfrac{q^2}{12}}$$

Substituting $q = \dfrac{A_q}{2^n}$ gives,

$$\text{SNR}_{\text{rms}} = \frac{12\,(V_{\text{rms}})^2}{\dfrac{(A_q)^2}{(2^n)^2}} = (12)(2^n)^2\left(\frac{V_{\text{rms}}}{A_q}\right)^2$$

Since $V_{\text{rms}} = V/2$ and $A_q = 2$ V (peak-to-peak),

$$\text{SNR}_{\text{rms}} = (12)(2^n)^2\left(\frac{\cancel{V}}{\sqrt{2}} \times \frac{1}{2\cancel{V}}\right)^2 = (12)(2^n)^2 \cdot \left(\frac{1}{2\sqrt{2}}\right)^2$$

In dB,

$$\text{SNR}_{\text{rms}} = 10\log(12) + 2(10\log 2^n) + 20\log\left(\frac{1}{2\sqrt{2}}\right) = 10.8 + 6n + (-9.03)$$

$$= 6n + 1.77$$

Since SNR = 46.6 dB (from step 2),

$$6n + 1.77 = 46.6 \text{ dB.}$$
$$6n = 46.6 - 1.77 = 44.83 \text{ dB.}$$
$$n = 8 \text{ bits}$$

COMPUTE VIDEO CHANNEL BIT RATE (f_{b_v})
Video channel bit rate is calculated as follows:

$$f_{b_v} \geq 2.4(\text{BW})(n)$$

where BW = 4.2 MHz

$$f_{b_v} = (2.4)(4.2 \text{ MHz})(8 \text{ bits}) = 80.64 \text{ Mb/s}$$

COMPUTE AUDIO CHANNEL BIT RATE f_{b_a}

$$f_{b_a} = (2.4)(\text{BW})(10)$$

(*Note:* For stereo hi-fidelity audio, 10 quantization bits are required.) Therefore,

$$f_{b_a} = (2.4)(20 \times 10^3)(10) = 480 \text{ Kb/s}$$

COMPUTE TOTAL CHANNEL BIT RATE (f_{b_c})
Channel bit rate is given by

$$f_{b_c} = f_{b_v} + f_{b_a} = 80.64 \text{ Mb/s} + 0.48 \text{ Mb/s} \cong 81 \text{ Mb/s}$$

COMPUTE SYSTEM BIT RATE (f_{b_s})
Total system bit rate is determined by

$$f_{b_s} = \frac{N(f_{b_c})}{F_{\text{TDM}}}$$

where
$$N = \text{number of video channels in the system (8)}$$
$$f_{b_c} = \text{channel bit rate (81 Mb/s)}$$
$$F_{\text{TDM}} = \text{efficiency of the TDM system (90\%–94\%).}$$

Assuming a TDM efficiency of 92%, then

$$f_{b_s} = \frac{8(81)}{0.92} = 704.3$$

Therefore, the system bit rate is established at $f_{b_s} = 705$ Mb/s.

System Characteristics

System capacity: $f_{b_s} = 705$ Mb/s.
Optical link distance: $d = 60$ km.
System performance: BER $= 10^{-9}$.

System Design

DETERMINE BANDWIDTH–LENGTH PRODUCT (BL)

$$BL = 705 \text{ Mb/s} \times 60 \text{ km} = 42.3 \text{ (Gb/s)km}$$

For the optical link distance of 60 km, a laser diode with the following operational characteristics is selected as the optical source.

Operating wavelength: 1300 nm
Optical power coupled into the fiber: 0 dBm
Spectral width: 2.5 nm
Rise time: 0.02 ns

SELECT FIBER
The relatively long distance of the optical link and the large BL product dictate the selection of a single-mode fiber. The selected single-mode optical fiber exhibits the following operational characteristics.

Operating wavelength window: 1300 nm
Attenuation: 0.4 dB/km
Material dispersion: 2.0 P_s/nm/km

SELECT OPTICAL DETECTOR
A PIN detector is utilized that exhibits the following operational characteristics.

Operating wavelength: 1300 nm
Sensitivity: -50 dBm
Gain: 1
Responsivity: 0.6
Rise time: 0.2 ns
Dark current: 1.0 nA
Coupling loss: 0.5 dB
Capacitance: 1.0 pF

SELECTED OPTICAL PREAMPLIFIER
The selected FET preamplifier is operating with the following device characteristics.

Transconductance, $g_m = 0.03$.
Input impedance, $R_i = 1$ MΩ.
Input capacitance, $C_i = 1$ pF.
Noise factor, NF $= 1.8$.

COMPUTE MAXIMUM ALLOWABLE SYSTEM LOSS $(L_{max_{dB}})$.
The maximum allowable system loss is given by.

$$L_{max_{dB}} = P_{o_{dB}} - P_{or_{dBm}}$$

Since $P_o = 0$ dBm and $P_{or} = -50$ dBm,

$$L_{max_{dB}} = 0 - (-50) = 50 \text{ dB}$$

COMPUTE SPLICE AND CONNECTOR LOSSES AND POWER MARGIN

Assuming 0.5 dB per splice and 4 km of optical link per splice, the required number of splices are as follows:

$$\text{Number of splices} = \frac{60 \text{ km}}{4 \text{ km/splice}} = 15$$

$$\text{Total splice losses} = L_s = 15 \times 0.5 = 7.5 \text{ dB}$$

Assume connector losses of 1.5 dB.

$$\therefore L_c = 1.5 \text{ dB}$$

Set a power margin of 12 dB, then

$$L_s + L_c + P_m = 7.5 + 1.5 + 12 = 21 \text{ dB}$$

COMPUTE FIBER LOSSES $L_{f_{dB}}$

$$L_f = L_{max} - (L_s + L_c + P_m) = 50 - 21 = 29 \text{ dB}$$

The maximum optical fiber loss in the link must not exceed 29 dB.

A preliminary investigation shows the following: Since the link distance is 60 km and the optical fiber attenuation is 0.4 dB/km, the total fiber attenuation is $60 \times 0.4 = 24$ dB. For the link to be functional, the optical fiber attenuation based on the manufacturer's specifications (24 dB) must be smaller than the maximum allowable system attenuation, established at 29 dB.

$$d_{max} > d$$

$$d_{max} = \frac{L_f}{L_{f_m}} = \frac{29}{0.4} = 72.5 \text{ km}$$

Therefore, $d_{max} > d$ (60 km), so the optical link is functional.

1. An optical fiber analog communications link is to be established covering a distance of 5 km. The link bandwidth is 35 MHz, and a 50-dB signal-to-noise ratio must be maintained. Determine all the necessary components and calculate receiver sensitivity.
2. Design an optical fiber link for the transmission of five video channels in a VSB/AM/FDM operating mode.
3. Design an optical fiber link for the transmission of six video channels in an FM/FDM mode of operation at a maximum distance while a 60-dB signal-to-noise ratio is constantly maintained.
4. Design an optical fiber link for the transmission of 50 Mb/s over a 10-km distance.
5. Design an optical fiber link for the transmission of 200 Mb/s over a 6.5-km distance.
6. Design an optical link for the transmission of 120 Mb/s of data over a distance of 150 km.
7. Design an optical fiber link for the transmission of five video channels in a PCM/TDM mode of operation and over a distance of 50 km.

System Measurements and Performance Evaluation

Objectives

- Define Fourier series, DFT, FFT, and **Z**-Transform.
- Describe the methods for modulator/demodulator sensitivity and differential gain measurements.
- Explain the importance of return-loss measurements.
- Identify the various methods of performance measurements for digital microwave links.
- Describe the various methods for a complete satellite performance test.
- Understand the principles involved in microwave link performance tests by the eye diagram and constellation display diagram.
- Describe the test procedures for noise-figure and jitter measurements.

Key Terms

Time domain
Frequency domain
Fourier series
DFT
FFT
Z-transform
Modulator sensitivity
Index of modulation
FM spectrum
Bessel function
Demodulator sensitivity
Path interference

Linearity
Differential gain
Microwave link analyzer
Amplitude response
Return loss
Reflector coefficient
Incident power
Return power
Intersymbol interference
Multipath fading
Adjacent channel
   interference (ACI)

The eye diagram
Cochannel interference
Modulator misalignment
Parity check
Timing jitter
Phase errors
Flat-fading
Symbol rate
C/I
CPI
Doppler shift
Constellation display

INTRODUCTION

During the design process or at the operational stages of a communication system, the display and analysis of the signals at key points along the system are absolutely necessary in determining system performance characteristics. The electric signal can be viewed in either the time or frequency domain. When a signal is viewed in the **time domain,** signal amplitude variations are observed in relation to time; when a signal is viewed in the **frequency domain,** the various frequency components can also be observed. Therefore, it is sometimes necessary to transform a signal from the time to the frequency domain or vice versa.

This signal transformation is necessary because it allows for an easier mathematical manipulation, the results of which are indicative of system performance. The mathematical tools required for such a transformation are the Fourier, Laplace and Z-transforms. As mentioned before, when a signal is displayed in the time domain, one is able to observe its amplitude variations as a function of time. This observation allows for certain conclusions to be derived relevant to system behavior. The basic instrument for displaying an electric signal in the time domain is the oscilloscope. In the frequency domain, the individual frequency components of an electric signal can be observed and analyzed, and appropriate conclusions can be drawn as to the system's properties and performance. The basic instrument required for displaying an electric signal in the frequency domain is the spectrum analyzer.

J. B. Fourier demonstrated that all complex signals are composed of a number of sine waves and that any complex signal can be constructed by a number of sine waves. A sine wave displayed in the time and frequency domains is shown in Figure 11.1. Two sine waves (Figure 11.2 a and b) added together produce the signal of Figure 11.2 (c), displayed in the time domain. Figure 11.3 displays the same complex signal in the frequency domain.

In digital communications, signal analysis is very important because in communications systems, the recovery of the carrier signal at the receiver end is of paramount importance. Therefore, the local oscillator's frequency purity and stability must be maintained within the predetermined tolerances during the system's entire operating life. The frequency purity and stability of such a circuit can only be determined by displaying the signal in the frequency domain through a spectrum analyzer.

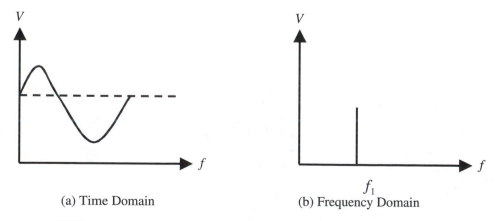

(a) Time Domain   (b) Frequency Domain

FIGURE 11.1 The construction of a complex signal of two sine waves.

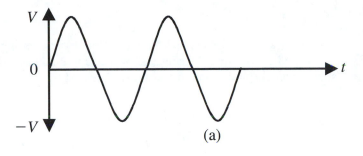

(a)

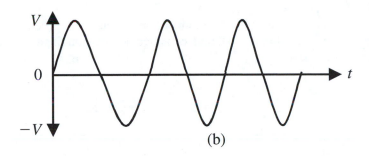

(b)

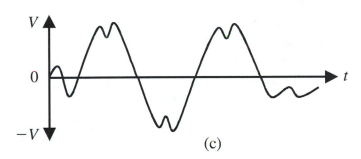

(c)

FIGURE 11.2

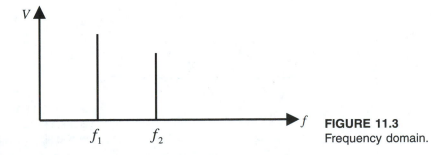

FIGURE 11.3
Frequency domain.

System Measurements and Performance Evaluation

387

11.1 FOURIER SERIES

The representation of a continuous signal by a **Fourier series** starting with a simple sine wave signal displayed in the time domain is shown in Figure 11.4. The mathematical expression of such a signal (pure, no harmonics) is expressed by

$$y = \sin(\omega t) \tag{11.1}$$

where $\omega = 2\pi f$

There is only one frequency component in this expression.

The signal of Figure 11.5 is a more complex signal composed of two sine waves, also displayed in the time domain. The mathematical expression of the signal in Figure 11.5 is given by

$$y = \sin(\omega t) + \frac{1}{3}\sin(3\omega t) \tag{11.2}$$

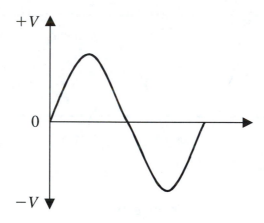

FIGURE 11.4   Sine wave displayed in the time domain.

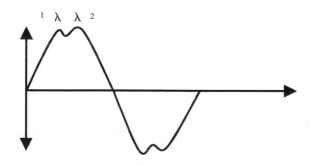

FIGURE 11.5   A continuous signal composed of two sine waves.

This expression indicates that the signal is composed of two signals, a fundamental and one harmonic. Similarly, the signal of Figure 11.6 is composed of a fundamental and two harmonics. The mathematical expression is given by

$$y = \sin(\omega t) + \frac{1}{3}\sin(3\omega t) + \frac{1}{5}\sin(5\omega t) \tag{11.3}$$

The general expression of a continuous signal with one fundamental and n harmonics is shown in Figure 11.7 and given by

$$y = \sin(\omega t) + \frac{1}{3}\sin(3\omega t) + \frac{1}{5}\sin(\omega t) + \cdots + \frac{1}{n}\sin(n\omega t) \tag{11.4}$$

Equation 11.4 represents the mathematical expression of a complex signal composed of a fundamental sine wave and n harmonics. If the number of harmonics is made very large, it is possible to construct a new signal very close to that of a square wave. For that matter, a square wave can be decomposed into a fundamental sine wave and an infinite number of harmonics.

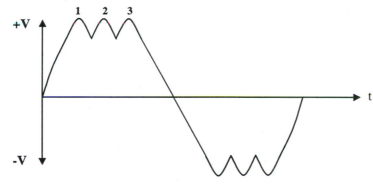

FIGURE 11.6 A continuous signal composed of three sine waves.

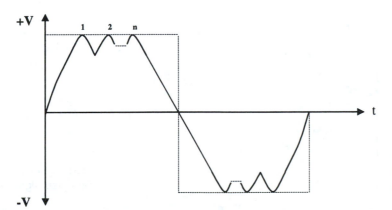

FIGURE 11.7 A continuous signal composed of n sine waves.

11.1.2 Fourier Transform (FT)

The mathematical transformation of a signal from the time domain to the frequency domain, or vice versa, is given by the Fourier transform:

$$x(\omega) = \int_{-\infty}^{+\infty} x(t)\exp(-j\omega t)\, dt \tag{11.5}$$

From the time domain to the frequency domain,

$$x(t) = \int_{-\infty}^{+\infty} x(\omega)\exp(-j\omega t)\, dt \tag{11.6}$$

From the frequency domain to the time domain. Equations (11.5) and (11.6) are both applicable to periodic and nonperiodic waveforms.

Discrete Fourier Transform (DFT)

The Fourier transform (FT) given earlier applies to continuous signals. For a sampled signal (Figure 11.8), a modified version of the Fourier-transform, the discrete-Fourier-transform (**DFT**), is applied as follows:
From Eq. (11.5)

$$x\omega = \int_{-\infty}^{+\infty} x(t)\exp(-j\omega t)\, dt \qquad \text{(continuous)}$$

or

$$x(m\Delta\omega) = \int_{-\infty}^{+\infty} x(t)\exp(-jm\Delta\omega t)\, dt \qquad \text{(discrete)}$$

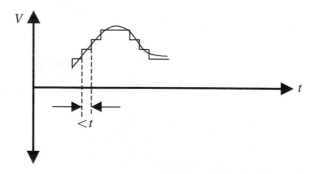

FIGURE 11.8
Sampled signal.

where $\quad m = 0, \pm1, \pm2, \pm3,...$
$\Delta\omega = 2\pi(\Delta f)$
Δf = frequency corresponding to Δt (sample spacing)

Computing the area under the continuous curve, we have

$$x(m\Delta\omega) = \Delta t \sum_{n=-\infty}^{+\infty} x(n\Delta t)\exp(-j2\pi m\Delta f \cdot n \cdot \Delta t) \qquad (11.7)$$

For a reasonable computation time and for approximate results, Eq. 11.7 is limited to a finite time given by

$$x(m\Delta\omega) = \Delta t \sum_{n=0}^{N-1} x(n\Delta t)\exp(-j2\pi m\Delta f \cdot n \cdot \Delta t) \qquad (11.8)$$

Finally, since

$$\Delta t = \frac{T}{N}$$

where $\quad T$ = period
N = number of samples

Then

$$x(m\Delta\omega) = \frac{T}{N}\sum_{n=0}^{N-1} x(n\Delta t)\exp\left(-j2\pi m\frac{n}{N}\right) \qquad (11.9)$$

Equation (11.9) expresses the discrete Fourier transform of a sampled continuous signal.

Fast Fourier Transform (FFT)

For a large value of N, the discrete Fourier transform (DFT) requires a lengthy computation time. For this reason, a simplified version is employed when the values of N are multiples of two. This method of evaluating the DFT is referred to as the "fast Fourier transform" (FFT). A more formal derivation of the DFT is as follows: Any function can be represented by a Mclaurin power series in x:

$$f(x) = C_0 + C_1(x) + C_2(x)^2 + C_3(x)^3 + \cdots + C_n(x)^n \qquad (11.10)$$

or by a Taylor power series

$$f(x) = C_0 + C_1(x - a) + C_2(x - a)^2 + C_3(x - a)^3 + \cdots + C_n(x - a)^n \quad \text{(11.11)}$$

where a_n = coefficient ($n = 0, \pm1, \pm2, ...$)
         x^n = variable

A Taylor series is used for numerical approximation calculations.

11.2 THE Z-TRANSFORM

The Z-transform is derived as follows: The variable (Z) is expressed by

$$Z = \text{Re}^{j\omega} \quad \text{(11.12)}$$

where   Z = complex number illustrated in Figure 11.9.
Recall the Taylor series expression

$$f(x) = \sum_{n=0}^{\infty} a_n x^n \quad \text{(11.13)}$$

Here again, the function $f(x)$ is expressed in terms of a variable (x) to the power of n, and the coefficient a_n. If (Z) is substituted for x, then $f(x)$ is given by

$$f(x) = \sum_{n=0}^{\infty} a_n Z^n \quad \text{(11.14)}$$

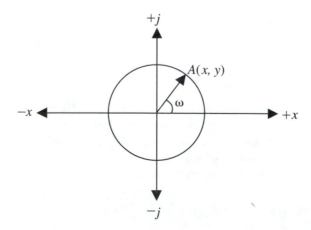

FIGURE 11.9

Applying the shifting theorem, 11.13 becomes

$$f(x) = \sum_{n=-\infty}^{+\infty} a_n Z^{-n} \qquad (11.15)$$

For radius $r = 1$,

$$f(x) = \sum_{n=-\infty}^{+\infty} a_n e^{-j\omega t} \qquad (11.16)$$

Finally, Eq. (11.15) can be expressed as

$$f(x) = \sum_{n=-\infty}^{+\infty} a_n(n) \qquad (11.17)$$

where $-\pi < \omega < +\pi$.

Equation 11.17 is referred to as the "discrete Fourier transform" (DFT).

11.3 MODULATOR/DEMODULATOR SENSITIVITY MEASUREMENTS

11.3.1 The Modulator

Modulator sensitivity is defined as the ratio of the carrier frequency deviation f_c to the input signal power (P):

$$\text{Modulator sensitivity} = \frac{\Delta f_c}{P} \qquad (11.18)$$

Therefore, if both components (Δf_c and P) are known, the modulator sensitivity can be determined.

The power of the input signal can be measured with a power meter, while the modulator output frequency deviation can be established indirectly from the **index of modulation** (m_f):

$$\text{Index of modulation } (m_f) = \frac{\Delta f_c}{f_m} \qquad (11.19)$$

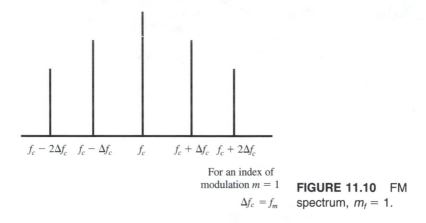

$$f_c - 2\Delta f_c \quad f_c - \Delta f_c \quad\quad f_c \quad\quad f_c + \Delta f_c \quad f_c + 2\Delta f_c$$

For an index of
modulation $m = 1$

$\Delta f_c = f_m$

FIGURE 11.10 FM spectrum, $m_f = 1$.

Since the modulation frequency (f_m) is known, the index of modulation can be established from **Bessel-function** curves. If a pure sinusoidal wave is applied at the input of the modulator and the output is displayed by a spectrum analyzer, the **FM spectrum** will be shown as in Figures 11.10 and 11.11.

This can also be expressed by Bessel curves. That is, the amplitude of the carrier and the amplitudes of the side bands can be plotted in relation to modulation index (Figure 11.12). From these curves, it is evident that the amplitude of the carrier signal crosses the zero point at specific values of the index of modulation. What is required in order to determine the carrier frequency deviation Δf_c is the first zero crossing at a particular index of modulation. This concept is utilized by the **microwave link analyzer** employed here to measure modulator/demodulator sensitivity (Figure 11.13).

When sweeping the local-oscillator frequency for the entire IF bandwidth, there is a point at which the local-oscillator frequency becomes equal to the IF frequency. At that point, a zero beat will be produced. The local oscillator and the time base of the oscilloscope are driven by the same source, that of the sweep generator. Varying the amplitude of the modulating signal (BB) until the first null is obtained and reading the corresponding index of modulation from Bessel curves, the carrier frequency deviation can now be established:

$$\Delta f_c = m_f \cdot f_m \tag{11.20}$$

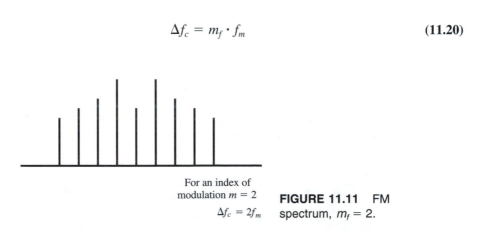

For an index of
modulation $m = 2$

$\Delta f_c = 2f_m$

FIGURE 11.11 FM spectrum, $m_f = 2$.

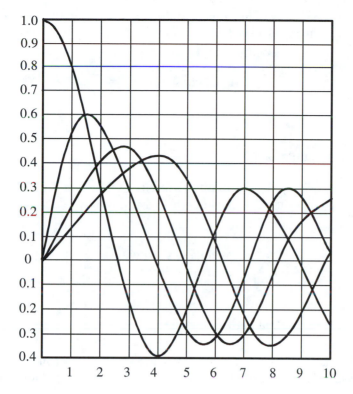

FIGURE 11.12 Bessel curves.

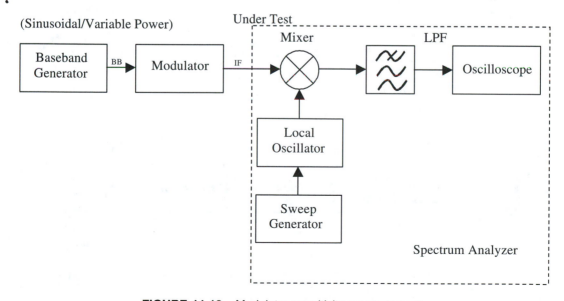

FIGURE 11.13 Modulator sensitivity measurement.

Having obtained with this indirect method the value of P and the value of Δf_c, the modulator sensitivity is then calculated.

$$\text{Modulator sensitivity} = \frac{\Delta f_c}{P} \tag{11.21}$$

11.3.2 The Demodulator

The setup for measuring **demodulator sensitivity** is shown in Figure 11.14. Demodulator sensitivity is the opposite of modulator sensitivity, given by

$$\text{Demodulator sensitivity} = \frac{P}{\Delta f_c} \tag{11.22}$$

where P = power of the the baseband signal at a specific value of carrier deviation Δf_c

Setting a specific value of Δf_c, the corresponding value of baseband power is measured by the power meter. Therefore, by taking the Δf_c ratio, the demodulator sensitivity can now be established.

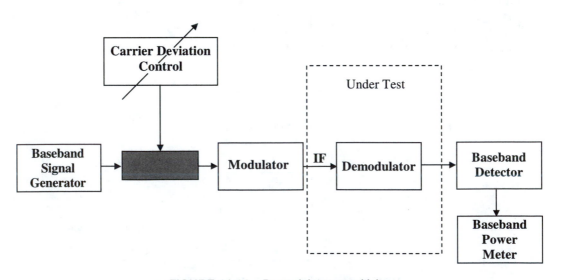

FIGURE 11.14   Demodulator sensitivity.

11.3.3 Modulator/Demodulator Linearity and Differential Gain

In line-of-sight or satellite microwave links employing FDM/FM modulation techniques, the degree of modulator/demodulator nonlinearities must be known in order to perform an accurate system evaluation. In the ideal case, the modulator output frequency deviation Δf_c of the carrier must be proportional to the input signal amplitude variation. In practice, this is not always true. Likewise, the output signal amplitude of a FM demodulator must be proportional to the input frequency derivation Δf_c. In practice, a certain degree of nonlinearity does exist between the output signal and the input frequency deviation, especially at a relatively large index of modulation. The errors induced into the system by modulator/demodulator nonlinearity are detrimental to the system's performance. Modulator/demodulator nonlinearity increases the overall modulator/demodulator impairments. These impairments are grouped under the term **differential gain.**

Differential gain is the measure of intermodulation of two frequencies. One is a high-frequency and low-power signal, the other a low-frequency, high-power signal. Therefore, the degree of nonlinearity and differential gain must be measured and its impact incorporated into the system's performance (Figure 11.15).

In Figure 11.15, the baseband generator provides a precise low-level, high-frequency signal and the sweep generator provides a high-level, low-frequency signal through the combiner circuit. The sweep and baseband signals are applied to the input of the demodulator circuit. At the modulator circuit, the IF signal is modulated by the low-level, high-frequency baseband signal while it is swept through its entire bandwidth by the low-frequency, high-level sweep generator. At the receiver end, the IF signal is fed into the demodulator. The

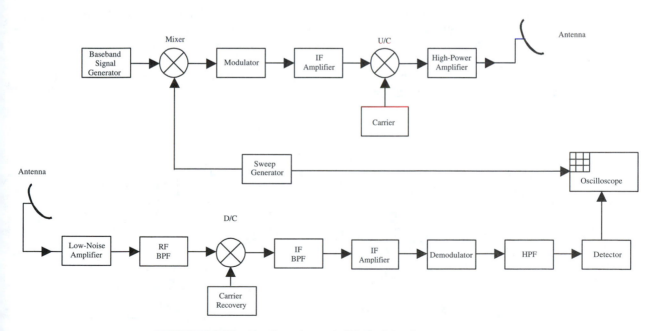

FIGURE 11.15 Nonlinearity and differential gain measurement.

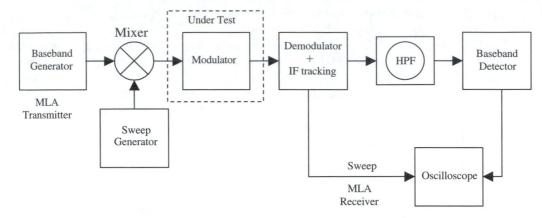

FIGURE 11.16 Modulator nonlinearity measurement using an MLA.

demodulator output of both baseband and sweep signals is fed into the input of a high-pass filter. Therefore, only the high-frequency component (baseband) will pass through and be detected by the detector circuit.

The output of the detector circuit is fed into the oscilloscope input, and the sweep signal is connected to the oscilloscope time base. The oscilloscope displays amplitude variations of the baseband signal in relation to a particular measure of the IF signal $\Delta V/\Delta f_c$. Thus, a measure of the differential gain of the modulator circuit is obtained. This type of differential gain measurement requires a direct connection between the transmitter and receiver, because the oscilloscope time base is connected to the sweep generator at the transmitter end. Normally, the distances separating the transmitter and receiver are quite long, thus presenting a rather difficult logistical problem. A better approach to differential gain and **linearity** measurement is to employ the microwave link analyzer (MLA) (Figure 11.16).

In Figure 11.17, the IF signal is modulated by the baseband signal while the sweep generator sweeps the IF signal through the entire bandwidth. At the receiver end, the high-pass filter allows only the baseband signal to be applied at the input of the detector circuit, while the recovered sweep signal is applied at the base of oscilloscope. Any variations in

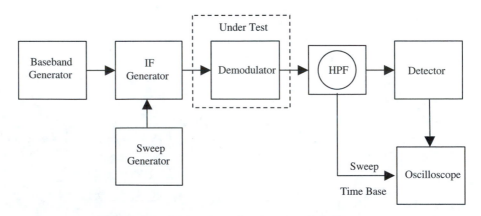

FIGURE 11.17 Demodulator nonlinearity measurement.

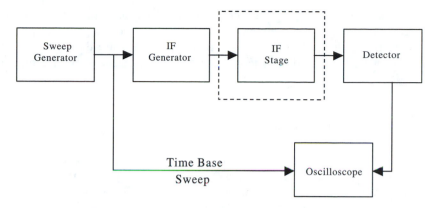

FIGURE 11.18 IF amplitude response measurement.

the baseband signal as the IF is swept through its entire bandwidth, will be detected and displayed at the oscilloscope.

11.3.4 IF Amplitude Response

The IF stages in a line-of-sight or satellite microwave link must be absolutely linear. That is, any amplitude variations of the output signal as a result of nonlinear response will result in an unwanted amplitude-to-frequency modulation. This irregularity contributes to the overall system degradation. An IF amplitude response measurement setup is shown in Figure 11.18.

The sweep generator slowly sweeps through the entire IF bandwidth while maintaining a constant signal amplitude. The output of the IF stage under test is fed into the detector while the oscilloscope time base is connected to the sweep generator. Any variations of the IF output signal will be detected and displayed by the oscilloscope. In this measurement, the potential nonlinearities of the IF generator must be known and the necessary allowances must be made for the accurate determination of the IF circuit amplitude response.

A more convenient way of measuring IF frequency response is shown in Figure 11.19. Here the use of the microwave-link analyzer can eliminate the logistical problem of

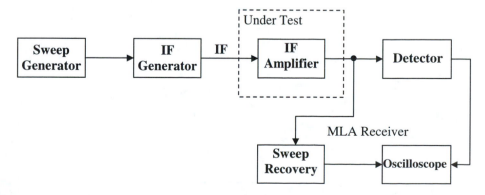

FIGURE 11.19 IF amplitude response measurement (MLA).

connecting the sweep generator to the oscilloscope time base. It was shown in previous tests that the MLA is capable of recovering the sweep signal required for the oscilloscope time base. The rest of the measurement procedures are the same as earlier. Therefore, the utilization of the MLA is more practical and more effective for measuring IF amplitude response.

11.3.5 Return-Loss Measurements

Return loss is the ratio of the microwave power reflected back to the source, to the power generated by the source resulting from a mismatched load. In vector form, this ratio, called the **reflector coefficient,** is shown in Figure 11.20 and is expressed by

$$\rho = \frac{E_r}{E_i} \angle \theta \qquad (11.23)$$

where
ρ = reflector coefficient
E_i = incident signal (vector)
E_r = return signal (vector)
θ = phase difference (degrees)

In dB,

$$L_{loss_{dB}} = -20 \log(E_r/E_i) = -20 \log(\rho) \qquad (11.24)$$

or

$$\text{return loss} = 10 \log(P_r/P_i) \qquad (11.25)$$

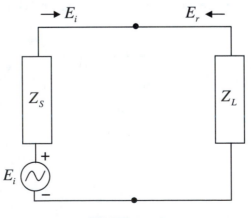

FIGURE 11.20

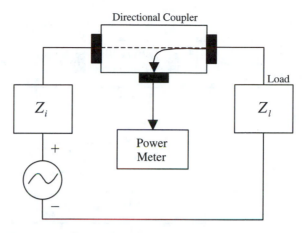

Directional Coupler

Load

Z_i

Z_l

Power
Meter

+

−

FIGURE 11.21   Return-loss measurement by mismatch method.

where P_i = **incident power**
P_r = **return power**

The concept of return loss is very important, especially in microwave link systems design, for two reasons: It inhibits maximum power transfer, and it induces group delay distortion into the system.

There are two basic ways of measuring return loss: the "standard mismatch method" and the "long cable method." The standard mismatch is the preferred method and is briefly presented here (see Figure 11.21).

The Standard Mismatch Method

The standard mismatch method of return loss measurement is shown in Figure 11.21. Microwave power from the source is applied to the load through the directional coupler. Ideally, if $(Z_i = Z_l)$, all the source power must be transmitted to the load. In the event of unequal loads, a fraction of the power will be reflected and measured by the microwave power meter. The amount of the reflected microwave power is proportional to the impedance mismatch. If the load impedance is known, the mismatch power meter can then be calibrated for a meaningful return loss measurement.

A more precise method of return loss measurement across the entire bandwidth is often required. Such a measurement setup is shown in Figure 11.22. The return loss measurement method of Figure 11.22 requires the power meter and the oscilloscope to be calibrated to a load of known mismatch. This standard load is replaced with the load to be tested. By sweeping through the entire bandwidth, the power meter will read the average return loss, while the oscilloscope (ac coupled) will display the return loss variations.

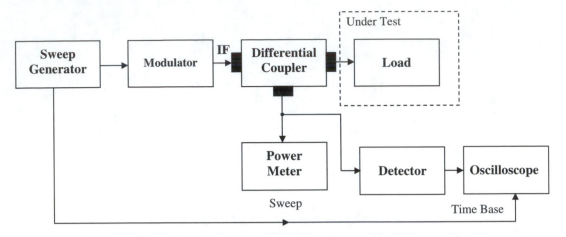

FIGURE 11.22   Return loss measurement: standard mismatch method.

11.4 DIGITAL MICROWAVE LINK MEASUREMENTS AND PERFORMANCE EVALUATION

Digital microwave links must constantly maintain pre-established performance character-istics during their entire operating lives. The most important performance indicator for a digital microwave link is the BER. System bit error rate must not exceed a maximum level at a specified carrier-to-noise ratio (C/N). The main causes for system performance degra-dation expressed by the BER are:

1. **Intersymbol Interference** (ISI)
2. **Multipath fading**
3. Jitter
4. AM/PM power amplifier conversion
5. Adjacent-channel and cochannel interference
6. Digital **modulator misalignment**

11.4.1. Error Performance

Tests performed when the link is either in service or off service can establish system error performance. Off-service error performance measurements are conducted by removing the normal traffic from the link and replacing it with a pseudo-random binary sequence (PRBS) signal (Figure 11.23). The pattern generator generates the appropriate PRBS signal and

FIGURE 11.23   Off-service error-performance measurement.

TABLE 11.1

F_b(Mb/s)	PRBS bit length (n)	Δf
T_1: 1.544	($n = 15$) 32,767	47.12
CEPT: 2.048	($n = 15$) 32,767	62.5
2: 6.312	($n = 15$) 32,767	192.6
8.448	($n = 12$) 32,767	257.82
T_3: 32.064	($n = 15$) 32,767	978.54
34.368	($n = 23$) 838,8607	4.1
T_4: 44.236	($n = 15$) 32,767	1365.3
139.264	($n = 23$) 838,8607	16.6

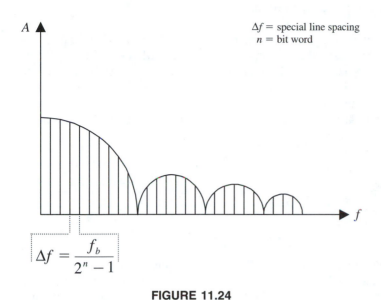

Δf = special line spacing
n = bit word

$$\Delta f = \frac{f_b}{2^n - 1}$$

FIGURE 11.24

transmits it through the link. At the receiver end, the error-detector unit compares the received signal with the exact replica of the PRBS signal stored in its memory and displays the difference between the two.

The CCITT has established the PRBS signal for various levels of TDM (Table 11.1 and Figure 11.24). The two basic instruments required for off-service error-performance evaluation are the pattern generator and the error detector. In-service error-performance measurements are performed without traffic interruption by a **parity check.** In this method of error-performance measurement, the parity bit of the incoming frame is compared to the parity bit of the previous frame. If an error bit has occurred, this error will be detected and analyzed by an error analyzer. This is one of several methods of in-service error-performance testing (Figure 11.25).

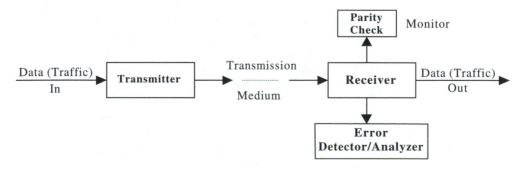

FIGURE 11.25 In-service error-performance measurements.

11.4.2 Noise and Interference Measurements

Digital radio link performance is determined by the bit error rate (BER) and the carrier-to-noise ratio (C/N); for example, in satellite transmission, a system BER equal to 10^{-8} corresponds to a specific carrier-to-noise level. If an abrupt reduction of the C/N occurs in the system (flat-fading conditions), the BER will also increase to a level beyond the set maximum for effective link operation.

Several internal or external sources of noise and interference directly contribute to system C/N and BER performance degradation:

1. **Timing jitter**
2. Modulator/demodulator **phase errors**
3. Amplifier nonlinearities
4. Clock phase errors

It is therefore imperative that an appropriate test be performed during the manufacturing, installation, and operating phases of a digital radio microwave link in order to identify and consequently rectify problems inhibiting proper link operation. The block diagram of a digital radio transceiver is shown in Figure 11.26. In this diagram, certain critical points are identified from which the processed signals can be interrupted directly or indirectly, and be displayed and analyzed. The results of such tests will determine the operational state of the link.

At the transmitter end, the digital interface CCITT signal modulates the IF signal through the modulator circuit. The modulated IF signal is then band limited through an IF bandpass filter and up-converted to the designed RF carrier frequency. The RF output signal is also band limited through an RF BPF filter, amplified by a traveling-wave tube (TWT) or solid-state microwave power amplifier, then fed into the antenna assembly for final transmission.

From this description of the operation of the digital radio transmitter, it is obvious that the signal has gone through a frequency translation and several signal amplification stages. After transmission, the signal suffers a significant attenuation of the order of 200 dB, while propagating through space. At the satellite receiver, the weak signal is amplified, frequency-translated to down-link carrier frequency, reamplified, and transmitted to the earth station with an attenuation almost equal to that of the up-link ($\cong 200$ dB).

The very weak signal received by the earth station antenna is amplified by the low-noise amplifier (LNA), usually located at the focal point of the receiver antenna (parabolic reflector). Low-noise amplifiers are designed with a gain of between 40 dB and 50 dB and a noise figure (NF) of approximately 1.5 dB. The LNA RF output signal is then processed through an RF BPF. This filter eliminates the frequency components outside the set bandwidth allowed for satellite transmission of (500 MHz) or for transponder frequency (36 MHz).

The RF BPF output is fed into the down-converter. This circuit performs a second frequency translation by converting the RF signal into IF. The IF bandpass filter further eliminates unwanted frequency components outside the IF bandwidth, thus positively contributing to a substantial decrease in noise power. The automatic-gain-control (AGC) stage between the IF BPF and the digital demodulator is intended to provide the constant-

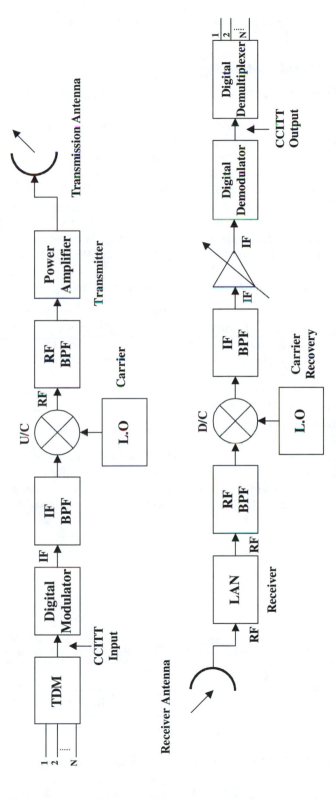

FIGURE 11.26 Digital radio transmitter.

amplitude IF signal level as required at the demodulator input. This signal level is usually between 1 mW and 1.5 mW.

Effects of Signal Flat-Fading

Signal **flat-fading** refers to the reduction of the carrier power at the input of the demodulator circuits while the noise power density remains at a constant level. Similarly, signal flat-fading is perceived when the noise power density (N_o) is substantially increased while the carrier signal power remains constant. Both interpretations lead to the same conclusion. That is, signal flat-fading implies that a substantial decrease of the carrier-to-noise ratio (C/N) has occurred at the input of the receiver demodulator stage. This C/N ratio decrease reflects a corresponding increase of the bit error rate (BER). If the BER level reaches a level above the set maximum, the microwave link can be declared nonoperational. It is imperative to note that the C/N at the input of the RF amplifier is the same as the C/N at the input of the receiver demodulator input. This is true because both the carrier signal and noise are simultaneously and equally amplified by the RF and IF receiver stages.

BER versus C/N Measurements with Flat-Fading

During normal operating conditions, a digital microwave link maintains a constant BER of between 10^{-7} and 10^{-9}. For proper link operation, performance tests must be periodically conducted to ensure proper system operation. If an effective test is to be performed in order to establish link performance, it must include the measurements of the carrier and of the noise power levels at the input of the demodulator circuit. At the same time, the corresponding BER levels for each C/N level must constantly be monitored. By comparing these measurements with the available theoretical curves, the link's performance state can easily be established. A basic setup for digital radio performance evaluation is shown in Figure 11.27.

To perform this measurement, an RF attenuator is placed between the low-noise amplifier and the down-converter. By varying the attenuator, different carrier signal levels can be obtained at the input of the demodulator. These carrier levels can now be compared with the spectral noise density (N_o). Spectral noise density is expressed by

$$N_o = KT_0(\text{NF}) \tag{11.26}$$

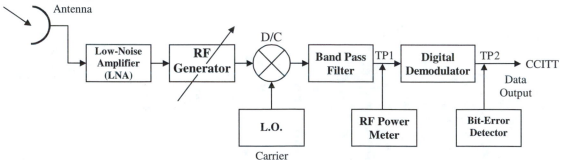

FIGURE 11.27   BER versus C/N measurement setup.

where K = Boltzmann's constant $(1.38 \times 10^{-23}$J-K$)$
          T_0 = 290 K
          NF = receiver noise figure (dB)

Several sets of C/N_o values are obtained from this process. These C/N_o values are further used to establish the carrier-to-noise ratio (C/N) as follows: The relationship between C/N_o and C/N is given by

$$\frac{C}{N} = \frac{C}{N_o} - 10 \log(\text{BW}) \qquad (11.27)$$

where   BW = receiver noise bandwidth (Hz)

The set values of C/N are used to compare the measured BER levels with the BER extrapolated from the theoretical curves.

The symbol-error-rate (SER) and carrier-to-noise ratio (C/N) can also determine the performance of a digital microwave link. The symbol rate (SR) can be calculated if the link capacity and modulation scheme employed are known. For example, for a digital microwave radio with 90 Mb/s capacity and an 8-PSK modulation scheme the system **symbol rate** (SR) is given by

$$\text{SR} = \frac{f_b}{N} \text{b/s} \qquad (11.28)$$

where $N = \dfrac{\ln(M)}{\ln(2)}$
          M = number of phase states

For 8-PSK, $(N = 3)$, so

$$\text{SR} = \frac{90}{3} \text{Mb/s} = 30 \text{ MB/s}$$

A set of theoretical curves reflecting the symbol error rate (SER) and carrier-to-noise ratio (C/N) are shown in Figure 11.28. The symbol rate (SR) is an important component defining the minimum theoretical system bandwidth required. However, in practical systems, the bandwidth required is much higher than the theoretical minimum. This excessive bandwidth is necessary in order to compensate for system filter α.

Since it is practically impossible to design system filters with a filter α equal to zero $(0 < \alpha < 1)$, a practical approach is to multiply the minimum theoretical system bandwidth by filter α. For example, a system requiring a theoretical minimum bandwidth of 30 MHz and incorporating filters with filter α equal to 0.2 requires a practical bandwidth of 36 MHz (1.2×30 MHz).

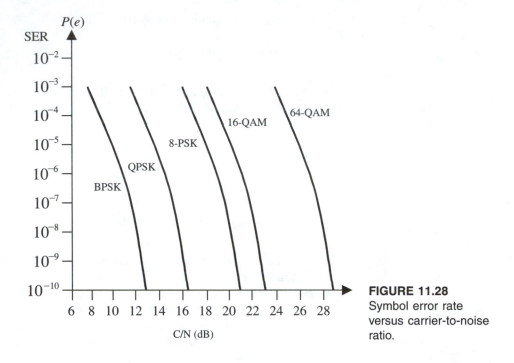

FIGURE 11.28
Symbol error rate versus carrier-to-noise ratio.

Carrier-to-Interference Ratio (C/I)

The carrier-to-interference ratio is as important as the carrier-to-noise ratio (C/N) in determining system performance. An interference signal detected by the receiver antenna will be processed through the IF and RF stages at the digital radio receiver, provided the spectral constant of the interference signal is the same as that of the bandpass filter (BPF). In essence, at the input of the receiver demodulator two components determine system performance: residual noise and interference.

For an effective operation of a digital microwave link, exact knowledge of the BER resulting from residual noise and interference is of critical importance. Several sources of interference contribute to system performance degradation.

1. *Adjacent Channel Interference (ACI).* In satellite communications links, 12 transponders share the 500-MHz available bandwidth (36 MHz per transponder), whereas 24 transponders can utilize the same 500-MHz bandwidth (36 MHz per transponder) in a cross-polarization mode of operation.

It is not uncommon for spectral power from one channel to spill over the adjacent channel, thus altering the amplitude of the home channel. This unwanted amplitude variation may shift the operation point of the power amplifier from a linear to a nonlinear mode of operation. This shift of the operating point may induce a signal distortion referred to as **adjacent channel interference.** Adjacent channel interference also occurs as a result of TWT imperfections on board the satellite, or when such amplifiers are driven to a nonlinear mode of operation.

2. *Cross-Polar Interference (CPI).* It was mentioned earlier that in satellite transmission, cross polarization is used in order to increase system capacity. Such systems employ two sets of antennas, one for vertical and the other for horizontal polarization. Under ideal conditions, the waves should not interfere with each other. In practice, unwanted information might be transmitted from one channel through the horizontal plane to a channel transmitted through the vertical plane of propagation, thus inducing an unwanted signal power into the home channel. This unwanted power is referred to as cross-polar interference.

3. *Doppler Shift.* Communications satellites are geosynchronous. That is, under ideal conditions the position of a satellite relative to a point on the earth's surface is absolutely constant. In practice, this is very difficult to maintain. Located approximately 36,000 km above the equator, the spacecraft is subject to various gravitational forces such as the moon, the planets, and the sun. Although the satellites are equipped with sophisticated orbital adjustment mechanisms, minor deviations are not uncommon.

These periodic changes in spacecraft height, although small, are still compatible with the satellite operating wavelengths. Therefore, the earth station receiver can detect a slight variation in the received carrier frequency. This phenomenon is called the **Doppler shift.** That is, when the satellite is moving away from the earth, a slight decrease of the transmitter frequency seems to be detected at a level proportional to the drifts. Likewise, when the satellite moves closer to the earth, the carrier frequency seems to increase, also proportional to the drift. These carrier-frequency changes are interpreted as interfering signals by the receiver demodulator circuit.

4. *Path Interference.* In a down-link satellite transmission, the carrier signal follows different propagation paths through the earth's atmosphere. The atmospheric conditions around the earth receiver station will determine the degree of refraction and consequently the total distance the signal has traveled away from its main path. Signals arriving at the input of the receiver at different times are perceived to be arriving with different amplitude and phases from the main signal. Here, the combined effect is interpreted as **path interference.** This combined interference effects the overall system performance to various degrees based on combined levels of intensity. Therefore, the combined level of residual noise and interference must be known and controlled to acceptable levels for an efficient link operation.

P(e) versus C/N and C/I

Assuming thermal noise and ISI are approximate to Gaussian noise and interference is a pure sine wave, the probability density function (PDF) for Gaussian noise is given by

$$P(u_n) = \frac{1}{\sigma\sqrt{2\pi}} e^{-(u_n)^2/2\sigma} \tag{11.29}$$

where σ = noise voltage (rms)
             u_n = noise voltage (instantaneous)

Assuming $\sigma = 1$ V (normalized), then Eq. (11.29) becomes

$$P(u_n) = \frac{1}{\sqrt{2\pi}} e^{-(u_n)^2/2} \qquad (11.30)$$

The area obtained by the multiplication of the PDF by an infinitesimal width (V_n) is ΔV_n (Figure 11.29). The probability that the value of a noise sample (V_n) is less than a pre-determined numerical value is known as the "cumulative probability density function" (CPDF). This function represents the probability that a signal in this case has a value V_n equal to V_{th}, where V_{th} is the threshold voltage. The normalized CPDF (V_{th}) of Gaussian noise with no dc component is given by

$$F(V_{th}) = \frac{1}{2\pi} \int_{-\infty}^{+V_{th}} e^{-v^2/e^2} dV \qquad (11.31)$$

The normalized complimentary probability density function is

$$Q(u_{th}) = 1 - F(V_{th}) \qquad (11.32)$$

or

$$Q(u_{th}) = \frac{1}{2\pi} \int_{V_{th}}^{\infty} e^{-V^2/2} dV \qquad (11.33)$$

Equation (11.33) cannot be evaluated explicitly, but there is an equivalent polynomial approximation that can be used instead:

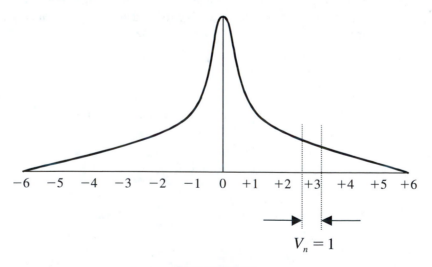

FIGURE 11.29

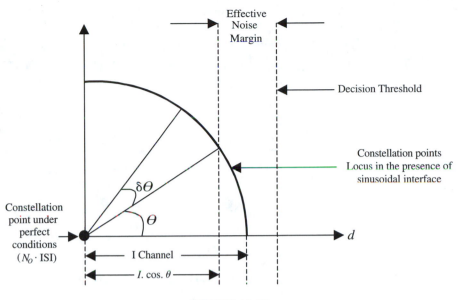

FIGURE 11.30

$$Q(u_{\text{th}}) = \frac{1}{2\pi} e^{-(V_{\text{th}})^2/2}(b_1 t + b_2 t^2 + b_3 t^3 + b_4 t^4 + b_5 t^5 \qquad \textbf{(11.34)}$$

where
$$t = \frac{1}{1 + P(V_{\text{th}})}$$
p = constant (0.231641)
b_1 = 0.31938153
b_2 = −0.35686782
b_3 = 1.781477937
b_4 = −1.3821255978
b_5 = 1.330274429

This approximation provides an acceptable degree of accuracy when calculating the error probability of a digital radio employing various modulation schemes at different C/N and C/I ratios.

For example, consider the probability of the in-phase (I-channel) signal exceeding the signal decision boundary (Figures 11.30 and 11.31). From Figure 11.30 the effective noise margin is equal to $d - I\cos\theta$. The threshold voltage is

$$V_{\text{th}} = \frac{d - I\cos\theta}{\delta} \qquad \textbf{(11.35)}$$

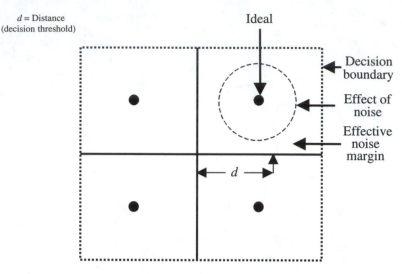

FIGURE 11.31   Constellation display.

For a small change of angle θ, the normalized complementary probability density function (CPDF) is

$$QV_{th} = \frac{d\theta}{2\pi} \qquad (11.36)$$

Therefore, error probability $P(e)$ is given as follows:

$$P(e) = QV_{th}\frac{d\theta}{2\pi} \qquad (11.37)$$

If N is the number of $d\theta$ increments, then the total error probability becomes

$$P(e) = \sum_{N=1}^{N} Q(V_{th})\frac{d\theta}{2\pi} \qquad (11.38)$$

Angle θ is given by the expression

$$\theta = \eta\delta\theta - 0.5\delta\theta$$

where $\qquad \delta\theta = \dfrac{2\pi}{N} \qquad (11.39)$

Since different modulation schemes have different decision boundaries (d), the final equation expressing error probability in reference to C/N, and C/I is given by

$$P(e) = \rho\frac{1}{\sqrt{2\pi}}\sum_{n=1}^{N}\frac{1}{N}e^{-(V_{th})^2/2}(b_1 t^1 + b_2 t^2 + b_3 t^3 + b_4 t^4 + b_5 t^5) \qquad \textbf{(11.40)}$$

and δ and ρ adjustment factors for different modulation schemes are given in Table 11.2.

$$I = \sqrt{2} \cdot d \cdot 10^{\frac{-[(C/I)-\delta]}{20}} \qquad \textbf{(11.41)}$$

$$\delta = d \cdot 10^{\frac{-[(C/I)-\delta]}{20}} \qquad \textbf{(11.42)}$$

where C/N and C/I are in dB.

Manipulating Eqs. (11.41), (11.42), and (11.35), the value of V_{th} can be established. Using V_{th} in (11.40), the error probability for different modulation schemes can be determined for various C/N and C/I levels. A set of graphs relating error probability with C/N and C/I levels for different modulation schemes is shown in (Figures 11.32 to 11.36).

    It is evident from the foregoing discussion that residual noise and interference are two key components contributing to system performance degradation in both line-of-sight and satellite transmission. For proper satellite transmission, the error probability, $P(e)$ or BER in relation to E_b/N_o is set as shown in Table 11.3.

INTELSAT (TDMA)

In order for noise and interference tests of a satellite link to be performed, the type of system and its performance characteristics must be known. For the INTELSAT-V series, having a 60Mb/sTDMA capacity, the link budget characteristics are shown in Tables 11.4 and 11.5.

    Satellite system testing is performed during the installation of the equipment (off-service testing) or during regular system operation (on-service testing). Such tests involve the three basic components of the system: the modem, the IF/RF, and the earth station–to–earth station link.

TABLE 11.2

Modulation Scheme	ρ	δ
Q-PSK	2	0
8-PSK	2	5.33
16-PSK	2	11.19
16-QAM	1.5	6.39
64-QAM	1.75	13.22

TABLE 11.3

E_b/N_o (dB)	BER
6.30	5×10^{-3}
10.00	10^{-4}
12.60	10^{-6}
14.00	10^{-7}

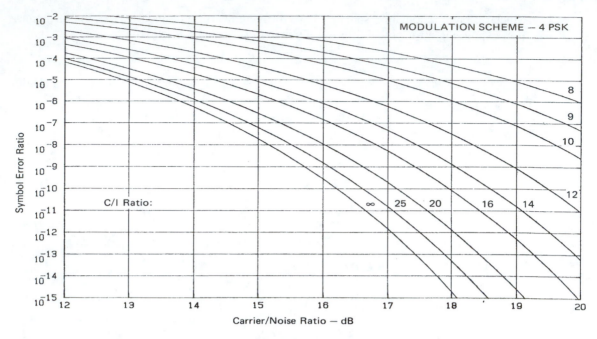

FIGURE 11.32
Source: Reproduced by permission of HP Canada.

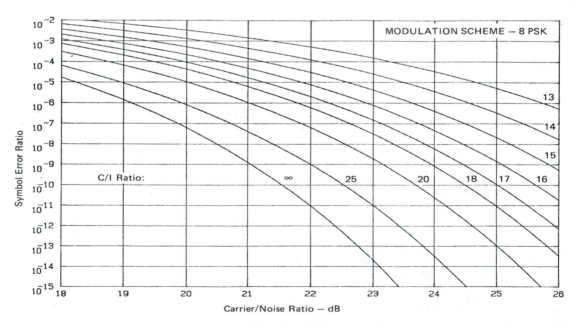

FIGURE 11.33
Source: Reproduced by permission of HP Canada.

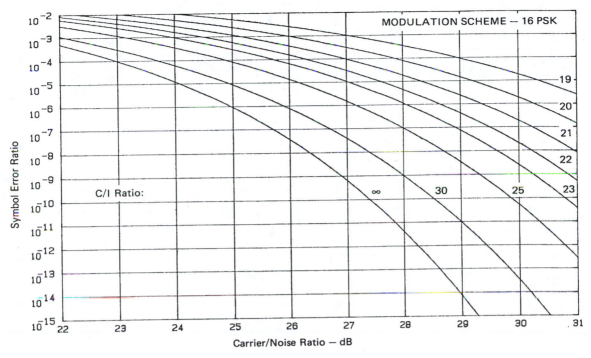

FIGURE 11.34
Source: Reproduced by permission of HP Canada.

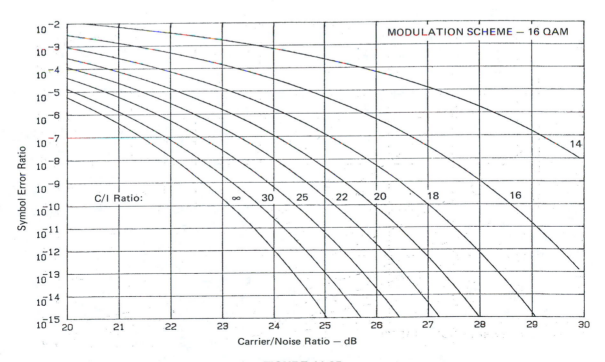

FIGURE 11.35
Source: Reproduced by permission of HP Canada.

System Measurements and Performance Evaluation

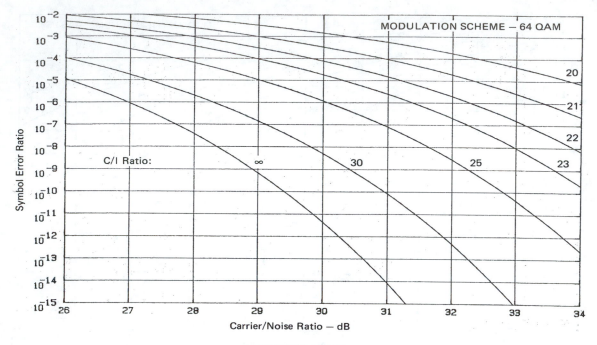

FIGURE 11.36

Source: Reproduced by permission of HP Canada.

Modem Testing

BER and E_b/N_o
BER and E_b/N_o with forward error correction (FEC)
IF frequency
Modulation ON/OFF ratio

TABLE 11.4 Up-Link

INTELSAT-V, 60 Mb/s

Operating frequency (GHz)	6/4
Bandwidth (MHz)	60
EIRP (earth station) (dB)	88.5
G/T: satellite (dB/K)	−11.6
Path losses (dB)	200.5
C/N (dB)	27.2
C/I (dB)	21.6
($C/N + I$) (dB)	20.5
Input backoff (dB)	3.0
Saturation flux density (dBw/m^2)	−72.0

TABLE 11.5 Down-Link

INTELSAT-V, 60 Mb/s

EIRP (dB)	29.0
G/T Earth station (dB/K)	40.7
Path losses	197.0
C/N (dB)	23.0
C/I (dB)	21.6
($C/N + I$) (dB)	19.2
E_b/N_o (dB)	9.6
C/I (External) (dB)	2.9
($C/N + I$) (dB)	16.5
EIRP variations (dB)	1.0
Available ($C/N + I$) dB: total	15.5
Available ($E_b/N_o + I$) dB: total	12.5

IF/RF *Testing*

Effective isotropic radiated power (EIRP)

Transmitter amplitude and group delay

Receiver amplitude and group delay

HPA output spectrum

BER loop test

Receiver frequency tolerance

Earth Station–to–Earth Station Testing

EIRP

BER measurement

Transmitter equalizer

Receiver equalizer

Modem IF Testing   Modem testing is required to determine baseband (BB) and IF performance. In such testing, the addition of noise or interference to the carrier is necessary in order to establish system BER (performance indicator). This additive noise or interference is also subject to carrier signal variations. It is therefore imperative that the instrument used as a source of noise and interference must be capable of continuously tracking carrier signal variation and of adjusting accordingly. In this way discrete levels of *C/N* or *C/I* ratios are generated and applied to the input of the demodulator circuit of the receiver, while the corresponding BER levels are detected by the bit-error detector. A very reliable instrument used to provide the noise and interference sources in such a test is the HP-3708A noise and interference test set (Figure 11.37).

In this setting, a binary signal closely resembling normal link traffic is generated by the pattern generator. This pseudo-signal is fed into the modulator stage. The output of the modulator is the IF signal modulated by the pseudo-signal with a constant amplitude. The noise and interference test set provides the noise or interference levels required to generate the discrete level of the *C/N* or *C/I* ratios, and the BER detector displays the various BER levels for the corresponding *C/N* or *C/I* ratios.

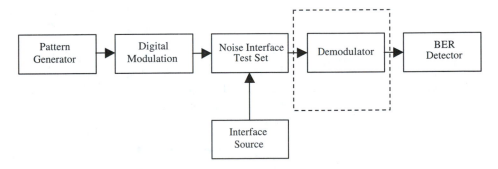

FIGURE 11.37   Modem test setup.

Modem and RF testing   This test involves the modem and RF section of a digital radio (Figure 11.38). In Figure 11.38, the pattern generator is set to generate a pseudo-signal closely re-sampling the normal traffic signal fed into the modulator input. The digitally modulated IF signal generated by the modulator stage is up-converted to the required 6-GHz RF signal for C-band satellite transmission. The RF bandpass filter provides the required bandwidth restrictions, and the high-power amplifier (HPA), usually a traveling-wave tube (TWT), amplifies the RF signal to specified levels.

The frequency translator stage is used to convert the 6-GHz up-link carrier frequency to the 4-GHz down-link frequency, and the RF attenuator simulates space losses down to approximately 200 dB. The low-noise amplifier, with a typical gain of between 45 to 50 dB and a low noise figure, typically 1.5 dB, are used to amplify the very weak signal to a level required by the down-converter. The RF bandpass filter eliminates out-of-band spectral noise. The down-converter translates the RF to IF frequency, and the IF BPF also eliminates noise outside the IF bandwidth. The noise and interference test set is connected between the IF BPF and the receiver demodulator stage, and the bit-error detector is connected to the output of the demodulator stage.

By setting the noise and interference test set to various interference levels applied at the demodulator input, these levels can now be compared to the carrier levels. If any variation of the carrier input level does occur, the test set automatically adjusts to the required levels, a necessary requirement for accurate *C/I* measurement. The BER for each particular *C/I* ratio is displayed by the error-detector set connected at the output of the demodulator stage.

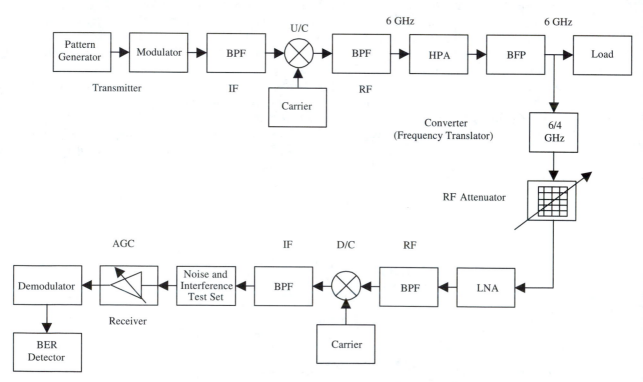

FIGURE 11.38   IF/RF performance measurement setup.

Complete Satellite Performance Test   The following test setup incorporates all three subsections of a satellite communications link; the modem, the IF/RF, and the ground station. This can be accomplished through the satellite link (Figure 11.39). Again, the carrier signal level must be kept constant at the input of the demodulator stage for accurate BER measurements.

This is accomplished by the automatic-gain-control (AGC) circuit at the receiver end. The noise and interference test set provides discrete levels of noise and interference at the input of the demodulator, which are compared with the constant carrier power present at the same input. The corresponding BER levels are detected and displayed by the BER detector.

The Eye Diagram

One way of determining digital radio link performance is through **the eye diagram.** This method provides a visual evaluation of the state of the link. Eye diagrams are formed by superimposing the *I* or *Q* binary streams on top of each other (Figure 11.40). Closure of the eye diagram in the horizontal axis is indicative of system degradations related to jitter and sampling circuit misalignments; closure in the vertical axis is the result of such degradations as ISI and AM/PM (Figures 11.41 and 11.42). The slope of the eye diagram curve reflects system error sensitivity. That is, the timing error probability increases with a slope decrease. An asymmetric eye diagram is the result of channel nonlinearities.

Ideally, the horizontal full opening of the eye diagram symbolizes the time required for the incoming symbol to be sampled in the absence of intersymbol interference (ISI), and the vertical opening of the eye symbolizes the amplitude of the incoming symbol. Maximum opening is indicative of an absence of amplitude distortion. The results from the test methods can now be compared to the theoretical curves shown in Figures 11.33 to 11.37, and the performance states of the satellite link can be determined.

Constellation Analysis

Constellation analysis is a very powerful tool for identifying digital radio impairments. Through this method, the sampled states of the I channel and Q channel at the input of the demodulator stage are superimposed to form a constellation display as shown in Figure 11.43. For a QPSK modulation scheme employed in a digital radio, the constellation display consists of four very fine points. Constellation displays for 16-QAM and 64-QAM are illustrated in Figures 11.44 and 11.45.

Theoretically, these points must be very fine and reflect a carrier-to-noise ratio (*C/N*) and a carrier-to-interference ratio (*C/I*) well above the minimum required. In practice, their size, even at the absence of interference, depends upon the system's residual noise (Figures 11.46 and 11.47). The performance of a digital radio is based on the ability of the demodulator to determine correctly whether the incoming symbol is a zero or one in the presence of noise. If the noise and interference levels are excessive, it is possible that the demodulator circuit will not be able to decide correctly and an error will occur.

The constellation method of performance evaluation is an in-service visual method of establishing a digital radio's operating state. System performance degradation occurs due to noise and interference effects on any or all of the three components of a digital radio link: the transmitter, the propagation path, and the receiver.

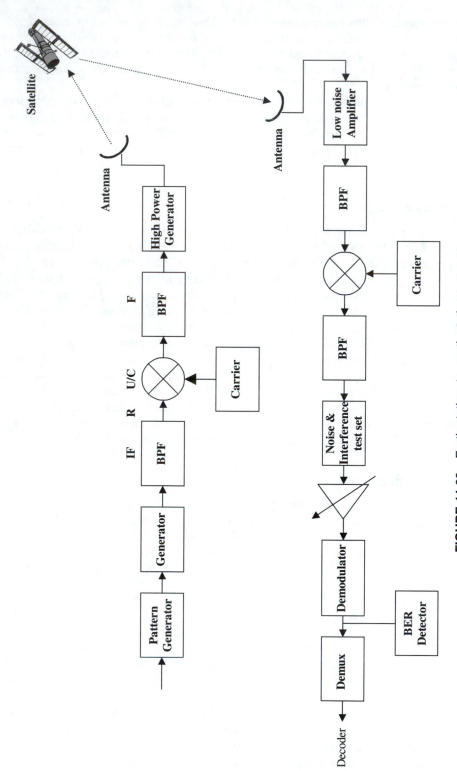

FIGURE 11.39 Earth station–to–earth station test setup.

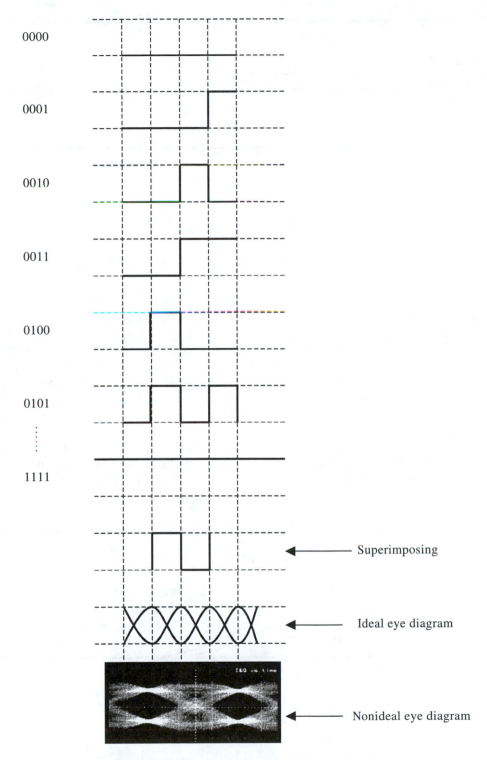

0000

0001

0010

0011

0100

0101

⋮

1111

Superimposing

Ideal eye diagram

Nonideal eye diagram

FIGURE 11.40 Formation of a 4-bit eye diagram.

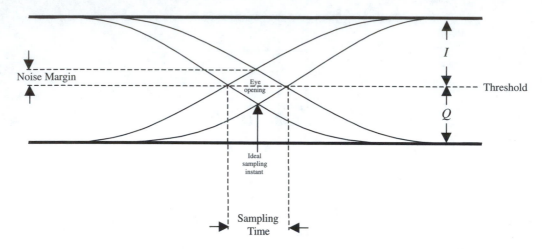

FIGURE 11.41 Eye diagram without impairments.

A constellation analyzer such as the HP-3709 B connects the appropriate inputs of the analyzer to the I channel, the Q channel, and the clock recovery of the demodulator in order to perform constellation analysis (Figure 11.48). The shape and position of the points displayed by the constellation analyzer will indicate impairments affecting both the I and Q channels, as well as the clock recovery circuit. Such sources of impairments are the RF and IF amplifiers, filters, the propagation path, up-converters and down-converters. Sources affecting the I or Q channels are modulator and demodulator circuits, baseband filters, and baseband amplifiers.

In-service performance evaluation of a digital radio link must be continuous and accurate. The constellation method of evaluation satisfies both these requirements. To take advantage of the constellation analysis method, link impairments are grouped into three

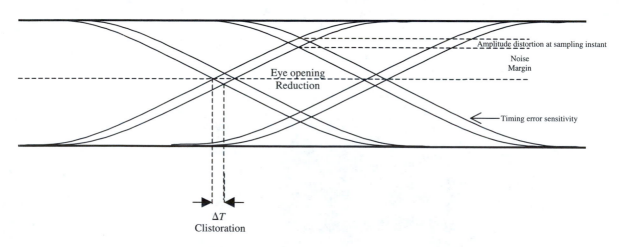

FIGURE 11.42 Eye diagram with impairment.

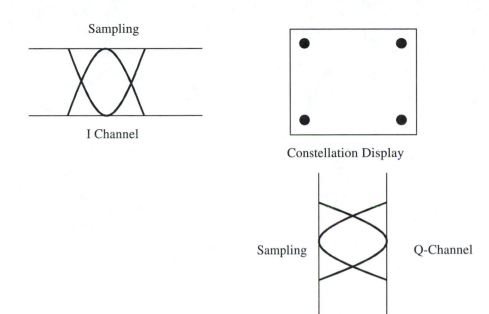

Sampling

I Channel

Constellation Display

Sampling   Q-Channel

FIGURE 11.43   Constellation display for QPSK modulation.

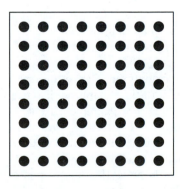

FIGURE 11.44
Constellation display
(16-QAM).

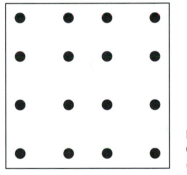

FIGURE 11.45
Constellation display
(64-QAM).

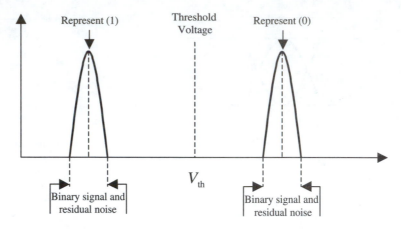

FIGURE 11.46

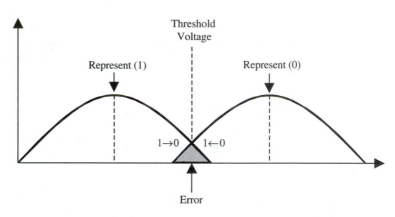

FIGURE 11.47

categories: (a) impairments induced by the transmitter, (b) impairments induced by the transmission path, and (c) impairments induced by the receiver. *Transmitter impairments are attributed to* modulator level misadjustment, phase noise, and AM/AM and AM/PM nonlinear distortion; *propagation path impairments are attributed to* low *C/N* and interference; and *receiver impairments* are attributed to lock angle error and quadrature angle error.

Measurements taken by the analyzer (display indicator) are as follows: *Closure* compares the cluster size (rms) to half of the separation distance between two consecutive clusters and is expressed in percentage. For example 12% of closure is considered to be normal under ideal operating conditions. *Lock angle error* is indicative of the average rotation of the *I* and *Q* axes in reference to the cluster, and is displayed in degrees. For example, 1° lock angle error is considered normal under ideal operating conditions. Both the I channel and the Q channel must be orthogonal (90° out of phase). Under normal operating con-

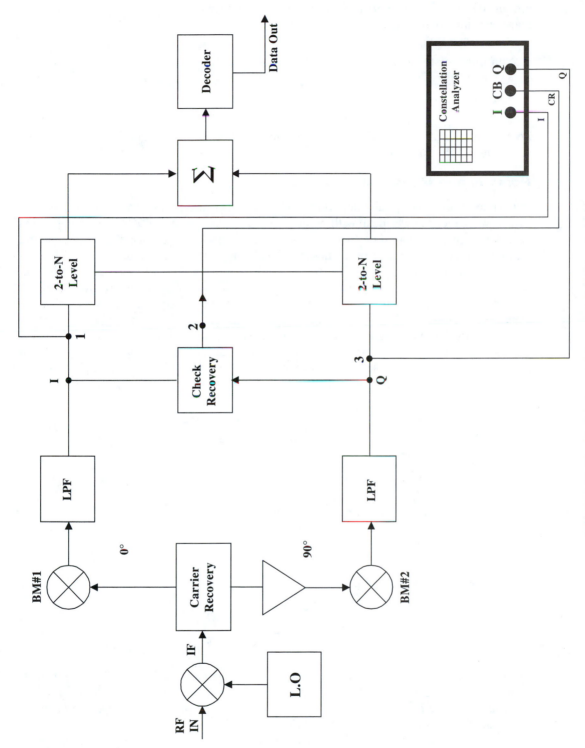

FIGURE 11.48 Digital demodulator constellation analysis.

ditions, the *quadrature angle error* is set at $-0.3°$. Other constellation displays under normal operating conditions are

Non-linear distortion: 2.2%
AM/AM: 0.6%
AM/PM: 0.3%

Some examples of constellation analyzer displays and impairment identifications are shown in Figures 11.49 through 11.51.

Noise Figure Measurements

In Chapter 1, the concept and importance of noise figure as a fundamental characteristic of a microwave receiver was dealt with in some detail. Again, the ability of a receiver to process very weak signals is based on two key factors: sensitivity and the noise figure (NF). Noise figure is a unique parameter reflecting individual component and circuit characteristics incorporated into the receiver design. Noise figure is more important when considering receiver systems in satellite applications. In such systems, the level of microwave power at the input of the receiver is in the order of -110 dBm and is very competitive with equivalent noise power.

When describing digital radio systems, it was mentioned that the two fundamental parameters determining system performance are the carrier-to-noise ratio and the bit error rate (BER). Both important parameters relate to receiver noise figure. The general expression of the noise figure (NF) is given by

$$NF_{(dB)} = 10 \log\left(\frac{N_a + KT_e\text{BW}G}{KT_0\text{BW}G}\right)$$ **(11.43)**

where
$$\begin{aligned}
N_a &= \text{output noise power } (KT_0\text{BW}G) \\
T_e &= \text{equivalent noise temperature (K)} \\
\text{BW} &= \text{operating system bandwidth (Hz)} \\
T_0 &= \text{ambient temperature (290 K)} \\
G &= \text{receiver gain (dB)}
\end{aligned}$$

Substituting for N_a in Eq. (11.43), gives Eq. (11.44).

$$\text{NF}_{dB} = 10 \log\left(\frac{T_e}{T_0} + 1\right)$$ **(11.44)**

Equation 11.44 shows that the noise figure of a microwave receiver is independent of its gain and operating bandwidth, and is expressed in terms of equivalent noise temperature in relation to $290°K(T_0)$. Equivalent noise temperature (T_e) expresses the sum total of external noise and internally generated noise (RTI).

In essence, the noise figure is a measure of signal deterioration when processed through the receiver. The carrier-to-noise ratio and the bit error rate are indicative of a system's op-

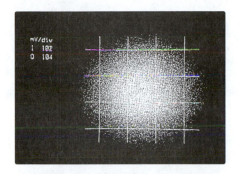

NO RF SIGNAL

This display indicates a complete loss of carrier signal. Sources of such impairment are the transmitter or the receiver (transmitter power loss, or receiver failure).

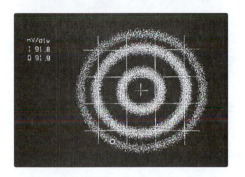

CARRIER RECOVERY LOOP OUT OF LOCK

This display also indicates the absence of a constellation pattern. Possible sources of impairment are:
1. No carrier recovery due to a very low *C/N* or *C/I*
2. Out-of-lock carrier recovery oscillator

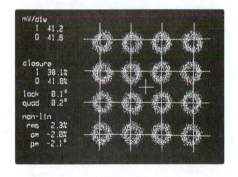

INTERFERING TONE

This display indicates the presence of an interference tone influencing cluster ringing. Possible source of impairment:
Cochannel interference.

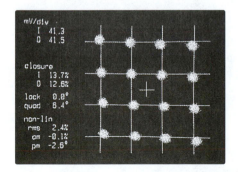

QUADRATURE ERROR

It is evident from this display that the clusters form a parallelogram instead of a square. Possible impairment causes:
Misalignments of the I channel and Q channel (out of 90° phase difference), due to either modulator or demodulator malfunctions.

FIGURE 11.49
Source: Reproduced by permission of HP Canada.

System Measurements and Performance Evaluation

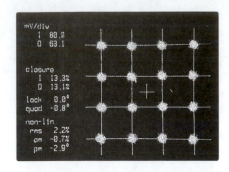

AMPLITUDE IMBALANCE

This constellation display appears to be normal with the exception that there exists an amplitude imbalance between the I and Q channel induced by the 2-to-N level converter.

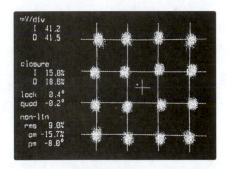

AM-AM/AM-CONVERSION

In this display a concentric displacement of the outermost clusters is observed. This displacement is attributed to TWT overdrive.

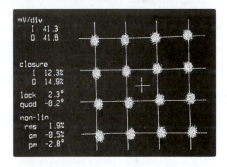

AM/PM CONVERSION

In this display, a rotation exists for the outermost clusters in reference to the innermost clusters. This cluster distortion is due to AM/PM conversion nonlinear distortion.

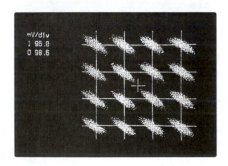

ADJUSTED TIMING SETUP

This elliptical display of the cluster is due to cross-coupling between the I and Q channels resulting from a frequency-dependent amplitude slope of the IF or RF stages of the digital radio.

FIGURE 11.50
Source: Reproduced by permission of HP Canada.

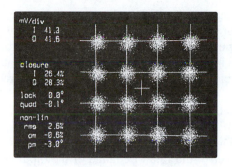

LOW *C/N*

In this constellation display the clusters appear to be larger than normal due to a relatively low *C/N* ratio. This may occur because of a low carrier signal, higher residual noise, or intersymbol interference (ISI).

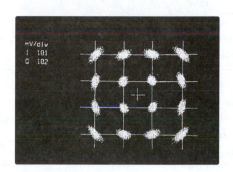

PHASE NOISE

This elliptical shape and rotation of the clusters is indicative of oscillator phase noise and/or carrier recovery circuit.

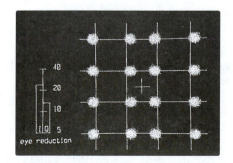

MISADJUSTED MODULATOR

The unequal distance-separating cluster is attributed to the erroneous amplitudes generated during the binary to *N*-level conversion process.

FIGURE 11.51
Source: Reproduced by permission of HP Canada.

erational conditions, the only difference being that the BER indicates whether the system is operational or not, whereas the noise figure indicates operational efficiency. For example, a decrease of the carrier signal at the input of the receiver will increase the error probability or BER. A progressive decrease of the input carrier power will reach a point at which the BER will cross the threshold of acceptable level and the system will be declared non-functional.

On the other hand, a receiver with smaller noise figure will compensate for a lower carrier input power, thus maintaining the BER at an acceptable level for normal system operation. It is assumed that the operating characteristic of a microwave amplifier is linear. That is, the rate of change of the output power to the rate of change of the input power is constant $(\rho P_o / \rho P_i)$. If the signal at the input of the receiver is removed, the power measured at the output under ideal conditions (noiseless receiver) must be zero. In practice, the power measured at the output of the receiver is not zero. This is attributed to the internally generated noise power, given by

$$P_n = KT_e \text{BW} \tag{11.45}$$

The characteristic curve relating receiver power output (P_o) and equivalent noise temperature (T_e) is shown in Figure 11.52. To effectively measure the receiver noise figure, the setup shown in Figure 11.53 is required.

To perform this noise-figure measurement of a particular microwave receiver, the following are required: hot and cold sources, a microwave power meter, and a noise-figure meter. Avalanche semiconductor diodes are usually employed as noise sources, covering the spectral range between 10 MHz and 18 GHz and capable of providing hot noise temperatures of approximately 400 K and cold temperatures of approximately 80 K. If P_{n_1} is the noise power output corresponding to the cold temperature source (T_c), and P_{n_2} the noise power corresponding to the hot temperature source (T_h), then the ratio of hot to cold power, called the "Y-factor," is given by

$$Y = \frac{P_{n_2}}{P_{n_1}} \tag{11.46}$$

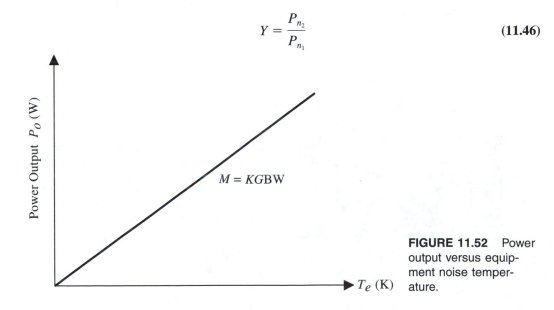

FIGURE 11.52 Power output versus equipment noise temperature.

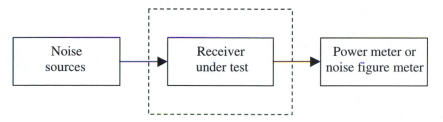

FIGURE 11.53 Receiver noise figure.

From the set-up of Figure 11.54, the values of P_{n_2} and P_{n_1} are obtained. The Y-factor is then used to calculate the equivalent noise temperature (T_e) expressed by

$$T_e = \frac{T_h - YT_c}{Y - 1} \tag{11.47}$$

Therefore,

$$\text{Noise figure NF} = 10 \log\left(\frac{T_e}{T_0} + 1\right) \tag{11.48}$$

A modern approach to noise-figure measurement is the utilization of the noise-figure meter. This sophisticated instrument is used instead of the power meter. It is capable of performing all the necessary mathematical calculations and displaying the noise figure in dB. Such an instrument is the HP-8970 noise figure meter. Low-noise amplifiers (LNAs) used in satellite systems (C band), such as the Avantek, exhibit a gain of 48 dB over the noise figure of 1.5 dB(120 K).

Jitter Measurements

One of the principal causes of digital radio system performance degradation is called **timing-jitter**. This impairment source must be identified, its magnitude measured, and its effect totally eliminated or at least minimized to an acceptable level in order to ensure effective system operation. There are two types of jitter: systematic and nonsystematic. Systematic jitter is pattern dependent. That is, jitter is introduced when certain irregularities are induced to the input data at the regenerative timing recovery circuit. The significant impact of this type of jitter on system performance is based on the fact that it is accumulative throughout the entire system. Nonsystematic jitter is related mainly to noise and is not as accumulative as is systematic jitter. Therefore, its impact on overall system performance is not so great.

Since jitter is so critical to system performance, it must be confined to acceptable limits. It is therefore very important that jitter levels be monitored and measured and that corrective action be taken when these levels exceed preset maxima. The required jitter measurements are

- Maximum tolerance to input jitter
- Maximum intrinsic output jitter
- Maximum output jitter
- Jitter transfer function

Maximum Tolerance to Input Jitter   Maximum jitter level at the input of a digital system or equipment is that level capable of triggering an error bit at the output of the system. To determine this maximum jitter level, the following test is required (Figure 11.54). The equipment necessary for such a test is

- A pattern generator
- An error detector

Jitter tolerance measurements are performed at the manufacturing stage and also at the installation point. During these tests, the amplitude of the jitter at the jitter generator is progressively increased while the error-detector display is constantly monitored. At the instant when an error hit is detected, the corresponding jitter amplitude is recorded. This jitter amplitude is referred to as "maximum tolerance to input jitter" or "upper limit." In this setup, the pattern generator is usually a PRBS generator simulating real data in (traffic). The results from the test are compared with pre-established standards set forth by the CCITT (G.823 and G.824) recommendations for various capacity digital systems (Table 11.6.)

It is imperative to note that in the case in which more than one input port exists in the system, all must comply with CCITT or AT&T requirements. It is also important to note that the measured maximum tolerance of the input jitter must be higher than the CCITT recommendations.

Maximum Intrinsic Output Jitter   This jitter is inherent to equipment or systems under test, and it must be measured and compared with pre-established standards. Similar to maximum tolerance input jitter, this internally generated jitter, if it is in excess of the specific limits, will result in a bit error, which will be detected and displayed at the output by the

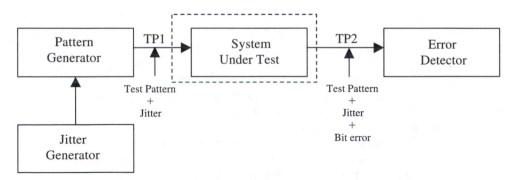

FIGURE 11.54   Maximum tolerance of input jitter.

TABLE 11.6   Maximum Tolerance of Input Jitter, CCITT Low Limits

Bit Rate (MHz)	Pattern	A_1 (UI)	A_2 (UI)	f_1 (Hz)	f_2 (Hz)	f_3 (kHz)	f_4 (kHz)
1.544	QRSS $2^{20} - 1$	5.0	0.1	10	120	6.0	40
6.312	QRSS $2^{20} - 1$	5.0	0.1	10	50	2.5	60
32.064	QRSS $2^{20} - 1$	2.0	0.1	10	400	8.0	400
44.736	PRBS $2^{15} - 1$	5.0	0.1	10	600	30.0	400
2.048	PRBS $2^{15} - 1$	1.5	0.2	20	2400	18.0	100
8.448	PRBS $2^{15} - 1$	1.5	0.2	20	400	3.0	400
34.368	PRBS $2^{23} - 1$	1.5	0.15	100	1000	10.0	800
139.264	PRBS $2^{23} - 1$	1.5	0.075	200	500	10.0	3500

error detector. The setup for the measurement of maximum intrinsic output jitter is shown in Figure 11.55.

Testing for intrinsic output jitter is performed during the manufacturing and installation of the digital equipment. The data obtained during these tests is documented for comparison with future tests. During the entire operating life of the system, monitoring the intrinsic output jitter is necessary in order to detect the progressive deterioration of the system's various components. This observation will eventually lead to the identification of specific faulty subsystems or components, the replacement of which will restore the system to its operational performance level.

Maximum Output Jitter   This jitter is measured at the output of the equipment under test and is related to the total accumulated jitter through the equipment stages, including intrinsic and extrinsic jitter. The testing setup for maximum output jitter is shown in Figure 11.56. If an increase of the output jitter beyond the maximum allowed is encountered, and if this results in a corresponding increase of the system's bit error rate, then the source or sources of that jitter must be traced and rectified, and the system restored to its normal operating mode.

Jitter Transfer Function   The jitter transfer function relates jitter gain, expressed as 20 $\log(J_o/J_{in})$, to frequency (Figure 11.57). Figure 11.58 illustrates the setup for jitter gain

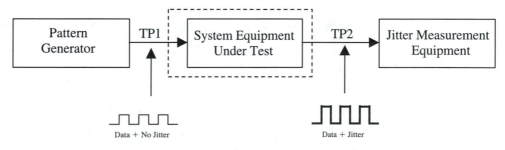

FIGURE 11.55   Testing setup for intrinsic output jitter.

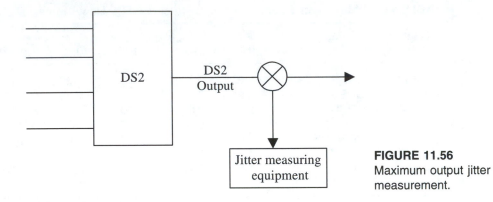

FIGURE 11.56
Maximum output jitter measurement.

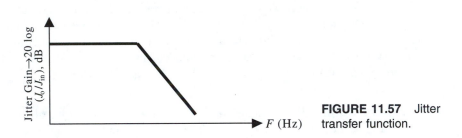

FIGURE 11.57 Jitter transfer function.

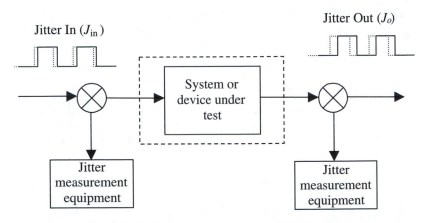

FIGURE 11.58 Jitter gain measurement setup.

measurements. For an effective monitoring of all the parameters indicative of digital radio system performance including jitter, a computerized central monitoring system is highly recommended. Such a system allows for economical and efficient system performance monitoring in real time.

QUESTIONS

1. Describe why sometimes signals must be transformed from the time domain to the frequency domain and vice versa.
2. Define *modulator/demodulator sensitivity*.
3. List the equipment required for modulation sensitivity measurement. With the assistance of a block diagram, briefly explain the operating procedures.
4. Define the terms *modulator/demodulator linearity* and *differential gain*.
5. With the assistance of a block diagram, explain the procedures for nonlinearity and differential gain measurements.
6. Define *return loss*. With the assistance of a block diagram, explain the procedure for return-loss measurements.
7. Describe the two basic methods for error-performance measurements for digital microwave links.
8. Sketch the block diagram, indicate the required instruments, and explain the necessary procedures for noise and interference measurements for digital radio systems.
9. Sketch the block diagram for bit error rate (BER) and carrier-to-noise ratio (CNR) measurements. Briefly explain the operating procedures.
10. Define the following terms: (a) *adjacent channel interference* (ACI), (b) *cross-polar interference* (CPI), and (c) *Doppler shift* (DS).
11. List the equipment necessary for Modem/IF testing. With the assistance of a block diagram, explain the operating procedures.
12. Describe in detail the operating procedures for Modem/RF testing.
13. Sketch the block diagram of a complete satellite (earth station–to–earth station) test and explain in detail its function of operation.
14. Define and compare both the eye diagram and constellation display methods of radio link performance evaluation.
15. Describe the equipment required for receiver noise-figure measurements and briefly explain the operating procedures.
16. Define *jitter*. Sketch the block diagram and describe the operating steps required for jitter measurements.

12

Elements of High-Definition TV

Objectives

- Define high-definition-TV (HDTV).
- Explain the difference between NTSC and HDTV.
- Identify the challenges facing the implementation of HDTV.
- Describe the system specifications.
- Recognize the importance of video compression.
- Describe the principle of operation of video encoders/decoders.
- Understand the basic concept of video signal transportation or packet delivery.

Key Terms

High definition TV
NTSC
Grand Alliance
Interlaced scanning
Dynamic resolution
Target specifications
Static luminance resolution

1080-I format
720-P format
Video compression
MPEG format standard
Discrete cosine transform
Huffman coding

Motion-compensated
 predictive coding
Bydirectional predictive
 coding
Coarse-motion simulator
Panel-encoding processors

INTRODUCTION

The concept of **High-definition TV** (HDTV) was conceived in Japan in June 1969. At that time, the Japanese government, through its own broadcasting system and in cooperation with other industrial groups from the electronics manufacturing sector, initiated research into the development of a worldwide high-definition TV system. In 1981, the Japanese were able to successfully demonstrate the first HDTV to the world. This system was based on 1125/60 lines, and it was hoped that the rest of the world would endorse it. This of course was not the case. Although the U.S. government did endorse the Japanese system, the Europeans rejected it and in 1987, came up with their own system of 1250 scanning lines and 60 Hz. This European action scuttled the Japanese attempt to promote a universal single-standard HDTV system.

Since system universality had failed to materialize, the U.S. government felt that it was the right time to take the initiative and to develop their own all-digital HDTV. At the same time, various industrial and academic sectors thought that HDTV confined within the 6-MHz channel bandwidth would be realized no later than 1990. The race for HDTV in the United States was underway; such companies and academic institutions as NBC, GI, MIT, Phillips, and others were involved. The next five years were marked by an intense competition among the participants to come up with the best all-digital HDTV system. To facilitate this effort and to evaluate the findings of all participants, a specially designed center was built in Virginia. In 1992, the committee responsible for the selection of the best system was unable to reach a decision and instead, recommended that all the industrial competitors get together and come up with one acceptable unified system. Thus the **Grand Alliance** came into being on May 24, 1993. Within two years from its birth in late 1995, the participants assembled the first prototype as follows:

The video encoder was designed and built by General Instruments and AT&T.

The digital modulator was built by Zenith.

The video decoder was built by Phillips.

The transport module was built by Thompson and Sarnoff.

The sound system was built by Dolby labs.

12.2 Challenges Facing the Implementation of HDTV

1. Efficient utilization of the available frequency spectrum.
2. Maintenance of the 6-MHz channel bandwidth.
3. Availability of HDTV receivers.
4. Maintenance of the current **NTSC** system for a specific time period.
5. HDTV must equal or surpass the same SNR (28 dB) as NTSC, a challenging task when HDTV is operating below the CNR threshold level.
6. HDTV must provide the same area coverage as NTSC with an acceptable level of cochannel interference.
7. Similarly, HDTV must be capable of sustaining cochannel interference from the NTSC system.

8. HDTV must be capable of providing system flexibility for current and future integrated systems.
9. HDTV must adapt to the same compression standards as the computer industry for both audio and video signals.
10. HDTV must be capable of transmitting through wideband optical fibers.

12.1 SYSTEM SPECIFICATIONS

The block diagram of the HDTV system is shown in Figure 12.1. The formats adapted by the select committee were (a) 720 vertical lines per frame and 1280 horizontal pixels or active samples per line with a 16:9 ratio, and (b) 1080 vertical lines per frame and 1920 horizontal pixels or active samples per line with a 16:9 ratio, 60-Hz **interlaced scanning,** and 50-Hz and 24-Hz progressive scanning for both formats.

12.1.1 Resolution

The Grand Alliance resolution was measured and compared with the **target specifications** set forth by the advisory committee. The test results showed the Grand Alliance data to exceed those of the target specifications. For example, for the 1080 by 1920 format, a **static luminance resolution** of 460 for horizontal compares to the 430 of target specifications. Similarly, chroma 250 exceeds the 215 target. For **dynamic resolution,** 500 horizontal, 200 vertical were obtained from Grand Alliance measurements, a favorable comparison to the 345 and 195 of the target specifications. Similar favorable results were obtained for the 720/1280 format.

In summary, static luminance resolution for the 1280-I format was found to be better than the specifications set forth by the advisory committee for both the horizontal and vertical, and a slight reduction was measured for the diagonal. For the same **1080-I format,** static chromatic resolution was measured to be higher than the set specifications for horizontal, and a substantial decrease of approximately 20% was measured for diagonal and vertical. For the **720-P format,** the results surpassed the specifications for static resolution (chromatic) by 20% for horizontal, 11% for vertical, and 60% for diagonal. For static luma, a 10% increase for horizontal, a 15% decrease for vertical, and an 8% decrease for diagonal were measured. For dynamic luma, a 13% increase for horizontal, an almost 19% decrease for vertical, and a 4% increase for diagonal was also measured. For dynamic chroma the horizontal gain was measured at 30%, the vertical gain at 19%, and the diagonal at 7%.

12.1.2 Format Comparison

When a 1080-I transmitting format is displayed on a 720-P format monitor, or when a 720-P transmitting format is displayed on a 1080-I monitor, the picture quality loss must not exceed a −1.0 grade loss. Tests conducted for both cases revealed the following: For

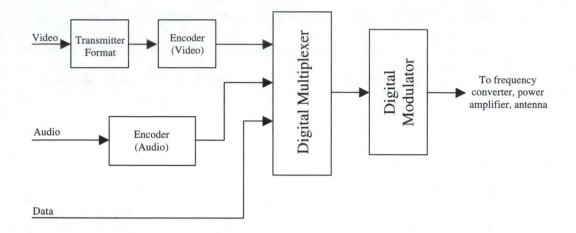

(a) Transmitter

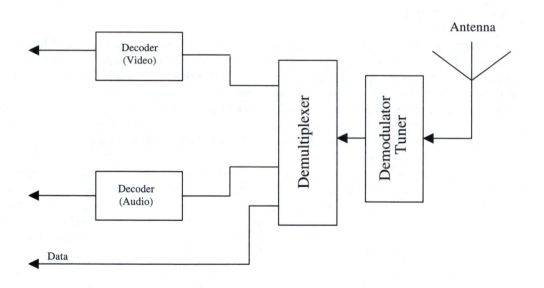

(b) Receiver

FIGURE 12.1 Transmitter and receiver block diagrams.

the 1080-I to 720-P format conversion, a -0.58 grade loss was measured. This value was favorable when compared to the -0.54 grade loss achieved without format conversion, thus revealing a total picture quality of -0.04 grade.

Similarly, transmitting with 720-P format and displaying on a 1280-I format monitor, a measured value of -0.69 grade loss was obtained. This value, compared with the -0.51 grade loss obtained without conversion, revealed an average picture loss of -0.18 grade. It is therefore evident from these tests that the picture quality loss based on format conversion was found to be much less than the target grade loss of -1.0 set by the advisor committee.

12.1.3 Video Compression

Video compression is referred to as the process whereby the data size is substantially reduced before transmission through the encoder module. At the receiver end, the compressed data must be received through a decoder module without any loss of quality. Video compression for HDTV is based on the MPEG standard.

12.1.4 The MPEG Profile

The Moving Picture Experts Group (MPEG) sets standards for video and audio compression. The evolution of the **MPEG format standard** began with MPEG-1, which was primarily designed to facilitate video compact disc and multimedia CD technology. This profile was capable of encoding progressive video with a 1.5 Mb/s rate of transmission. Following the development of MPEG-1, the MPEG-2 profile was developed to facilitate the encoding of progressive video transmission.

This format is capable of encoding interlaced pictures with a transmission rate of approximately 4 Mb/s. It was thought at that time that a separate compression standard was required for the emerging HDTV. Hence the MPEG-3 standard was proposed. Further study revealed that the MPEG-2 modified standards were sufficient for HDTV requirements to encompass future technological demands such as computer visualization, video and speech synthesis, artificial intelligence, and more. Therefore, the MPEG-4 standards are now under development.

The MPEG-2 standards adopted by the Grand Alliance to be used for HDTV video compression are based on the following technologies:

> **Discrete cosine transform** (DCT)
> **Huffman coding**
> **Quantization**
> **Motion-compensated predictive coding**
> **Bidirectional prediction**

The video encoder/decoder block diagram is shown in Figure 12.2.

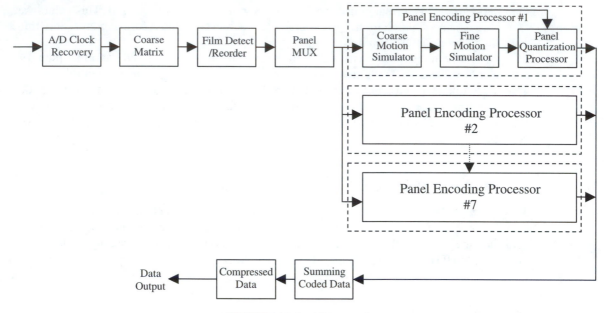

FIGURE 12.2 Video coder.

12.1.5 Video Encoder

The clock recovery and analog-to-digital converter converts the video analog input into digital for the three fundamental colors green, blue, and red, while at the same time generating the 27-MHz clock frequency required by the transport system. The function of the film detector/recorder is to detect 24-frame film. It also provides a progressive and gradual two-way transition of the 24-frame processing code. The function of the panel multiplexer is to vertically sample the signals for color difference, divide the obtained data, and feed it into the six **panel-encoding processors,** composed of a coarse-motion simulator, a fine-motion estimator, and a panel-quantization processor.

The **coarse-motion simulator** is a specially designed chip capable of performing motion vector calculations per macroblock over the entire horizontal range of ±63 pixels (B-Picture) with an accuracy of one pixel, and ±127 pixels (P-Picture) with the same one-pixel accuracy. It also performs similar calculations of ±15 pixels per vector and ±31 pixels per frame over the vertical range. In addition, the panel-encoding processor determines field vectors and discrete-cosine-transform-coding mean absolute errors for either picture or frame. Furthermore, the discrete-cosine-transform/fine-motion estimator module is designed to compute the half-pixel motion vector, based again on mean absolute errors. The discrete-cosine-transform coefficients are fed into the panel quantization processor for quantization. The macroblock layer data, picture data, group of pictures, and slice and sequence layer are ready to be applied to the transport system in a compressed and formatted packet.

12.1.6 Video Decoder

The block diagram of a video decoder is shown in Figure 12.3. To a great extent, the video decoder performs the opposite function to that of the encoder. The compressed data is rearranged into video packets and then fed into the rate buffer. Simultaneously, the video packet is fed into the packetized elementary-stream-layer-adaption field. This module is designed to recover the control and timing data. The recovered data is then fed into the system and display controller, and the output is fed into the rate buffer. The output of the rate buffer is applied to the input of the eight decoder engines. Each decoder engine performs the function of decoding based on the inverse discrete cosine transform, inverse quantization, and motion compensation. The outputs of the decoder engines (decoder video) are fed into the input of the video buffer, whose main function is to sort them into video frames. At the final stage the video frames are converted to the video signal through the postfilter digital matrix and digital-to-analog converter module.

12.1.7 Transportation (Packet Delivery)

The main function of the transport system is to multiplex, synchronize and efficiently deliver the compressed and packetized audio, video, and related data of a specific length based on the MPEG standard adopted by the Grand Alliance.

The transport packet is composed of a 4-byte header and 184 bytes of information (Figure 12.4). The 4-byte packet header includes synchronization, packet loss protection, and encryption data; the variable-length adaption header incorporates random access flag, time synchronization, media synchronization, and bit-stream point flag data. The adaption header is part of the 184-byte payload. The standard 4-byte packet header is very useful when the system operates under unfavorable transmission conditions. In such a case, when a complete failure of the error protection scheme occurs, the header informs the elementary stream decoders. These decoders reject the packet and prevent it from appearing on the receiver set. If the system requires access to the compressed data stream for periodic video and audio firing synchronization, an additional adaption header of variable length is incorporated into the payload. The transport data stream is classified as single program and multiprogram, and its composition is shown in Figure 12.5. Tests were conducted on a prototype transport system developed by the Grand Alliance. During these tests, transmission of a multiple programs utilizing the standard 6-MHz channel bandwidth and data rates of 4.738 Mb/s, 5.744 Mb/s, 3.747 Mb/s, and 4.717 Mb/s exhibited an impressive error-free signal of -28 dB.

12.1.8 Audio

For high-definition TV, the Grand Alliance has proposed a discrete multichannel digital audio system. The audio system is based on an AC-3 compressor system developed by Dolby. It encodes into a 384 Kb/s data stream for right, left, and center, as well as right surround, left surround, and four frequency-enhancement channels through a time-domain to frequency-domain coding process.

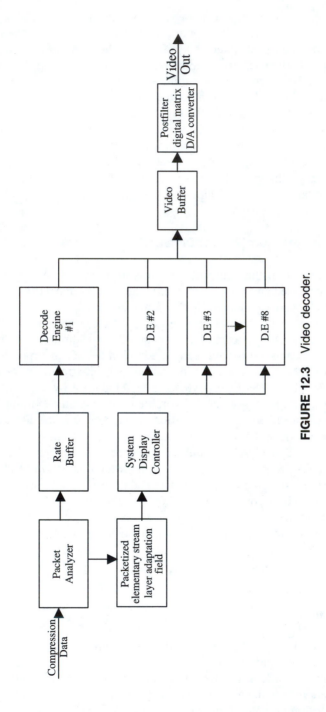

FIGURE 12.3 Video decoder.

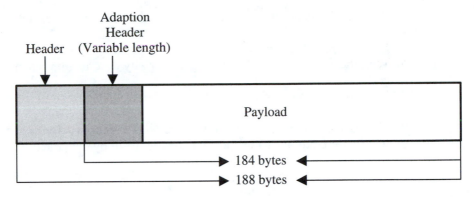

FIGURE 12.4 The transportation packet.

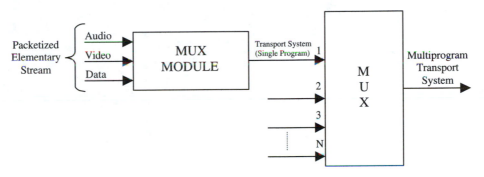

FIGURE 12.5 Grand Alliance prototype transport system.

The continuously analyzed audio signal spectrum is divided into significant and insignificant components. The significant components are quantized and encoded to generate the target signal-to-noise ratio; the insignificant spectral components are ignored. A major concern with HDTV is the possibility of an induced time delay between voice and video signals attributed to voice and video signal digitalization. Tests conducted by the Grand Alliance measuring the time delay between video and audio revealed the following: For the 720-P format, the video was lagging the audio by 36–40 ms. This time delay was considered acceptable. For the 1080-I format, the video signal was leading the audio by 9–13 ms. Again, this delay time was also considered acceptable.

QUESTIONS

1. List and briefly explain the difficulties facing the implementation of HDTV.
2. Sketch the block diagram of a transmitter HDTV system and explain its operation.
3. Sketch the block diagram of a receiver HDTV system and briefly explain its operation.

4. Describe the concept of resolution and format compression as defined by the Grand Alliance.
5. What is the MPEG profile?
6. With the assistance of a block diagram, briefly describe the function of a video encoder.
7. With the assistance of a block diagram, briefly explain the function of a video decoder.
8. Sketch and describe the Grand Alliance HDTV packet delivery system.
9. Briefly describe the audio specifications set forth by the Grand Alliance for HDTV.

Glossary

A/D Analog-to-digital conversion.

A-law The European companding law used for PCM encoders/decoders.

Aliasing Distortion occurring when the Nyquist criteria have not been met.

AM Amplitude modulation.

APD Avalanche photodiode.

Apogee The furthest point from the earth of an elliptical orbit.

Asynchronous transmission When carrier recovery at the receiver end is not required.

Attenuation The magnitude relationship between output and input signals is less than one.

Attenuation (optical) The progressive amplitude reduction of the light ray traveling through a fiber.

Bandwidth The frequency range within a carrier signal able to carry analog or digital information.

Base band A normally low-frequency information-carrying signal before modulating an RF carrier.

Bending losses Losses induced due to fiber bending.

Bit rate The number of bits a transmission channel is capable of carrying.

C/T Carrier-to-temperature ratio.

Carrier An RF signal with constant frequency and amplitude. This signal is served as the transport of the information-carrying baseband signal.

C Band The 6/4-GHz satellite system.

CCITT Consultative Committee for International Telephone and Telegraph.

CMR Common-mode rejection. The process whereby an undesirable signal is simultaneously applied to a differential two-input circuit in order to reduce its overall negative effect.

CNR Quantitative measure of a modulated carrier signal in relation to the noise level.

Coder An encoder/decoder is the analog-to-digital or digital-to-analog conversion module in a digital transmission system.

Companding A logarithmic compression and expansion of a dynamic signal before the encoding or after the decoding process.

Cross-Talk The unintentional energy transfer from one information-carrying channel (usually voice) to the adjacent channel.

CVSDM Continuous variable-slope delta modulation (a modified version of delta modulation).

D/A conversion Digital to analog conversion.

D/C Down-converter. The module that converts an RF to IF frequencies. Mainly applied to microwave and satellite links.

Dark current The leakage current of a photodetector device.

Data compression A method of compressing the channel data for more efficient bandwidth utilization in a communications system without compromising quality of reception.

Decoding The process through which a varying-amplitude sampled signal is reconstructed from a binary stream of information.

Delta modulation A method of converting an analog signal into a digital signal.

Demand assignment A method by which a transponder is assigned a channel from a pool of channels upon request.

De-spun Refers to parabolic reflector antennas of a spin satellite. Although the satellite is spinning, the antenna maintains a fixed position in reference to a point on the surface of the earth.

Deviation Mainly applied to frequency modulation, and refers to frequency change above and below a center frequency.

Dispersion The spreading of the pulse traveling through an optical fiber.

Distortion The inability of the transmitted signal to be accurately reproduced at the receiver end.

Domestic satellites The satellite system that provides communications to a particular country.

Down-link The signal path between satellite and earth.

DPSK Differential phase-shift keying. A method of digital modulation.

Duplex The simultaneous and independent two-way transmission of data or voice signals.

Earth station The facility encompassing the transmitter and receiver sections of a satellite communications link at a particular geographic location.

Echo The phenomenon whereby some signal power is reflected back to the source because of time delays or impedance mismatch.

Echo suppression The circuit designed to eliminate the echo signals.

EIRP The product of the transmitter power and transmitter antenna gain in reference to isotropic propagation.

Elevation The angle between earth and satellite at a particular point on the surface of the earth.

Equalizer The electric circuit designed to compensate for undesirable phase and amplitude changes of a signal propagating through a transmission medium.

Equatorial orbit A satellite placed in the orbit above the equator with a 0° inclination.

Extrinsic absorption Based on optical fiber impurities unintentionally injected during the fabrication process.

F Noise Factor. Relating to equivalent noise temperature and ambient noise temperature.

FCC Federal Communications Commission.

FDM Frequency-division multiplexing: The process of combining several voice signals into a composite baseband signal before modulation.

FM Frequency modulation.

FM (IF) Improvement factor. The method used to improve the signal-to-noise ratio at the receiver end.

Frame A time slot identifiable by a number of digits and synchronized with frame alignment.

Frequency domain The display of an electrical signal in reference to frequency.

Frequency reuse The utilization of one carrier frequency in both vertical and horizontal polarization mode of transmission through the use of two sets of appropriately polarized parabolic reflector antennas.

FSK Frequency-shift keying. A form of digital modulation.

G/T Figure of merit. Refers to the ratio of the transmitter antenna gain to system noise temperature.

Geosynchronous Refers to a satellite placed in an equatorial orbit and performing one full orbit in 24 hours.

HPA High-power amplifier, operating at microwave frequencies usually referred to as "TWTs."

Intermodulation The modulation of various components of the same complex signal processed through a nonlinear device or circuit.

Intrinsic absorption The result of interaction between free electrons and light wavelength traveling through the fiber.

Jitter The undesirable variation of a signal from its reference point.

Ku band The 14/12-GHz communications satellite band.

Laser Light amplification by stimulation emission of radiation. A light-generating device based on stimulation emission of radiation.

LED Light-emitting diode. An electroluminescent device composed of semiconductor materials, and capable of generating light.

LNA Low-noise amplifier.

LNB The combination of low-noise amplifier and down-converter in a single module.

Mie loss Mie loss occurs when the diameter of the obstacle is larger than the wavelength of the propagating wave.

Multimode Optical fiber capable of carrying more than one signal of the same wavelength following different traveling paths.

Multiple access A substantial number of customers having access to the same satellite channel pool.

Multiplexing Combining several input signals into one composite signal before modulation.

Noise figure (NF) A quantitative method of establishing receiver performance in terms of equivalent noise temperature.

Numerical aperture (NA) The angle between the axis of the core and the impeding light ray in an optical fiber.

Optical fiber A dielectric cylindrical system capable of converging electromagnetic waves occupying the visible spectrum.

PCM Pulse-code modulation. A process by which analog signals are quantified in order to be digitally transmitted.

Photodetection The process whereby optical power is detected and converted to electrical power.

Photodetector capacitance A parasitic component affecting the response time of the device.

Photodetector noise The result of the ionization and photocurrent multiplication process.

PIN photodetector A P-region, Intrinsic, N-region semiconductor photodetector.

Polar orbit An orbit with a 90° angle of inclination.

Propagation delay The time required for a signal to propagate through a device or circuit.

QPSK Quadrature phase-shift keying. A type of digital modulation scheme (four phases).

Quantization noise An equivalent noise generated through the process of voice signal quantization.

Quantum efficiency The ratio of photon generation, to the number of induced electrons.

Rayleigh-scattering loss Rayleigh loss occurs when the diameter of the obstacle is smaller than the wavelength of the propagating wave.

Repeater A complete transreceiver unit employed in line-of-sight microwave links. (The communications satellite is also referred as an "active repeater."

Response time The time required by the generated carriers to travel the absorption region under reverse-bias conditions.

RTI Return to input.

Sampling rate The clock frequency used to sample an analog signal.

Satcom Communications satellite.

Scattering The result of interaction between the traveling light wave and small variations of material density within the fiber.

Single-mode fibers Optical fibers capable of conveying only one signal of a specific wavelength.

SLED Surface-emitting LEDs.

SNR Signal-to-noise ratio. Indicative of signal quality.

SPADE Single-channel-per-carrier, pulse-code-modulation, multiple-access, demand-assignment equipment.

Step-index fibers Optical fibers with gradual change of their diffractive index from core to cladding.

TDM Time-division multiplexing. A digital multiplexing method combining a number of digitally coded analog signals into one single stream of data.

Transponder The communications equipment on board the satellite.

TWA Traveling-wave tube. A high-power amplifier operating at microwave frequencies.

U/C Up-converter. The circuit designed to convert the IF to RF or microwave frequencies in a line-of-sight or satellite transmission.

Up-link The distance the signal travels from the earth station to a communications satellite.

WAGN White additive Gaussian noise.

References

Allard, F. C., *Fiber Optics Handbook*. McGraw-Hill, 1990.

Allen, J. L., "Array Antennas," *IEEE Spectrum*, November 1964.

Bellamy, J., *Digital Telephony*. John Wiley and Sons, 1982.

Berg, D., "An Introduction to Satellite Television," *CQ*, February 1982.

Carlson, A. B., *Communication Systems*. McGraw-Hill, 1986.

Collette, R. C., "The European Communications Satellite Program," AIAA 4th Communications Satellite Conference, Washington DC, April 1972.

Cooper, R. B., "Home Reception via Satellite," *Radio Electronics*, August 1979.

Elliott, R. S., *Antenna Theory and Design*. Prentice-Hall, 1981.

Feher, K., *Advanced Digital Communications*. Prentice-Hall, 1987.

Feher, K., *Digital Communications with Microwave Applications*. Prentice-Hall, 1981.

Feher, K., *Telecommunication Measurements Analysis and Instrumentation*. Prentice-Hall, 1987.

Freeman R. L., *Telecommunication Transmission Handbook*. John Wiley and Sons, 1981.

Fthenakis, E., *Manual of Satellite Communications*. McGraw-Hill, 1984.

Gibson, R., "Satellite Communications I," *IEEE Spectrum*, March 1980.

Ha, Tri T., *Digital Satellite Communications*. McGraw-Hill, 1990.

Haykin, S., *Communication Systems*. McGraw-Hill, 1983.

HP-3709B, "Constellation Analyzer Training Manual."

HP-3730-1, "Product Note."

HP-3785 A/B, "Training Manual."

HP-3785A-1, "Practical Jitter Measurements."

HP Application Note 57-1.

HP Application Note 207.

HP Application Note 243.

HP Application Note 355.

HP Application Note 922.

HP Product Note 3708-1.

HP Product Note 3708-3.

HP Product Note 3708-5.

HP-360 Application Note.

IEEE MTT-S Digest, 1988, pp. 255–258, 499–502, 823–830, 941–943.

IEEE Proceedings, 1977, pp. 294–307, 332–341.

IEEE Transactions on Microwave Theory and Techniques, Vol. MTT-27, No. 5, May 1979, pp. 45–451, 434–441, 442–449, 483–491.

J. Luminescence, Vol. 7, pp. 390–414. North-Holland, 1973.

Jong, M. T., *Methods of Discrete Signals and Systems Analysis*. McGraw-Hill, 1982.

Keiser, G., *Optical Fiber Communications*. McGraw-Hill, 1983.

Killen, H., *Digital Communications with Fiber Optics and Satellite Applications*. Prentice-Hall, 1988.

Kraus J. D., *Electromagnetics*. McGraw-Hill, 1984.

Kraus J. D., "Antennas: Our Electronic Eyes and Ears," *Microwave J*, 1989.

Ladbrook, P., "MMIC Design." ARTECH House Inc. Norwood, MA 1989.

Leopold, R. J., and Miller, A., "The Iridium Communications System," *IEEE Potentials*, 1993.

Maral, G., *Satellite Communications System*. John Wiley and Sons, 1986.

Martin, J., *Future Development in Telecommunications*. Prentice-Hall, 1977.

Meredith C. O., "Lessons Learned from Intelsat-III Satellite Program." TRW Systems, Redondo Beach CA, 1972. AIAA 4th Communications Satellite Conference, Washington DC, April 1972.

Microwave Journal, June 1973, pp. 29–36.

Microwave Journal, June 1986, pp. 121–132.

Microwave Journal, September 1988, pp. 197–210.

Nyguist, H., "Certain Factors Affecting Telegraph Speed." *BSTJ*, Vol. 3, April 1924, pp. 324–326.

Owen, F., *PCM and Digital Transmission Systems*. McGraw-Hill, 1982.

Kate, P. P., and Nickelson, R. L., "Design for INSAT," AIAA 4th Communications Satellite Conference, Washington DC, April 1972.

Pettai R, *Noise in Receiving Systems*. John Wiley and Sons, 1984.

Proakis, J., *Digital Communications*. McGraw-Hill, 1983.

Proc. IEEE, December 1973, pp. 1703–1751.

Proc. IEEE, Vol. 65, No. 3, March 1977, pp. 317–330, 457–473, 480–481.

Ricardi, L. J., "Communication Satellite Antennas," *Proc. IEEE,* Vol. 65, No. 3, March 1977.

Rowden, C., *Speech Processing*. McGraw-Hill, 1992.

Schwatrz, M., *Information Transmission Modulation Noise*. McGraw-Hill, 1970.

Shanmugam, K. Sam., *Digital and Analog Communication Systems*. John Wiley and Sons, 1979.

Sielman, P. F., Schwartz, L., and Noji, T. T., "Multiple Beam Communications Satellites with Remote Beam Steering and Beam Shaping," AIAA 4th Communications Satellite Conference, Washington DC, April 1972.

Skitek, G. G., *Electromagnetic Concepts and Applications*. Prentice-Hall, 1982.

Spilker, J., *Digital Communication by Satellite*. Prentice-Hall, 1977.

Stremler, F. G., *Introduction to Communication Systems*. Addison-Wesley, 1982.

Taub, H., and Schilling, D. L., *Principles of Communication Systems*. McGraw-Hill, 1971.

Trans. Microwave Theory, Vol. MTT-23, January 1975, pp. 20–30.

Tsui, J. B-Y., "Digital Microwave Receivers ARTECH House, Inc., Norwood MA, 1989.

Varian, K. R., "DROs at 4, 6 and 11 GHz" *Microwave J*, October 1986.

Viterbi, A. J., *Principles of Digital Communications and Coding*. McGraw-Hill, 1979.

Watkins, E. T., Post, B. M., Knust-Graichen, and Schellenberg, J. M., "Low-Noise, 20MHz Receiver Front-End," *Microwave J*, May 1986.

Wheeler C. A., and Livingston, S.M., "A Solid-State Amplifier for Satellite Communications" *Microwave J* (technical note), July 1975.

Wu, W. W., *Elements of Digital Communications*. Computer Science Press. Rockville, MD, 1985.

Index